# THE WILEY GUIDE TO PROJECT TECHNOLOGY, SUPPLY CHAIN & PROCUREMENT MANAGEMENT

# THE WILEY GUIDES TO THE MANAGEMENT OF PROJECTS

Edited by

Peter W. G. Morris and Jeffrey K. Pinto

The Wiley Guide to Project, Program & Portfolio Management
978-0-470-22685-8

The Wiley Guide to Project Control
978-0-470-22684-1

The Wiley Guide to Project Organization & Project
Management Competencies
978-0-470-22683-4

The Wiley Guide to Project Technology, Supply Chain &
Procurement Management
978-0-470-22682-7

# THE WILEY GUIDE TO PROJECT TECHNOLOGY, SUPPLY CHAIN & PROCUREMENT MANAGEMENT

Edited by

Peter W. G. Morris and Jeffrey K. Pinto

JOHN WILEY & SONS, INC.

*Library of Congress Cataloging-in-Publication Data:*

ISBN: 978-0-470-22682-7

Printed in the United States of America

10 9 8 7 6 5 4 3 2 1

# CONTENTS

# THE WILEY GUIDE TO PROJECT TECHNOLOGY, SUPPLY CHAIN & PROCUREMENT MANAGEMENT: PREFACE AND INTRODUCTION

Peter W. G. Morris and Jeffrey Pinto

In 1983, Dave Cleland and William King produced for Van Nostrand Reinhold (now John Wiley & Sons) the *Project Management Handbook*, a book that rapidly became a classic. Now over twenty years later, Wiley is bringing this landmark publication up to date with a new series *The Wiley Guides to the Management of Projects*, comprising four separate, but linked, books.

Why the new title—indeed, why the need to update the original work?

That is a big question, one that goes to the heart of much of the debate in project management today and which is central to the architecture and content of these books. First, why "the management of projects" instead of "project management"?

Project management has moved a long way since 1983. If we mark the founding of project management to be somewhere between about 1955 (when the first uses of modern project management terms and techniques began being applied in the management of the U.S. missile programs) and 1969/70 (when project management professional associations were established in the United States and Europe) (Morris, 1997), then Cleland and King's book reflected the thinking that had been developed in the field for about the first twenty years of this young discipline's life. Well, over another twenty years has since elapsed. During this time there has been an explosive growth in project management. The professional project management associations around the world now have thousands of members—the Project Management Institute (PMI) itself having well over 200,000—and membership continues to grow! Every year there are dozens of conferences; books, journals, and electronic publications abound; companies continue to recognize project management as a core business discipline and work to improve company performance through it; and, increasingly, there is more formal educational work carried out in university teaching and research programs, both at the undergraduate, and particularly graduate, levels.

Yet, in many ways, all this activity has led to some confusion over concepts and applications. For example, the basic American, European, and Japanese professional models of project management are different. The most influential, PMI, not least due to its size, is the most limiting, reflecting an essentially execution, or delivery, orientation, evident both in its *Guide to the Project Management Body of Knowledge, PMBOK Guide, 3rd Edition* (PMI, 2004) and its *Organizational Project Management Maturity Model, OPM3* (PMI, 2003). This approach tends to under-emphasize the front-end, definitional stages of the project, the stages that are so crucial to successful accomplishment (the European and Japanese models, as we shall see, give much greater prominence to these stages). An execution emphasis is obviously essential, but managing the definition of the project, in a way that best fits with the business, technical, and other organizational needs of the sponsors, is critical in determining how well the project will deliver business benefits and in establishing the overall strategy for the project.

It was this insight, developed through research conducted independently by the current authors shortly after the publication of the Cleland and King *Handbook* (Morris and Hough, 1987; Pinto and Slevin, 1988), that led to Morris coining the term "the management of projects" in 1994 to reflect the need to focus on managing the definition and delivery of *the project itself* to deliver a successful outcome.

These at any rate are the themes that we shall be exploring in this book (and to which we shall revert in a moment). Our aim, frankly, is to better center the discipline by defining more clearly what is involved in managing projects successfully and, in doing so, to expand the discipline's focus.

So second, why is this endeavor so big that it takes four books? Well, first, it was both the publisher's desire and our own to produce something substantial—something that could be used by both practitioners and scholars, hopefully for the next 10 to 20 years, like the Cleland and King book—as a reference for the best-thinking in the discipline. But why are there so many chapters that it needs four books? Quite simply, the size reflects the growth of knowledge within the field. The "management of projects" philosophy forces us (i.e., members of the discipline) to expand our frame of reference regarding what projects truly *are* beyond the traditional *PMBOK/OPM3* model.

These, then, are not a set of short "how to" management books, but very intentionally, resource books. We see our readership not as casual business readers, but as people who are genuinely interested in the discipline, and who is seek further insight and information—the thinking managers of projects. Specifically, the books are intended for both the general practitioner and the student (typically working at the graduate level). For both, we seek to show where and how practice and innovative thinking is shaping the discipline. We are deliberately pushing the envelope, giving practical examples, and providing references to others' work. The books should, in short, be a real resource, allowing the reader to understand how the key "management of projects" practices are being applied in different contexts and pointing to where further information can be obtained.

To achieve this aim, we have assembled and worked, at times intensively, with a group of authors who collectively provide truly outstanding experience and insight. Some are, by any standard, among the leading researchers, writers, and speakers in the field, whether as academics or consultants. Others write directly from senior positions in industry, offering their practical experience. In every case, each has worked hard with us to furnish the relevance, the references, and the examples that the books, as a whole, aim to provide.

What one undoubtedly gets as a result is a range that is far greater than any individual alone can bring (one simply cannot be working in all these different areas so deeply as all these authors, combined, are). What one does not always get, though, are all the angles that any one mind might think is important. This is inevitable, if a little regrettable. But to a larger extent, we feel, it is beneficial for two reasons. One, this is not a discipline that is now done and finished—far from it. There are many examples where there is need and opportunity for further research and for alternative ways of looking at things. Rodney Turner and Anne Keegan, for example, in their chapter on managing innovation (*The Wiley Guide to Project Technology, Supply Chain & Procurement Management,* Chapter 8) ended up positioning the discussion very much in terms of learning and maturity. If we had gone to Harvard, to Wheelwright and Clark (1992) or Christensen (1999) for example, we would almost certainly have received something that focused more on the structural processes linking technology, innovation, and strategy. This divergence is healthy for the discipline, and is, in fact, inevitable in a subject that is so context-dependent as management. Second, it is also beneficial, because seeing a topic from a different viewpoint can be stimulating and lead the reader to fresh insights. Hence we have Steve Simister giving an outstandingly lucid and comprehensive treatment in *The Wiley Guide to Project Control,* Chapter 5 on risk management; but later we have Stephen Ward and Chris Chapman coming at the same subject (*The Wiley Guide to Project Control,* Chapter 6) from a different perspective and offering a penetrating treatment of it. There are many similar instances, particularly where the topic is complicated, or may vary in application, as in strategy, program management, finance, procurement, knowledge management, performance management, scheduling, competence, quality, and maturity.

In short, the breadth and diversity of this collection of work (and authors) is, we believe, one of the books' most fertile qualities. Together, they represent a set of approximately sixty authors from different discipline perspectives (e.g., construction, new product development, information technology, defense / aerospace) whose common bond is their commitment to improving the management of projects, and who provide a range of insights from around the globe. Thus, the North American reader can gain insight into processes that, while common in Europe, have yet to make significant inroads in other locations, and vice versa. IT project managers can likewise gather information from the wealth of knowledge built up through decades of practice in the construction industry, and vice versa. The settings may change; the key principals are remarkably resilient.

But these are big topics, and it is perhaps time to return to the question of what we mean by project management and the management of projects, and to the structure of the book.

## Project Management

There are several levels at which the subject of project management can be approached. We have already indicated one of them in reference to the PMI model. As we and several other of the *Guides*' authors indicate later, this is a wholly valid, but essentially delivery, or execution-oriented perspective of the discipline: what the project manager needs to do in order to deliver the project "on time, in budget, to scope." If project management professionals cannot do this effectively, they are failing at the first fence. Mastering these skills is the *sine qua non*—the 'without which nothing'—of the discipline. Volume 1 addresses this basic view of the discipline—though by no means exhaustively (there are dozens of other books on the market that do this excellently—including some outstanding textbooks: Meredith and Mantel, 2003; Gray and Larson, 2003; Pinto, 2004).

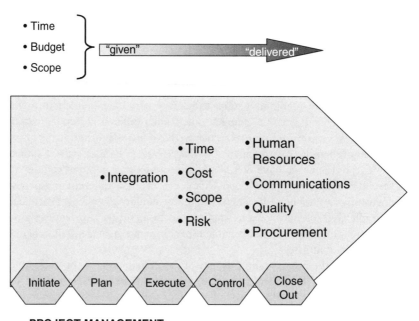

**PROJECT MANAGEMENT:**
**"On time, in budget, to scope" execution/delivery**

The overriding paradigm of project management at this level is a control one (in the cybernetic sense of control involving planning, measuring, comparing, and then adjusting performance to meet planned objectives, or adjusting the plans). Interestingly, even this model—for us, the foundation stone of the discipline—is often more than many in other disciplines think of as project management: many, for example, see it as predominantly oriented around scheduling (or even as a subset, in some management textbooks, of operations management). In fact, even in some sectors of industry, this has only recently begun to change, as can be seen towards the end of the book in the chapter on project management in the pharmaceutical industry. It is more than just scheduling of course: there is a whole range of cost, scope, quality and other control activities. But there are other important topics too.

Managing project risks, for example, is an absolutely fundamental skill even at this basic level of project management. Projects, by definition, are unique: doing the work necessary to initiate, plan, execute, control, and close-out the project will inevitably entail risks. These need to be managed.

Both these areas are mainstream and generally pretty well understood within the traditional project management community (as represented by the PMI *PMBOK*® '*Guide*' (PMI, 2004) for example). What is less well covered, perhaps, is the people-side of managing projects. Clearly people are absolutely central to effective project management; without people projects simply could not be managed. There is a huge amount of work that has been done on how organizations and people behave and perform, and much that has been written on this within a project management context (that so little of this finds its way into *PMBOK* is almost certainly due to its concentration on material that is said in *PMBOK* to be "unique" to project management). A lot of this information we have positioned in Volume 3, which deals more with the area of competencies, but some we have kept in the other volumes, deliberately to make the point that people issues are essential in project delivery.

It is thus important to provide the necessary balance to our building blocks of the discipline. For example, among the key contextual elements that set the stage for future activity is the organization's structure—so pivotal in influencing how effectively projects may be run. But organizational structure has to fit within the larger social context of the organization—its culture, values, and operating philosophy; stakeholder expectations, socioeconomic, and business context; behavioural norms, power, and informal influence processes, and so on. This takes us to our larger theme: looking at the project in its environment and managing its definition and delivery for stakeholder success: "the management of projects."

## The Management of Projects

The thrust of the books is, as we have said, to expand the field of project management. This is quite deliberate. For as Morris and Hough showed in *The Anatomy of Major Projects* (1987), in a survey of the then-existing data on project overruns (drawing on over 3,600 projects as well as eight specially prepared case studies), neither poor scheduling nor even lack of teamwork figured crucially among the factors leading to the large number of unsuccessful projects in this data set. What instead were typically important were items such as client changes, poor technology management, and poor change control; changing social, economic, and environmental factors; labor issues, poor contract management, etc. Basically, the message was that while traditional project management skills are important, they are often not *sufficient* to ensure project success: what is needed is to broaden the focus to cover the management of external and front-end issues, not least technology. Similarly, at about the same time, and subsequently, Pinto and his coauthors, in their studies on project success (Pinto and Slevin, 1988; Kharbanda and Pinto, 1997), showed the importance of client issues and technology, as well as the more traditional areas of project control and people.

The result of both works has been to change the way we look at the discipline. No longer is the focus so much just on the processes and practices needed to deliver projects "to scope, in budget, on schedule," but rather on how we set up and define the project to deliver stakeholder success—on how to manage projects. In one sense, this almost makes the subject impossibly large, for now the only thing differentiating this form of management from other sorts is "the project." We need, therefore, to understand the characteristics of the project development life cycle, but also the nature of projects in organizations. This becomes the kernel of the new discipline, and there is much in this book on this.

Morris articulated this idea in *The Management of Projects* (1994, 97), and it significantly influenced the development of the Association for Project Management's Body of Knowledge as well as the International Project Management Association's Competence Baseline (Morris, 2001; Morris, Jamieson, and Shepherd, 2006; Morris, Crawford, Hodgson, Shepherd, and Thomas, 2006). As a generic term, we feel "the management of projects" still works, but it is interesting to note how the rising interest in program management and portfolio management fits comfortably into this schema. Program management is now strongly seen as the management of multiple projects connected to a shared business objective—see, for example, the chapter by Michel Thiry (*The Wiley Guide to Project, Program & Portfolio Management,* Chapter 6.) The emphasis on managing for business benefit, and on managing projects, is exactly the same as in "the management of projects." Similarly, the recently launched *Japanese Body of Knowledge, P2M (Program and Project Management),* discussed *inter alia* in Lynn Crawford's chapter on project management standards (*The Wiley Guide to Project Organization & Project Management Competencies,* Chapter 10), is explicitly oriented around managing programs and projects to create, and optimize, business value. Systems management, strategy, value management, finance, and relations management for example are all major elements in *P2M:* few, if any, appear in *PMBOK.*

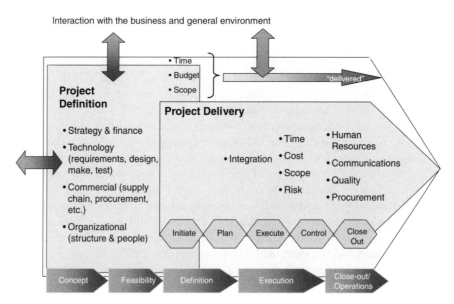

THE MANAGEMENT OF PROJECTS involves managing the definition and delivery of the project for stakeholder success. The focus is on the project in its context. Project and program management – and portfolio management, though this is less managerial – sit within this framework.

("The management of projects" model is also more relevant to the single project situation than *PMBOK* incidentally, not just because of the emphasis on value, but via the inclusion of design, technology, and definition. There are many single project management situations, such as Design & Build contracts for example, where the project management team has responsibility for elements of the project design and definition).

## Structure of *The Wiley Guide to Project Technology, Supply Chain & Procurement Management*

*The Wiley Guides to the Management of Projects* series consists of four distinct, but interrelated, volumes:

- *The Wiley Guide to Project, Program & Portfolio Management*
- *The Wiley Guide to Project Control*
- *The Wiley Guide to Project Organization & Project Management Competencies*
- *The Wiley Guide to Project Technology, Supply Chain & Procurement Management*

This book, *The Wiley Guide to Project Technology, Supply Chain & Procurement Management,* addresses two specialized, but absolutely key areas in certain sectors: the management of technical issues and the procurement of third party resources.

It is a well-documented fact that technology represents a major issue in the effective management of projects. Technology can be broadly (and often confusingly) defined to evoke a wide range of meanings—some helpful and others not. For our purposes, a working definition of technology within the context of the management of projects involves not so much actually *doing* the "technical" elements of the project as managing the processes and practices needed to ensure the technical issues by which projects are transformed from concepts into actual entities—doing this effectively within the time, cost, strategic, and other constraints on the project. In this regard, this section includes a number of chapters that guide us through the key lifecycle issues that define the project, ensure its viability, manage requirements, and track changes; in short, that highlight the key steps in transforming and realizing the technical definition of the project.

1. In Chapter 1 Al Davis, Ann Hickey and Ann Zweig take us carefully and systematically through the different types of requirements, basically from a strong IT/systems perspective, showing how requirements have to be elicited and selected (triage), and how this leads to specification. The sequencing of the requirements management process is then examined for different types of (systems) projects, and different types of requirements management tools are discussed. The chapter concludes by looking at trends in requirements management research and practice.
2. From the elicited and triaged requirements and the resulting specifications, the project design can be elaborated. Peter Harpum, in Chapter 2, discusses design management

in another comprehensive overview covering the nature of design and how designers design, systems engineering, and whole-life design (life cycle management); and design management "techniques" such as Design-for-Manufacturability, Concurrent Engineering, CAD/CAM, risk management, as well as several for scheduling, cost, and quality design management.

3. Hans Thamhain extends this discussion with his treatment, in Chapter 3, of concurrent engineering. Concurrent Engineering (CE) is one of those terms whose meaning varies from industry to industry, and firm to firm. It is in reality a combination of several things; as Hans says, CE "is a systematic approach to integrated project execution that emphasizes parallel, integrated execution of project phases, replacing the traditional linear process of serial engineering." Hans lists the following characteristics as typifying good CE: a uniform process model; Integrated Product Development; gate functions; standard project management process; Quality Function Deployment; early testing; and organizational involvement and transparency. He concludes with an extensive list of recommended practices for different phases of the project development cycle.

4. In Chapter 4 Rachel Cooper, Ghassan Aouad, Angela Lee, and Song Wu broaden the discussion into process and product modeling and the management of projects. They begin with a formal review of process modeling techniques, showing how process models can be used to represent the development of new product development and construction projects. Product modeling is then introduced with particular emphasis on object modeling. It is shown how product models might ultimately be integrated with activity models so that product and project information could be drawn off the same integrated model.

5. The emphasis on information modeling is taken further in Chapter 5 by Callum Kidd and Tom Burgess on configuration management (CM). They begin by explaining what CM is, showing that it has application in a wide spectrum of industries ranging from power plants to new product development as well as systems/IT projects. The heart of their chapter centers around the way contemporary information management practices are shaping CM. The chapter concludes on a somewhat ambivalent note with research data from the authors suggesting that often Tier 2 and 3 suppliers (in aerospace) use CM more to be compliance with Tier 1 requirements than for real business benefit.

The next two chapters may both seem rather industry-specific, though in fact both have wide-ranging and important implications. Both are extremely authoritative.

6. In Chapter 6 Alistair Gibb discusses the management of Safety, Health and Environmental (SHE) issues. As in so many of the chapters, Alistair argues the need to consider these issues from the very early stages of a project. He identifies the requirements for early definition work on SHE policy and objectives, risk assessment, designer actions, sustainability assessment, and SHE plan preparation, and goes on to discuss the implications of method statements, procurement strategy, and resources and competence development. He discusses key issues in implementation throughout the project lifecycle. Though the chapter is written from a predominantly construction perspective, nearly

all the points raised apply more broadly to most manufacturing situations and even, albeit to a lesser extent, in many IT projects.

7. Hal Mooz, in Chapter 7, discusses verification—what in old language would probably have been termed testing. (Today verification and validation have quite specific meanings: Hal begins by defining these along with several other key terms.) The chapter is set in a predominantly systems development context, but systems being defined here quite broadly (one of Hal's memorable examples is from Harley Davidson). Verification and validation have to be understood in terms of their position in the systems development/integration process. Here Hal introduces the Vee++ systems development cycle that he and his colleagues Kevin Forsberg and Howard Cotterman introduced in *Visualizing Project Management* (1996). Validation and verification techniques are described together with some key insights developed over years of experience in managing verification. Again, while the terminology may seem a little strange to some people, particularly those not working in manufacturing or systems, the applicability of Hal's chapter is instructive and relevant.

8. Rodney Turner and Anne Keegan give some insights, in Chapter 8, on managing innovation. They do so from an organizational learning perspective, emphasizing the importance of creating an environment supportive of, and the management conditions most likely to be conducive to, innovation in a project context. They introduce four practices—systems and procedures, project reviews, benchmarking, project management communities—that have been adopted for the selection, retention, and distribution of technological developments and conclude by showing how these relate to different stages of organizational learning and of project management maturity. (Both topics that are discussed in Volume 3, in Chapters 8 by Bredillet, 9 by Morris, and 13 by Cooke-Davies.)

An extremely important development in the management of projects over the past decade or so has been the manner in which logistics and concern for supply chain functions has impacted on how we develop projects. We could reasonably argue that these have always been associated with project management and yet, as more and more organizations adopt project management as a principal method for operating their primary activities, they are discovering that the traditional models of procurement (lowest cost bidding, contract administration, supplier expediting, tracking, and so forth)—once regarded as the overriding concerns of the construction industry—have branched out and embraced most organizations managing projects today. Understanding and proactively managing the critical steps in a firm's supply chain have proven to directly contribute to a company's bottom line success. This phenomenon is particularly important in project-based industries. The following chapters of *The Wiley Guide to Project Technology, Supply Chain & Procurement Management* take us directly into the mainstream of supply chain logistics and procurement for the management of projects.

9. In Chapter 9 David Kirkpatrick, Steve McInally, and Daniela Pridie-Sale address Integrated Logistics Support (ILS). The emphasis is on looking at whole life operations:

these need to be planned and managed from the earliest stages of the project. They look at ILS largely from the defense sector's perspective, though with examples from civil manufacturing, medical equipment, and construction. ILS, though clearly centered within the acquisition process, involves significant interaction with the project's / program's technical functions, as can be seen in the discussion on Logistics Support Analysis. They also show how ILS integrates with systems engineering and with private sector finance initiatives (PFI etc.). They discuss CALS (Continuous Acquisition and Life-Cycle Support), largely from the data handling perspective (resonating with the parallel discussion in Chapter 5 on configuration management). They conclude, very usefully, with a review of the difficulties faced in implementing ILS: the quality of data available, the difficulties of forecasting over such long periods, changes in usage and organizational composition, managing stakeholders towards long-term objectives, etc.

10. Ray Venkataraman turns to a less specialized topic in Chapter 10 with his overview of supply chain management (SCM). Having described the general critical issues in SCM— value optimizing the way customers, suppliers, design & operations, logistics and inventory are effectively managed—he refines these in terms of the key challenges in projects. A three-stage project supply chain framework is proposed covering procurement, conversion, and delivery; ways value can be enhanced through integration are discussed. A model (the Supply Chain Council's SOR model) is proposed for tracking supply chain performance. This operates at three levels: overall structure and performance targets based on best-in-class performance; supply chain configuration; and operational metrics such as performance, tools, processes, and practices. Ray concludes by describing the upcoming issues in project SCM as he sees them.

11. In Chapter 11, Mark Nissen bears down more specifically on project procurement. Mark takes a broad, process perspective, illustrating the customer-buying and vendor-selling activities, and in particular the role of the project manager in optimizing key "hand-offs" (friction points) from his research in the U.S. high tech sector. The project manager is presented with a dilemma Mark believes—and he is right—in being torn on the one hand to be tough and firm, and on the other to be accommodating and build synergy through trust and cooperation etc. What advice do we have, therefore, for the project manager? Mark proffers several tips: do not tinker, manage the critical path, question the matrix, benchmark, and really watch IT and software.

12. Dave Langford and Mike Murray in Chapter 12 make the discussion more specific with their analysis of procurement trends in the UK construction industry and elsewhere. They show that there have been some major shifts in project procurement practice since the early '80s (the time of Cleland and King's *Project Management Handbook*). Again, they show how much of the key procurement activity happens in the early stages of the project—not just in acquisition planning, but in the whole involvement of construction/manufacturing in the early design and definition stages (as we saw in the chapters by Ray Venkataraman and Hans Thamhain, among others). These changes in procurement practices reflect a general trend away from simple transaction-based procurement to more long-lived, relational procurement where trust and value for money (whole life— see Ive Vol. 1, Chapter 14) count for more than simply lower capital cost; these trends are accentuated by (a) the increase in technical and organizational complexity on many

projects, and (b) the increasing sophistication and active involvement of clients in the management of their projects. New forms of procurement have arisen and become increasingly dominant, the most significant being partnering and performance-based contracting.

13. George Steel, in Chapter 13, takes us through one of Mark Nissen's "friction points"—tender management—in considerable detail showing how business benefit can be obtained by clearly following established processes and practices. The key, essentially, is to build a contracting and tender management strategy that reflects the organization's values and drivers; to recognize the difference between "hard money" and "soft" (actual costs versus estimated intangibles); and, however tough the bidding process may have been, to build the supply chain synergy (team spirit etc.) once the contract is awarded.

14. David Lowe in Chapter 14 similarly keeps us at a highly practical level in his expert review of contract management. This is a vitally important aspect in the management of any project that entails third party contracts (or in-house ones for that matter!). He uses the contract form of FIDIC as his reference. Though broadly construction-oriented, it is not exclusively so, and it is used in over sixty countries. David shows how contracts deal primarily with risk identification, apportionment, and management, and relationship management. He believes a project manager should have a thorough understanding of the procurement process and post-tender negotiation; of the assumptions made by the purchaser and the supplier; of the purchaser's expectations of the service relationship; of the contract terms and conditions; and of the legal implications of the contract. To this end, he takes us on a high-level tour of contractual issues; contract types and strategy; roles, relationships and responsibilities; time, payment and change provisions; remedies for breach; bonds, guarantees and insurance; claims; and dispute resolution. He concludes with two lists of best practice guidance drawn from the Association for Project Management.

15. Change Management is one of the most important and delicate areas of project management. Poorly handled, it will lead to the project getting out of control; however, sometimes change can be for the project's benefit. Kenneth Cooper and Kimberly Sklar Reichelt examine this in Chapter 15, culling in the results of interviews with dozens of managers and using simulation models to look at the potential disruptive impacts of changes. They show why it is better for changes to be handled quickly, before their effect can begin to become cumulative, and that disruption is reduced by less tight schedules. They conclude by looking at good practice in managing change claims and disputes.

# About the Authors

## Alan M. Davis

Al Davis is professor of information systems at University of Colorado at Colorado Springs and is president of The Davis Company, a consulting company. He has consulted for many corporations over the past twenty-seven years. Previously, he was a member of the board of directors of Requisite, Inc.; Chairman and CEO of Omni-Vista, Inc., a software company in Colorado Springs; Vice President of Engineering Services at BTG, Inc., a Virginia-based company that went public in 1995, and acquired by Titan in 2001; a Director of R&D at GTE Communication Systems in Phoenix, Arizona; Director of the Software Technology Center at GTE Laboratories in Waltham, Massachusetts. He has held academic positions at George Mason University, University of Tennessee, and University of Illinois at Champaign-Urbana. He was Editor-in-Chief of IEEE Software from 1994 to 1998. He is an editor for the *Journal of Systems and Software* (1987–present). He is the author of *Software Requirements: Objects, Functions and States* (Prentice Hall, 1990 and 1993) and the best-selling *201 Principles of Software Development* (McGraw Hill, 1995). He is founder of the IEEE International Conferences of Requirements Engineering. He has been a fellow of IEEE since 1994, and earned his Ph.D. in Computer Science from University of Illinois in 1975.

## Ann Hickey

Ann Hickey is an assistant professor of information systems at the University of Colorado at Colorado Springs. She worked for seventeen years as a program manager and senior systems analyst for the Department of Defense before beginning her academic career. She teaches graduate and undergraduate courses in systems analysis and design, enterprise systems, and information systems project management. Her research interests include requirements elicitation, elicitation technique selection, collaboration, and scenario and process modeling. Her work has been published in the *Journal of Management Information Systems*, the *Database for Advances in Information Systems*, the *Journal of Information Systems Education*, the *Requirements Engineering Journal* and national and international conference proceedings. She received her B.A. in mathematics from Dartmouth College and her M.S. and Ph.D. in MIS from the University of Arizona.

## Ann Zweig

In 1997, Ms. Zweig co-founded the startup software company, Omni-Vista, in Colorado Springs, Colorado. Omni-Vista provided products and services that assist software development companies in making informed business decisions about existing or planned products and projects by performing intelligent tradeoff analyses that incorporate critical business factors as well as technology. Most recently, Ms. Zweig served as President and COO. Before becoming a computer scientist, Ms. Zweig was a biologist with The Nature Conservancy and also at the Rocky Mountain Biological Laboratory in Gothic, Colorado. Ms. Zweig served two years as a Peace Corps Volunteer in the Kingdom of Tonga where she taught biology, chemistry, and physics. Ms. Zweig has an M.S. in Computer Science from the University of Colorado and a B.S. in Biology from the University of Kansas.

## Peter Harpum

Peter Harpum is a project management consultant with INDECO Ltd, with significant experience in the training and development of senior staff. He has consulted to companies in a wide variety of industries, including retail and merchant banking, insurance, pharmaceuticals, precision engineering, rail infrastructure, and construction. Assignments range from wholesale organizational restructuring and change management, through in-depth analysis and subsequent rebuilding of program and project processes, to development of individual persons' project management capability. Peter has a deep understanding of project management processes, systems, methodologies, and the 'soft' people issues that programs and projects depend on for success. Peter has published on: design management; project methodologies, control, and success factors; capability development; portfolio and program value management; and internationalization strategies of indigenous consultants. He is a Visiting Lecturer and examiner at UMIST on project management.

## Hans J. Thamhain

Dr. Hans J. Thamhain specializes in technology-based project management. Currently a Professor of Management and Director of Project Management Programs at Bentley College, Waltham / Boston, his industrial experience includes twenty years of high-technology management positions with GTE / Verizon, General Electric and ITT. Dr. Thamhain has PhD, MBA, MSEE, and BSEE degrees. He is well known for his research on technology-based project control and team leadership. He has written over seventy research papers and five professional reference books on project and technology management. Dr. Thamhain is the recipient of the Distinguished Contribution Award from the Project Management Institute in 1998 and the IEEE Engineering Manager of the Year 2000 Award. He is certified as New Product Development Professional, NPDP, and Project Management Professional, PMP.

## Rachel Cooper

Rachel Cooper is Professor of Design Management, at the University of Salford, where she is Director of the Adelphi Research Institute for Creative Arts and Sciences, and also co-director of the EPSRC Funded Salford Centre for Research and Innovation in the Built and Human Environment. She has been undertaking design research for the past twenty years. Her work covers design management; new product development; design in the built environment; design against crime; and socially responsible design. Projects includes: Engineering & Physical Sciences Research Council study of Requirements Capture; Cost and Benefits of Partnering; Generic Design & Construction Process Protocol; Future scenarios for Distributed Design Teams; three projects for the Design Council / Home Office on Design Against Crime; and for the Design Council and government, a study of the use of Design in Government Departments; an eighteen country study of New Product Development in High Technology Industries. Professor Cooper was Founding Chair of the European Academy of Design, and is also Founding Editor of *The Design Journal*. Professor Cooper has written over one-hundred papers and six books, her latest co-authored with Professor Mike Press, *The Design Experience*, was published June 2003.

## Ghassan Aouad

Professor Ghassan Aouad is Head of the School of Construction & Property Management, and director of the Centre for Research and Innovation in the Built and Human Environment at the University of Salford. He also leads the prestigious research (from 3D to nD modeling). Professor Aouad's research interests are in: modeling and visualization, development of information standards, process mapping and improvement, and virtual organizations. He has published extensively in these areas.

## Angela Lee

Dr. Angela Lee is a research fellow at the University of Salford, and has worked on numerous projects including '3D to nD Modelling,' 'PeBBu (EU Thematic Performance Based Building Network) and Process Protocol II. Her research interests include performance measurement, process modeling, process management and requirements capture, and has published extensively in these fields. She completed a BA (Hons) in Architecture at the University of Sheffield and her PhD at the University of Salford.

## Song Wu

Song Wu is a research fellow on the 3D to nD Modelling project and previously worked on the Process Protocol II project at the University of Salford. He was trained as civil engineer and worked as a Quantity Surveyor in Singapore and China for thee years. Song was awarded an M.Sc. in Information Technology in Construction in 2000, and is currently completing his PhD in IT support for construction process management at the University of Salford. His research interests include data modeling, database management, information management and IFC (industry foundation classes) implementation.

## Alistair G F Gibb

Alistair Gibb is program director at Loughborough University of the Construction Engineering Management program, sponsored by thirteen major construction organizations. His research work falls primarily into two main areas: off-site fabrication and health and safety. In off-site fabrication, Alistair has managed a string of major research projects and has recently secured the primary academic role in *prOSPa*, the prestigious UK Government funded Pii Programme on the subject. In the health and safety area, he is Director of the *APaCHe* partnership for construction health and safety, working closely with the HSE and industrial collaborators. He also has a leading role in both European and international networks in health and safety. Since 1995, he has been project director of the European Construction Institute's Safety, Health and Environment task force, and in 2001, he joined the main board of ECI-ACTIVE as a non-executive director. He has been a member of numerous committees and task groups, including the Association of Planning Supervisors (co-opted board member). He is a member of the Conseil International de Batiment (cib) working commission on health and safety; ICBEST (the international council for building envelope); ISSA (International Social Security Association); and WSIB (Canadian Workplace Safety & Insurance Board) Research Advisory Council.

## Callum Kidd

Having worked in managing configurations for over ten years in the process and then aerospace industries, Callum Kidd moved to UMIST to set up and run the Industrial Management Centre. He later moved to Leeds University where in 1994 he established the Configuration Management Research Group, and was awarded research grants from BT, Royal Mail, European Commission, Ericsson and Vickers Defence to research the future of CM in a variety of contexts. He moved back to UMIST in 2000 to manage the Project Management Professional Development Programme, where he is currently carrying out research into the synergy between CM and PM.

## Thomas F. Burgess

Thomas Burgess is Senior Lecturer in Operations and Technology Management at Leeds University Business School. After qualifying as an engineer, Tom worked in roles connected with operations management, information systems, and project management in a number of consulting and engineering companies prior to entering academia. His MBA (Bradford) is in Production Management and Ph.D. (Leeds) in Computer Studies. His research has focused on major management-related process innovations and their impact on organizations; and the use of modeling and simulation methods to assist in understanding and supporting these new innovations. Lately, his research has centered on improving the processes for development projects in the chemicals and pharmaceuticals industry.

## Hal Mooz

Hal Mooz is founder and CEO of the Center for Systems Management—a company dedicated to training, mentoring, consulting, and culture-building in project management and systems engineering and related disciplines. His experience covers being a Chief Systems Engineer, Program Manager, and Deputy Director of Programs for intelligence satellites at Lockheed Missiles and Space Company. He is co-author of *Visualizing Project Management* published 1996 by John Wiley and Sons and is a contributing author to *The Handbook of Managing Projects* to be published by Wiley and Sons in 2004. He is a member of PMI and presenter of papers and tutorials at PMI, ProjectWorld, and international project management conferences; he is a Certified Project Management Professional (PMP) by Project Management Institute (PMI). Recertified in 2001; a Member of International Council on Systems Engineering (INCOSE) and presenter of papers and tutorials at international INCOSE conferences. Several papers have been judged best of conference; Hal Mooz is a recipient of the CIA Seal Medallion for contributions in project management, and recipient of the INCOSE 2001 Pioneer Award for furthering the cause of Systems Engineering.

## Rodney Turner

Rodney Turner is Professor of Project Management at Erasmus University Rotterdam, in the Faculty of Economics. He is also an Adjunct Professor at the University of Technology

Sydney, and Visiting Professor at Henley Management College, where he was previously Professor of Project Management, and Director of the Masters program in Project Management. He studied engineering at Auckland University and did his doctorate at Oxford University, where he was also for two years a post-doctoral research fellow. He worked for six years for ICI as a mechanical engineer and project manager, on the design, construction and maintenance of heavy process plant, and for three years with Coopers and Lybrand as a management consultant. He joined Henley in 1989 and Erasmus in 1997. Rodney Turner is the author or editor of seven books, including *The Handbook of Project-based Management*, the best selling book published by McGraw-Hill, and the *Gower Handbook of Project Management*. He is editor of *The International Journal of Project Management*, and has written articles for journals, conferences and magazines. He lectures on and teaches project management world wide. From 1999 and 2000 he was President of the International Project Management Association, and Chairman for 2001–2002. He has also helped to establish the Benelux Region of the European Construction Institute as foundation Operations Director. He is also a Fellow of the Institution of Mechanical Engineers and the Association for Project Management.

## Anne Keegan

Anne Keegan is a University Lecturer in the Department of Marketing and Organisation, Rotterdam School of Economics, Erasmus University Rotterdam. She delivers courses in Human Resource Management, Organisation Theory and Behavioural Science in undergraduate, postgraduate and executive level courses. She has been a member of ERIM (Erasmus Research Institute for Management) since 2002. In addition, she undertakes research into the Project Based Organisation and is a partner in a European Wide Study into the Versatile Project Based Organisation. Her other research interests include HRM in Knowledge Intensive Firms, New Forms of Organising and Critical Management Theory. Dr. Keegan has published in *Long Range Planning* and *Management Learning* and is a reviewer for journals including the *Journal of Management Studies* and the International Journal of Project Management. She is a member of the American Academy of Management, the European Group for Organisation Studies (EGOS) and the Dutch HRM Network. Dr. Keegan studied management and business at the Department of Business Studies, Trinity College Dublin, and did her doctorate there on the topic of *Management Practices in Knowledge Intensive Firms*. Following three years post-doctoral research she now works as a university lecturer and researcher. Dr. Keegan has also worked as a consultant in the areas of Human Resource Management and Organizational Change to firms in the computer, food, export and voluntary sectors in Ireland and the Netherlands.

## David Kirkpatrick

Professor David Kirkpatrick was trained as an aeronautical engineer (and later an economist) and had a distinguished thirty-three year career in the scientific civil service of the UK Ministry of Defence (MoD). During this period, he did research on aerodynamics and aircraft design at the Royal aircraft Establishment Farnborough, military operational analysis for the chief scientist (RAF) cost analysis and forecasting in support of defence equipment procurement. He also served for three years as an attaché in Washington DC, promoting UK / U.S. collaboration in defense technology. He retired from the MoD in 1995 and joined the Defence Engineering Group at University College London to lecture in various aspects of defense equipment acquisition, and to undertake associated research consultancy work. He was appointed to a personal chair of Defence Analysis in 1999. In addition to many technical and official papers printed within MoD, he has written for external publication over sixty papers on aerodynamics and aircraft design, the cost and effectiveness of defense equipment, defense economies, and military history. He is an independent member of the Defence Scientific Advisory Council, and a specialist adviser to the House of Commons Defense Committee. He is a Fellow of the Royal Aeronautical Society and an Associated Fellow of the Royal United Services Institute.

## Steve McInally

Dr. McInally initially was trained as an electro-mechanical engineer, working in the telecommunications industry in the mid-1970s and early 1980s. Between 1985 and 1994, he was employed by Philips Medical Systems in a variety of engineering roles, installing, commissioning, and maintaining radiotherapy equipment, then later as a requirements elicitation specialist on radiotherapy system design projects. With Philips' sponsorship, he completed his B.Sc. in Business and Industrial Systems at Leicester Polytechnic in 1992, and his Ph.D. in Instrument System Design with UCL's Defense Engineering Group in 2001. He was appointed Research Fellow at the Defense Engineering Group in 2002. Recent publications include papers on requirements elicitation in the systems engineering process; the use of heuristics in systems engineering; and introspective learning models for advanced motorcycle riding. Dr. McInally also designs on-line teaching modules for systems engineering education, acts as Rapporteur for organizations such as Royal United Services Institute and the Royal Academy of Engineering in London.

## Daniela Pridie-Sale

Daniela Pridie-Sale has studied languages, geography, and management. She commenced her career in the leisure industry. After a period of teaching, she went on to work in the oil and finance industries. More recently, she was marketing manager for an international college before joining UCL's Defense Engineering Group. She provides teaching and research support in the application of management science within systems engineering, and is currently studying for an M.Sc. in international business at Birkbeck College, London.

## Ray Venkataraman

Dr. Ray Venkataraman is an Associate Professor of Operations Management at the School of Business in Penn State University, Erie, Pennsylvania. He received his PhD in Management Science from Illinois Institute of Technology. Dr.Venkataraman also has a B.S in Chemistry from the University Madras, India, an M.B.A in Information Systems and an

M.S in Accounting from DePaul University. Dr. Venkataraman has published in *The International Journal of Production Research, Omega, International Journal of Operations and Production Management, Production and Operations Management (POMS), Production and Inventory Management, Production Planning and Control, The International Journal of Quality and Reliability Management,* and other journals. His current research interests are in the areas of Manufacturing Planning and Control Systems, Supply Chain Management and Project Management. Dr. Venkataraman is also a member of The Decision Sciences Institute, Production and Operations Management Society (POMS) and American Production and Inventory Control Society (APICS). He has served on the editorial review board of *Production and Operations Management (POMS)* Journal.

## Mark Nissen

Mark Nissen is Associate Professor of Information Systems and Management at the Naval Postgraduate School and a Young Investigator. His research focuses on the study of knowledge and systems for innovation, and he approaches technology, work and organizations as an integrated design problem. Mark's publications span the information systems, project management, and related fields, and he received the Menneken Faculty Award for Excellence in Scientific Research, the top research award available to faculty at the Naval Postgraduate School. Before his information systems doctoral work at the University of Southern California, he acquired over a dozen years of management experience in the aerospace and electronics industries, and he spent a few years as a direct-commissioned officer in the Naval Reserve.

## David Langford

David Langford has published widely; books he has contributed to include: *Construction Management in Practice, Direct Labour Organisations in Construction, Construction Management Vol. I and Vol. II, Strategic Management in Construction, Human Resource Management in Construction and Managing Overseas Construction.* He has co-edited a history of government interventions in the UK construction industry since the war. He has contributed to seminars on the field of construction management on five continents. His interests are travel, cricket, and he plays golf with more enthusiasm than skill. David Langford holds the Barr Chair of Construction in the Department of Architecture and Building Science at the University of Strathclyde in Glasgow, UK. He has published widely in the field of Construction Management and has co-authored eight books and edited one and three volumes of construction research. He is a regular visiting lecturer at universities around the world.

## Michael Murray

Mike is a Lecturer in construction management within the Department of Architecture and Building Science at the University of Strathclyde. He completed his PhD research in January 2003 and also holds a 1st class honors degree and M.Sc. in construction management. He has lectured at three Scottish universities (The Robert Gordon University, Heriot Watt, and currently at Strathclyde) and has developed a pragmatic approach to both research and lecturing. He has delivered research papers to academics and practitioners at UK and overseas symposiums and workshops. He began his career in the construction industry with an apprenticeship in the building services sector and was later to lecture in this topic at several Further Education colleges. Mike is co-editor of two text-books, *Construction Industry Reports 1944–1998* (2003) and The RIBA *Architects Handbook of Construction Project Management* (2003).

## George Steel

George Steel is the founder and Managing Director of INDECO, a management consultancy specializing in Project and Contract Management. He has personally led many international corporate value improvement initiatives, and has been responsible for developing and negotiating many major contracts. Prior to founding INDECO, George was a partner of Booz Allen & Hamilton, New York, where he worked with a number of international oil and gas companies on the development of their organization, and on the management of major development programs. Earlier in his career, George was a Project Manager with an international engineering and construction contractor designing and constructing oil refineries and LNG Projects. He has an Honors degree in Engineering from the University of Edinburgh and is a Fellow of the Association of Project Management.

## David Lowe

David Lowe is a Chartered Surveyor and a member of the Project Management, Construction and Dispute Resolution Faculties of the Royal Institute of Chartered Surveyors. He is a lecturer in Project Management at the Manchester Centre for Civil and Construction Engineering, UMIST, where he is Programme Director for the M.Sc. in the Management of Projects. He is also joint program director for a distance-learning M.Sc. in Commercial Management, a bespoke program for a blue-chip telecommunications company. Consultancy work includes benchmarking the engineering and project management provision of an international pharmaceutical company. His PhD, completed at UMIST, investigates the development of professional expertise through experiential learning. Further research projects include the growth and development of project management in the United Kingdom construction industry; an investigation of the cost of different procurement systems and the development of a predictive model; and a project to assist medium sized construction companies develop strategic partnerships and diversify into new business opportunities offered by public and private sector clients. Dr. Lowe has over thirty refereed publications.

## Kenneth Cooper

Kenneth G. Cooper is a member of the Management Group of PA Consulting and leads the practice of system dynamics within PA. His management consulting career spans thirty years, specializing in the development and application of computer simulation models to a variety of strategic business issues. Mr. Cooper has directed over one-hundred and fifty consulting engagements, among them analyses of one-hundred major commercial and defense development projects. His group's office is in Cambridge, Massachusetts, USA. Mr. Cooper received his bachelor's and master's degrees from M.I.T. and Boston University, respectively.

## Kimberly Sklar Reichelt

Kimberly Sklar Reichelt is a managing consultant in PA Consulting Group's Decision Sciences Practice. For fifteen years, she has specialized in building and using system dynamics models to aid management decision-making. While her experience has been in a variety of industries, from sports to medical to financial, she has focused in particular on project management assignments for both commercial and defence contractors. Ms. Reichelt received her bachelor's and master's degrees in Management Science from the Massachusetts Institute of Technology.

*The Wiley Guides to the Management of Projects* series offers an opportunity to take a step back and evaluate the status of the field, particularly in terms of scholarship and intellectual contributions, some twenty-four years after Cleland and King's seminal *Handbook*. Much has changed in the interim. The discipline has broadened considerably—where once projects were the primary focus of a few industries, today they are literally the dominant way of organizing business in sectors as diverse as insurance and manufacturing, software engineering and utilities. But as projects have been recognized as primary, critical organizational forms, so has recognition that the range of practices, processes, and issues needed to manage them is substantially broader than was typically seen nearly a quarter of a century ago. The old project management "initiate, plan, execute, control, and close" model once considered the basis for the discipline is now increasingly recognized as insufficient and inadequate, as the many chapters of this book surely demonstrate.

The shift from "project management" to "the management of projects" is no mere linguistic sleight-of-hand: it represents a profound change in the manner in which we approach projects, organize, perform, and evaluate them.

On a personal note, we, the editors, have been both gratified and humbled by the willingness of the authors (very busy people all) to commit their time and labor to this project (and our thanks too to Gill Hypher for all her administrative assistance). Asking an internationally recognized set of experts to provide leading edge work in their respective fields, while ensuring that it is equally useful for scholars and practitioners alike, is a formidable challenge. The contributors rose to meet this challenge wonderfully, as we are sure you, our readers, will agree. In many ways, the *Wiley Guides* represent not only the current state of the art in the discipline; it also showcases the talents and insights of the field's top scholars, thinkers, practitioners, and consultants.

Cleland and King's original *Project Management Handbook* spawned many imitators; we hope with this book that it has acquired a worthy successor.

## References

Christensen, C. M. 1999. *Innovation and the General Manager*. Boston: Irwin McGraw-Hill.
Cleland, D. I. and King, W. R. 1983. *Project Management Handbook*. New York: Van Nostrand Reinhold.

Cleland, D. I. 1990. *Project Management: Strategic Design and Implementation*. Blue Ridge Summit, PA: TAB Books.

Gray, C. F., and E. W. Larson. 2003. *Project Management*. Burr Ridge, IL: McGraw-Hill.

Griseri, P. 2002. *Management Knowledge: a critical view*. London: Palgrave.

Kharbanda, O. P., and J. K. Pinto. 1997. *What Made Gertie Gallop?* New York: Van Nostrand Reinhold.

Meredith, J. R., and S. J. Mantel. *Project Management: A Managerial Approach*. 5th Edition. New York: Wiley.

Morris, P. W. G., and G. H. Hough. 1987. *The Anatomy of Major Projects*. Chichester: John Wiley & Sons Ltd.

Morris, P. W. G. 1994. *The Management of Projects*. London: Thomas Telford; distributed in the USA by The American Society of Civil Engineers; paperback edition 1997.

Morris, P. W. G. 2001. "Updating the Project Management Bodies Of Knowledge" *Project Management Journal* 32(3):21–30.

Morris, P. W. G., H. A. J. Jamieson, and M. M. Shepherd. 2006. "Research updating the APM Body of Knowledge 4th edition" *International Journal of Project Management* 24:461–473.

Morris, P. W. G., L. Crawford, D. Hodgson, M. M. Shepherd, and J. Thomas. 2006. "Exploring the Role of Formal Bodies of Knowledge in Defining a Profession—the case of Project Management" *International Journal of Project Management* 24:710–721.

Pinto, J. K., and D. P. Slevin. 1988. "Project success: definitions and measurement techniques," *Project Management Journal* 19(1):67–72.

Pinto, J. K. 2004. *Project Management*. Upper Saddle River, NJ: Prentice-Hall.

Project Management Institute. 2004. *Guide to the Project Management Body of Knowledge*. Newtown Square, PA: PMI.

Project Management Institute. 2003. *Organizational Project Management Maturity Model*. Newtown Square, PA: PMI.

Wheelwright, S. C. & Clark, K. B. 1992. *Revolutionizing New Product Development*. New York: The Free Press.

CHAPTER ONE

# REQUIREMENTS MANAGEMENT IN A PROJECT MANAGEMENT CONTEXT

Alan M. Davis, Ann M. Hickey, and Ann S. Zweig

Project success is the result of proper planning *and* proper execution. Fundamental to proper planning is making sure that the work to be performed by the project is well understood and that the amount of work is compatible with available resources. Requirements management is all about learning and documenting the work to be performed by the project, and ensuring compatibility with resources. A well-executed on-time project that does not meet customer needs is of no use to anybody.

## Requirements

Requirements define the desired behavior of a system[1] to be built by a development project. More formally, a *requirement* is an externally observable characteristic of a desired system. The two most important terms of this definition are *externally observable* and *desired*. Externally observable implies that a customer, user, or other stakeholder is able to determine if the eventual system meets the requirement by observing the system. Observation here could encompass using any of the five senses, as well as any kind of device or instrument. Next, a requirement must state something that is desired by some stakeholder of the system. Stakeholders include all classes of users, all classes of customers, development personnel, managers, marketing, product support personnel, and so on. It is not so easy to determine

---

[1]A *system* is any group of interacting elements that together perform one or more functions. The elements could be electronic hardware, mechanical devices, software, people, and/or any physical materials.

if a candidate requirement is a valid requirement from this perspective. In fact, the only way to make the determination is to ask the stakeholders. The word *desired* was chosen purposefully and is meant to encompass both wants and needs (see *Wants vs. Needs* later in the chapter).

## Requirements Management

This chapter is all about how project managers and analysts manage requirements. *Requirements management* is the discipline of

- learning what the candidate requirements are—the learning aspects of requirements management are generally called *elicitation*;
- selecting a subset of those candidate requirements that are compatible with the project's goals, budget, and schedule—the selecting aspects of requirements management are generally called *triage*;
- documenting the requirements in a fashion that optimizes communication and reduces risk—the documenting aspects of requirements management are generally called *requirements specification*; and
- managing the ongoing evolution of those requirements during the project's execution.

On large projects, the individuals who perform requirements management are generally called analysts, requirements analysts, requirements managers, requirements engineers, systems analysts, business analysts, problem analysts, or market analysts. In companies that mass-market the products of their development projects, these individuals are generally within the marketing organization of the company. In companies that build custom products for their customers, these individuals are generally within either the marketing or the development organizations of the company. In IT organizations where the products of development projects are used within the company, these individuals are within the IT organization itself and interface with the internal customers, or are within the internal customer organization and interface with the IT organization.

On smaller projects, the project manager often performs a majority of the requirements management activities because these strategic activities are so critical to project success.

## Requirements Management and Project Management

Much of requirements management can be thought of as part of (or preceding) project planning, because one goal of requirements management is the decision concerning *what* system is to be built. However, because needs of customers are often in constant flux, requirements must be addressed throughout the project. At project inception, the project manager is often intimately involved in defining requirements. Because any subsequent change to requirements affects project scope, the project manager tends to stay involved in the requirements management process throughout development.

Project management of requirements activities is unique among most project responsibilities because of two factors: (1) the strong customer focus and (2) the "softness" of the discipline. In most aspects of project management, the constraints upon the task are predefined, known, and finite. The project manager's job is to control the project in such a way that the short-term and long-term project goals are achieved. In the case of requirements, none of that is true. The stakeholders who are the source of the requirements may not be available when needed. Even worse, their needs are constantly in flux. The very act of asking the stakeholders for their needs induces the stakeholders to conceive of new requirements hitherto not thought of. Every time a requirement is stated, the stakeholders will think of many more. Every time a prototype is constructed and demonstrated to the stakeholders, they will think of dozens of additional requirements. The phenomenon is likened to a continuous application of Maslow's hierarchy of needs. Every time any need is satisfied, more needs appear. Thus, the actual performance of requirements management causes the project to expand in scope.

Most activities being planned, controlled, and monitored by project management tend to appeal to the left side of the brain. Everything is (or should be) well defined, concrete, measurable, and to a large degree controllable. Requirements management requires a large dose of both left-side and right-side brain function. For example, the skills required to perform requirements elicitation primarily reside in the right side of the brain. Such skills deal with communication, feeling, and listening. On the other hand, the skills needed to record and manage the changes to requirements (including the use of so-called requirements management tools) reside primarily in the left side of the brain. These skills deal with specification, attention to detail, and precision. For this reason, requirements management is more like project management than like the other tasks performed by the individuals reporting to the project manager. Requirements management, like project management, require a very diverse set of skills.

# Types of Requirements

We defined a requirement as an externally observable characteristic of a desired system. Although this sounds fairly specific, in practice requirements come in a wide variety of flavors and serve a wide variety of purposes. The following sections describe some of this richness.

## User/Customer vs. System (Problem vs. Solution)

Some authors demand that requirements describe a problem purely from the perspective of the customer and must omit any reference to any solution system. Other authors demand that requirements specifically describe the external behavior of the solution system itself (IEEE, 1993). We have found that most practitioners divorce themselves from either extreme and recognize that as the requirements process proceeds, requirements naturally evolve from descriptions of the problem to descriptions of the solution. When requirements are stated in terms of the problem without reference to a solution, they look like this:

We need to reduce billing errors by 50 percent.

When requirements are stated in terms of the external behavior of the solution, they look like this:

The system shall provide an "audit" command, which verifies the accuracy of bills.

There is only a fine line separating the problem and the solution. In the preceding examples, one *could* argue that the former is actually within the solution domain. After all, reducing billing errors is just one way of trying to accomplish some real goal, such as increasing collections, increasing revenue, or maximizing cash flow.

Lauesen (2002) differentiates between user requirements and system or software requirements. He states that user requirements are supposed to address just the needs of the user, and system or software requirements are supposed to address the expected behavior of the solution system. However, he also correctly points out that in practice, most requirements describe external behavior of the solution system anyway, and that the term user requirements is generally applied to any requirements that are written in a language that users can understand.

## Systems of Systems vs. Single Systems

By their very nature, systems are composed of other systems, as shown in Figure 1.1. For such systems, requirements are written for every system, usually starting with the top one. When requirements are written for the topmost system, they are written from a perspective outside that system, thus ensuring that all its requirements are externally observable. After these requirements are documented in a *system requirements specification*, system design (generally not considered part of requirements) is performed to decompose the system into its constituent subsystems and then to document those subsystems. Then requirements are written for

### FIGURE 1.1. SYSTEMS ARE COMPOSED OF SYSTEMS.

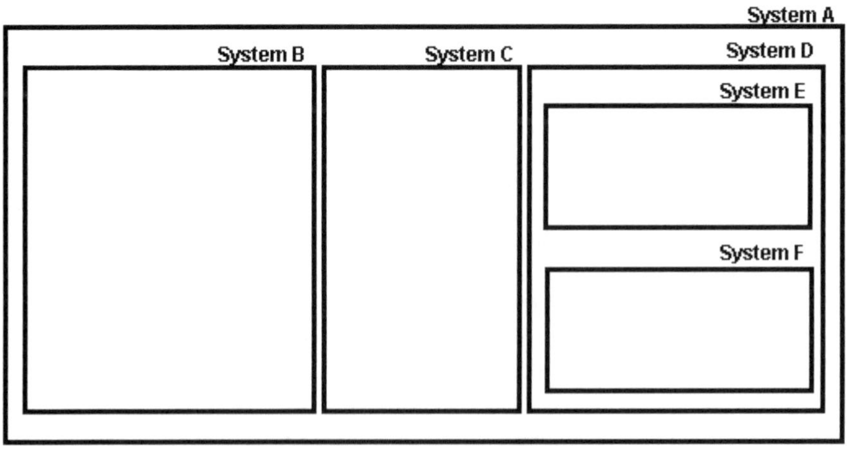

each subsystem, from a perspective outside each of those subsystems, and the process repeats itself. As we get toward the lower-level systems, the system requirements are often replaced with two documents, a *software requirements specification* and a *hardware requirements specification*, each of which defines the requirements for its part of the system.

When a system is simple enough to not require decomposition into subsystems, the most common approach is to write a *system requirements specification* for the overall system, allocate each of the requirements to either software or hardware or both, and then proceed to write a software requirements specification and a hardware requirements specification.

When a system is composed entirely of either software or hardware, just one document is usually written—either a software requirements specification or a hardware requirements specification.

## Primary vs. Derived

Thayer and Dorfman (1994) differentiate between requirements that are defined initially and requirements that are derived from those original requirements because of design decisions. For example, once the decision is made to include this requirement:

The system shall provide service $x$ to the customers.

it becomes evident that we must also include this requirement:

The system shall bill the customers for using service $x$.

## Project vs. Product

IEEE Standard 830 (1993) and Volere (Robertson and Robertson, 2000) make a clear distinction between requirements that constrain the solution system itself, for instance:

When the button is pressed, the system shall ignite the light.

and requirements that constrain the project responsible for creating the product, for instance:

The product must be available for commercial sale no later than April 2004.

IEEE Standard 830 calls the former *product requirements* and the latter *project requirements*. Volere differentiates between two types of *product* requirements: functional and nonfunctional; and three types of *project* requirements: project constraints, project drivers, and project issues.

Much agreement exists in the industry that product requirements are requirements, but little agreement exists concerning whether project requirements are really requirements. We happen to believe they are not requirements, but it is only a semantic issue. The fact is that during requirements activities, the team *will* need to perform trade-off analyses between both types of "requirements."

## Behavioral vs. Nonbehavioral

Some requirements describe the inputs into and the outputs from a system, and the relationships among the inputs and outputs. Others describe general characteristics of the system without defining inputs, outputs, and their interrelationships—that is, the functions that the system is intended to support. The former requirements are called *behavioral requirements*, although they have also been called *functional requirements* by the Robertsons (2000) and Davis (1993). The latter requirements are called *nonbehavioral requirements*, although they have also been called *developmental quality requirements* by Faulk (1997) and by the quite ambiguous and almost deceptive term *nonfunctional requirements*, by the Robertsons (2000) and Davis (1993).

Following are examples of behavioral requirements:

When the button is pressed, the system shall ignite the light. If the power is on and the on-off button is pressed, the system shall turn power off. When the user enters the command *xyz*, the system shall generate the report shown in Appendix H.

Examples of nonbehavioral requirements include all aspects of performance, reliability, adaptability, throughput, response time, safety, security, and usability, and they include such requirements as the following:

The system shall handle up to 25 simultaneous users. All reports shall be completely printed by the system within five minutes of the request by the user. The user interface shall conform to Microsoft standard *xxx*.

## Wants vs. Needs

Many requirements writings seem to imply that one of the responsibilities of the analyst is to remove from consideration any requirements that are deemed to be "wants" rather than "needs" of the customers/users (IEEE, 1983; Swartout and Balzer, 1982; Siddiqi and Shekaran, 1996). Common wisdom and experience contra indicates this. Marketing studies have shown that people decide to buy or use a system because it satisfies their wants as well as their needs.

## Requirements vs. Children of Those Requirements

When requirements are documented, they often are recorded more abstractly than is desirable, for example,

The system shall be easy for current system users to use.

This may be sufficient for early discussions, but it must be refined before the parties should agree to the effort. The most common way to do this is to document the refined requirements as subrequirements of the parent requirement, as in the following:

The system shall be easy for current system users to use.

(a) The system shall include conventional keyboard and mouse.
(b) The system shall exhibit the same "look and feel" of the existing legacy system.

Requirements should be refined whenever a discussion arises concerning the meaning or implications of a requirement.

### Original Requirements vs. Modified Requirements

According to Standish Group Reports (1995), 58 percent of all requirements defined for software-based systems will change during the development process. According to Reinertsen (1997), a similar rate of change occurs for all products in general. This constant flux requires us to recognize that requirements evolve not only toward increasing detail but also toward altered functionality. We must clearly differentiate between requirements that were originally documented and requirements that become apparent only after development began.

### Requirements in One Release vs. Requirements in Another

Almost all products evolve. Many requirements stated for, and implemented in, release $n$ will undergo change in subsequent releases. This observation makes it clear that we must record the relationship between specific requirements and specific product releases.

## Requirements Activities

Three distinct types of activities are performed under the auspices of requirements: elicitation, triage, and specification. The following subsections elaborate on these.

### Elicitation

The first major set of activities within requirements management is called *elicitation*. Elicitation is the process of determining who the stakeholders are and what that they need—in other words, what their requirements are. Some of these needs can be "gathered"—that is, they are known and understood by the stakeholders, and all the analyst needs to do is "pick them up" from the stakeholder. Others may surface only as the result of stimulating the stakeholders; this type of activity most closely corresponds to the dictionary definition of "elicitation." Other requirements need to be learned through study, experimentation, reading, or consultation with subject matter experts. Still others are discovered via observation. Regardless of the process used, and regardless of what the activity is called, the analysts must find out what the stakeholders needs are. Elicitation includes not just obtaining the needs but also analyzing and refining those needs to improve the team's understanding of them. Once elicited, analyzed, and refined, these needs should be recorded as a list of candidate requirements, as shown in Figure 1.2

## FIGURE 1.2. ELICITATION CREATES A LIST OF CANDIDATE REQUIREMENTS.

The user starts the RLM by placing it within the border of a defined lawn and pressing BEGIN MOWING from the Main Menu.

The RLM shall determine if it is in a defined lawn. If not, the RLM shall sound the error tones and display the message MOWER NOT IN RECOGNIZED LAWN on the first line and RETURN on the second line.

If correctly placed, the RLM shall beep once and wait for the user to step back beyond the safe distance range. After the user has moved beyond this range, the RLM shall move to a starting location within the lawn and begin mowing.

While mowing, the RLM's panel shall display nothing except in the event of an error condition, dump or refueling required, or an obstacle comes within the minimum safe distance.

The RLM shall check the grass height, grass type, grass density, and moisture of the lawn to determine the settings proper for cutting. Adjustments to the blade position and speed shall be made as required. When a swath is properly cut, the RLM shall move to an uncut area.

The cutting pattern shall begin with the perimeter of the lawn and work inward to the lawn's center. Each pass shall overlap the previous pass by a width less than or equal to 33% of the RLM's swath but greater than or equal to 25% of the RLM's swath.

This normal cutting pattern may be altered by obstacle avoidance maneuvers but shall resume when avoidance maneuvers are complete.

During avoidance maneuvers, the RLM may, for the sake of fuel efficiency, temporarily shut off its blades if over an area that has been properly cut. Obstacle avoidance is discussed in Requirement 510.

The RLM shall shut off the blades if fouling occurs to the degree that the RLM may damage itself. Should blade fouling occur, the RLM shall sound the error tones and display the message BLADES FOULED on the first line of the display. Should there be more than one blade . . .

The individual who conducts elicitation is generally called an *analyst*. An experienced analyst is adept at using a wide variety of elicitation techniques and possesses the sensitivities and skills necessary to assess the political, technical, and psychological characteristics of a situation to determine which elicitation technique to apply (Hickey and Davis, 2003; and Hickey and Davis, 2003a). Some of the classic techniques used during elicitation are as follows:

- *Interviewing* is the process of repeatedly prompting one or more stakeholders to verbalize their thoughts, opinions, concerns, and needs. The most effective prompts are open-ended questions, which force the stakeholder to think and respond in nontrivial ways. For example, prompts such as these are open-ended: "Would you please elaborate upon the problems you are experiencing now?" and "Why do you consider this a problem?" Other important aspects of effective interviewing include listening, taking notes, and playing back what you heard to verify that it was what was intended. Because over half of communication among individuals is nonverbal (Knapp and Hall, 1997), face-to-face interviewing is best. However, interviewing can also be performed over a telephone, though less efficiently. Gause and Weinberg (1989) provide a wealth of ideas on how to perform interviewing.
- *Brainstorming* is the process of gathering multiple stakeholders in a room, posing an issue or question, encouraging the stakeholders to express their ideas aloud, and having those ideas recorded somehow. The reason for demanding that ideas be expressed aloud is to encourage people to piggyback their own ideas on top of others' ideas. Criticism is generally discouraged. A wide range of variations of such meetings exists. Some variations

enforce anonymity via a tool; some have stakeholders record their own ideas, while others utilize a single scribe to record all ideas; and some discourage voicing the ideas aloud.

- *Conducting collaborative workshops* involves gathering multiple stakeholders together in structured, facilitated workshops to define the requirements for a system. Workshops may run from several hours to several days. During the workshops, facilitators lead stakeholders through a series of preplanned activities designed to produce the requirements deliverables needed. For example, participants may brainstorm on a variety of issues; create or review models, prototypes, or specifications; or negotiate and prioritize requirements. JAD (Wood and Silver, 1995) is probably the most widely known type of collaborative workshop, but there are many other variations, some of which use collaborative tools to increase efficiency (Dean et al., 1997). Gottesdeiner (2002) provides the best compendium of ideas on how to use collaborative group workshops for requirements elicitation.

- *Prototyping* is the process of creating a partial implementation of a system, demonstrating it to stakeholders, and perhaps allowing them to play with it. The bases for prototyping are that customers (a) can often think of new requirements only when they can visualize more basic requirements and (b) often can identify what they don't want more easily than what they do want. Davis (1995) provides the best overall summary of prototyping techniques and effects.

- *Questionnaires* are composed of series of questions that are then distributed to many stakeholders. Their responses are then collected, compiled, and analyzed to arrive at an understanding of general trends among the stakeholders' opinions. Unlike interviews and brainstorming, questionnaires assume that the relevant questions can be articulated in advance. For this reason, they are most effective at confirming well-formulated hypotheses concerning requirements, rather than assisting with the requirements synthesis process itself.

- *Observation* is an ethno-methodological technique where the analyst observes the users and customers performing their regular activities. In such cases, the analyst is passive and aims to not affect the activities in any way. It is the ideal technique for uncovering tacit knowledge possessed by the stakeholders. The best survey of techniques involving observation can be found in Goguen and Linde (1993).

- *Independent study* includes reading about problems and solutions, performing empirical studies, conducting archeological digs (Booch, 2002), or consulting with subject matter experts. Independent study is effective when others have addressed a similar problem before but the problem is relatively new to you.

- *Modeling* involves the creation of representations of the problem or its solutions in a notation that increases communication and provides fresh insights into the problem or solution. A wide range of modeling approaches exist, including object diagrams, data flow diagrams (DFD), the Unified Modeling Language (UML), Z, finite-state machines (FSMs), Petri nets, the System Description Language (SDL), statecharts, flowcharts, use cases, decision tables and trees, and so on. See (Davis, 1993; Kowal, 1992; Wieringa, 1996) for descriptions of most of these modeling notations. Although each provides the analyst with unique insights into the problem or its solution, the largest benefit often comes from using more than one. This is because each induces the analyst to ask (or answer) a certain class of questions, and the combination of multiple models induces more questions than the sum of using each one separately.

## Triage

It is a rare project that has sufficient resources to address all the candidate requirements. To overcome this problem, project managers or teams need to conduct a scoping exercise typically called *triage*. Triage is the process of determining the appropriate subset of candidate requirements to attempt to satisfy, given a desired schedule and budget (Davis and Zweig, 1990; Davis, 2003). It is an activity conducted for an individual project that is quite similar to the performance of portfolio management, in which a set of projects are competing for the same finite set of resources and the project manager must choose from among them. See Chapter 2 in Meredith and Mantel (2003).

Triage is conducted in a formal meeting, usually led by the project manager, product manager, or independent facilitator. The participants must include representatives of at least three groups:

- *Primary stakeholders* need to determine the relative priority of candidate requirements and ensure that the voices of all classes of users and customers are expressed. Ideally, these representatives should be customers and users themselves, but often they are composed of marketing personnel, analysts, or subject matter experts.
- *Development* needs to be present to ensure that the requirements selected for inclusion in any release are reasonable relative to the realities of schedule and budget demands.
- *Financial support* must also be present. Otherwise, it is too easy for the other two parties to solve the triage problem by simply increasing available budgets.

Triage can be conducted by viewing the problem as one of balancing a multiarmed seesaw (see Figure 1.3). The three arms are the selected candidate requirements, the available budget, and the desired schedule. These three variables must be repeatedly manipulated until they are in balance. In this case, balance implies that there is a reasonably acceptable probability that the selected requirements can be satisfied by the project within the budget and schedule. Although the traditional development project manager's goal is to ensure

### FIGURE 1.3. TRIAGE BALANCES A SEESAW.

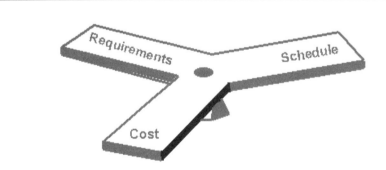

completion on schedule and within budget, an even more responsible project manager takes a larger view. Just because the selected requirements *can* be built within the budget and schedule constraints does not mean that the project *should* be undertaken. A responsible project manager thus considers additional arms of the seesaw, which capture the risks associated with and the effect on achievement of business goals of the selected requirements. Thus, if the product is to be sold externally, additional arms include aspects of marketing, finance, personnel, and other factors as shown in Figure 1.4, adapted from Chapter 2 of Meredith and Mantel (2003). If the product is to be used internally, fewer factors must be considered, as shown in Figure 1.5.

The result of triage is a pruned version of the list shown in Figure 1.4. Although most practitioners think of this as a pruned list, a more reasonable way to visualize it is as the full original list, with each requirement annotated by whether or not it is included in the next release, as shown in Figure 1.6.

## FIGURE 1.4. ADDITIONAL SEESAW ARMS.

- **Marketing Factors**
    - Size of Potential Market
    - Likely Market Share
    - Time Entering Market Window
    - Impact on Existing Products
    - Consumer Acceptance
    - Estimated Life of Product
    - Spin-Off Potential
    - Degree to Which We Understand Market
- **Financial Factors**
    - Revenue Expectation
    - Profitability (Net Present Value)
    - Cash Flow Impact
    - Payout Period
    - Cash Requirements
    - Time to Breakeven
- **Personnel Factors**
    - Training Needs
    - Labor Skill Needs
    - Level of Resistance

- **Other Factors**
    - Impact from Government Standards
    - Impact on Other IT Systems
    - Reaction from Stockholders (if a Corporation)
    - Reaction from Securities Markets (if Publicly Held Company)
    - Patent and Trade Secret Protection
    - Potential for New Patent Creation
    - Impact on Brand
    - Impact on Image with Customers and Competitors
    - Degree to Which We Understand New Technology
    - Ability to Direct and Control New Process
    - Experience We Gain from this Project to Be Applied to Future Projects
    - Average Order Size

*Source:* Adapted from Meredith and Mantel, 2003.

## FIGURE 1.5. ADDITIONAL SEESAW ARMS FOR INTERNAL DEVELOPMENT.

- **Demand Factors**
  - Size of Potential Use
  - Customer Acceptance
  - Estimated Life of Product
- **Financial Factors**
  - Increased Revenue Expectation
  - Decreased Cost Expectation
  - Cash Flow Impact
  - Payout Period
  - Cash Requirements
- **Personnel Factors**
  - Training Needs
  - Labor Skill Needs
  - Level of Resistance

- **Other Factors**
  - Impact on Other IT Systems
  - Degree to Which We Understand New Technology
  - Ability to Direct and Control New Process
  - Experience We Gain from this Project to Be Applied to Future Projects

## Specification

Once a subset of requirements is selected and agreed to by all parties, those requirements need to be refined and documented. This process is often called *requirements specification*.

***Forms of Specification.*** A variety of common practices exist in the industry for documenting requirements, including the following:

- *A polished word-processed document.* Such a document typically follows one of the many standards available in the industry (e.g., IEEE, 1993 and Robertson and Robertson 2000) and is typically called a *software requirements specification* (SRS). Like all technical documents, it is composed of chapters and paragraphs. The biggest advantage of this approach is that all parties can read the document with a minimum of training. On the other hand, the biggest disadvantages are that (a) often many resources are expended polishing the noncritical parts of the document, (b) triage is almost impossible, (c) natural language can prove to be ambiguous, and (d) it is awkward to annotate each requirement *in situ*. This is a popular approach for constructing large embedded real-time critical applications, where "critical" usually means *life*-critical, *financial*-critical, or *security*-critical.
- *A hierarchical list of requirements.* Whether the list is packaged within the constraints of a formal SRS or not, it appears as a two-dimensional table, with each row corresponding to a single requirement and each column corresponding to an attribute of that requirements, including a unique identifier, the text, the priority, estimated development cost, and so on. The biggest advantages of this approach are that (a) all parties can read the list with a minimum of training, (b) fewer words means less time spent polishing, (c) triage

# FIGURE 1.6. TRIAGE CREATES A LIST OF SELECTED CANDIDATE REQUIREMENTS.

The user starts the RLM by placing it within the border of a defined lawn and pressing BEGIN MOWING from the Main Menu.　NO／YES

The RLM shall determine if it is in a defined lawn. If not, the RLM shall sound the error tones and display the message MOWER NOT IN RECOGNIZED LAWN on the first line and RETURN on the second line.　NO／YES

If correctly placed, the RLM shall beep once and wait for the user to step back beyond the safe distance range. After the user has moved beyond this range, the RLM shall move to a starting location within the lawn and begin mowing.　NO／YES

While mowing, the RLM's panel shall display nothing except in the event of an error condition, dump or refueling required, or an obstacle comes within the minimum safe distance.　NO／YES

The RLM shall check the grass height, grass type, grass density, and moisture of the lawn to determine the settings proper for cutting. Adjustments to the blade position and speed shall be made as required. When a swath is properly cut, the RLM shall move to an uncut area.　NO／YES

The cutting pattern shall begin with the perimeter of the lawn and work inward to the lawn's center. Each pass shall overlap the previous pass by a width less than or equal to 33% of the RLM's swath but greater than or equal to 25% of the RLM's swath.　NO／YES

This normal cutting pattern may be altered by obstacle avoidance maneuvers but shall resume when avoidance maneuvers are complete.　NO／YES

During avoidance maneuvers, the RLM may, for the sake of fuel efficiency, temporarily shut off its blades if over an area that has been properly cut. Obstacle avoidance is discussed in Requirement 510.　NO／YES

The RLM shall shut off the blades if fouling occurs to the degree that the RLM may damage itself. Should blade fouling occur, the RLM shall sound the error tones and display the message BLADES FOULED on the first line of the display. Should there be more than one blade . . .　NO／YES

can be performed easily, and (d) it is trivial to annotate the requirements. On the other hand, the biggest disadvantage is that natural language can prove to be ambiguous.

- *Few or no documented requirements.* In this scenario, documentation of requirements is seen as a detractor from getting the product out. In effect, the code is the requirements, or more correctly, the code implies the requirements. The biggest advantage of this approach is that (in theory) no time is required to write or review the requirements, and thus total development time can be reduced by, say, 15 percent. However, this advantage does not come without the considerable risk of building the wrong product altogether. The proponents of this approach possess a variety of motivations. For example, some of those in the entrepreneurial world feel that getting to market fast with an innovative product is so critical to its market success they cannot afford to spend the time "investigating" the requirements—and they may be right! Meanwhile, those in the agile development community (Cockburn, 2002; Highsmith and Cockburn, 2001) claim that they build such small increments of the product, and if they make a mistake in such an iteration, it is easy to back it out and try again. Justification for recording requirements can be found in Hoffman and Lehner (2001).
- *The model is the requirements.* In some industries, requirements are not documented in natural language but are instead captured adequately in a model (see previous discussion of models). For example, in some business applications, a majority of the requirements can be captured using use cases, data flow diagrams, and entity relation diagrams. In some user-interface-intensive applications, a majority of requirements can be captured using use cases. And in some real-time systems, a majority of the requirements can be captured using Petri nets, finite-state machines, or statecharts. The unified modeling language (UML; Booch, 1999) is an attempt to capture all these models in one notation. The biggest advantage to this approach is that systems people (on the IT side and the customer side) can read the notations easily. The biggest disadvantages are that (a) nonsystems people on the customer side have difficulty understanding the notations; (b) no model is sufficient to represent *all* requirements, so they must be augmented in some way (for example, few of the aforementioned notations provide the ability to capture nonbehavioral requirements as described previously); (c) triage is likely to be difficult; and (d) it is almost impossible to annotate individual requirements.
- *The prototype is the requirements.* In this case, a prototype system is constructed and the customer likes it. Then the real system is constructed to mimic the behavior of the prototype. The biggest advantage to this approach is that customers can witness the intended system's behavior first hand. The biggest disadvantages are that, (a) by definition, a prototype does *not* exhibit all the behaviors of the real system, so it must be augmented in some way, (b) triage is likely to be difficult, and (c) it is almost impossible to annotate individual requirements.

All of the approaches can be followed in an incremental manner (i.e., document a little, build a little, validate a little, then repeat) or a full-scale manner (i.e., document a lot, build a lot, validate a lot). Table 1.1 summarizes the advantages and disadvantages of the five approaches. In this table, notice that just because a technique has more check marks in its

## TABLE 1.1. DISADVANTAGES OF VARIOUS REQUIREMENTS DOCUMENTATION APPROACHES.

| Disadvantages | Documentation Approach | | | | |
|---|---|---|---|---|---|
| | Document | List | Few/None | Model | Prototype |
| **Natural language is inherently ambiguous** | √ | √ | | | |
| **Challenging for multinational efforts** | √ | √ | | | |
| **Notation not already known by customer** | | | | √ | |
| **Difficult to annotate individual requirements** | √ | | √ | √ | √ |
| **Difficult to select subset of requirements for inclusion** | √ | | √ | √ | √ |
| **Insufficient to represent *all* requirements** | | | | √ | √ |
| **Could imply unintentional requirements** | | | | | √ |
| **High risk of building the wrong product** | | | √ | | |
| **Risk of incurring unnecessary up-front (perhaps nonrecoverable) costs** | √ | √ | | √ | √ |
| **Could be challenging to maintain** | √ | √ | | √ | √ |
| **Difficult to trace to origins and be traced from downstream entities** | √ | | | √ | √ |
| **Difficult to diagnose reasons for misunderstandings** | | | √ | | |

column does not necessarily make it a worse approach; each comes with its own inherent risks. Only the project manager can decide which risks are worth taking.

As requirements are documented using any of the precedinig approaches, disagreements will naturally arise concerning what individual requirements mean. In such cases, three solutions exist: (a) document the requirement in less ambiguous terms but using the same general approach, (b) supplement the requirement with another approach that has less ambiguity, and (c) refine the requirement into its constituent subrequirements, as described previously.

***Attributes of a Specification.*** As work proceeds on requirements, they should evolve toward increased value to the project team. That means they should become less ambiguous, more correct, more consistent, and more achievable. For a more complete list of attributes that requirements should exhibit see Davis (1995). The activities involved in determining if the requirements are evolving toward increased quality are generally called *validation and verification*, or V&V for short (Wallace, 1994). There appears to be some confusion within the industry concerning the differences between the two terms as applied to requirements, for example, see Christensen and Thayer (2001), Leffingwell and Widrig (2000), Wiegers (1999); and Young (2001). The confusion arises from the use of the terms in latter phases of system development. In later phases, *verification* of that phase's output is the process of ensuring that the output is correct relative to the outputs of the previous phase, and *validation* of that phase's output is the process of ensuring that the output is correct relative to the requirements (IEEE, 1986). Since requirements are usually considered the first phase of a system development life cycle, those definitions do not apply. However, if you consider that these words imply that verification ensures that the product is being built right and validation ensures that the right product is being built (Boehm, 1982), then we can extrapolate their meanings to requirements, as follows:

- *Requirements verification* ensures that the requirements themselves are written in a quality manner.
- *Requirements validation* ensures that the requirements as documented reflect the actual needs of the users/customers.

Then, to *verify* the quality of requirements, the following attributes must be addressed:

- *Ambiguity* is the condition in which multiple interpretations are possible given the identical requirement. Ambiguity is inherent to some degree in every natural-language statement. Thus, the parties can easily spend their entire project budget attempting to remove every bit of ambiguity. A more successful project will reword or refine a requirement only when the potential for adverse consequences is evident if the requirement stays as is. Another way to decide on whether a requirement statement is "good enough" is to determine if *reasonable, knowledgeable, and prudent* individuals would make different interpretations of the requirement.
- An SRS is *inconsistent* if it contains a subset of requirements that are mutually incompatible. For example, if two requirements are incompatible, or are in conflict with each other, then the SRS is inconsistent. Furthermore, an SRS should also be consistent with all other documents that have been previously agreed to by the parties.
- Requirements should also be *achievable*, which means it is possible to build a system with available technology, and within existing political, cultural, and financial constraints.

To *validate* requirements, the following attribute must be addressed:

- A requirement is *correct* if it helps to satisfy some stakeholder's need. Obviously, if a candidate requirement fails this test, it should be triaged out of the product.

# Variations of Requirements Management Practices

Requirements management practices vary based on many aspects of the project. Let's look at some of these aspects and see how they effect requirements management.

## Size of iterations

All product development efforts are iterative because as soon as customers start using any product, new requirements appear, thus driving another iteration. The differences lie in how big each iteration is and whether or not the team tries to satisfy "all the known requirements" in each iteration. As iterations increase in size (either in terms of elapsed time or sheer number of requirements), risks increase. In particular, the risks that increase include the likelihood of exceeding the budget, of completing after the desired delivery date, and of failing to meet customer needs. On the other hand, as iterations decrease in size, the effort for overhead tasks become a larger proportion of the total effort. With larger iterations, more effort must be expended during the requirements phases of each iteration.

## Relationship of Iterations to Planning

In some cases, an entire product's requirements are explored and documented at project inception, and a product rollout strategy is developed that incorporates successively larger subsets of requirements in each iteration. In other cases, limited requirements activity occurs up front. The initial product is released primarily to acquire requirements feedback. Each successive iteration's requirements are defined based on the feedback acquired from the previous iteration.

## Use of Throwaway Prototypes

Any iteration can be prefaced with the construction of a prototype. The purpose of the prototype is to remove the risk of building the wrong iteration. By seeing a prototype, stakeholders can provide valuable feedback concerning whether or not the development team is on the right track. Such an approach reduces the risk of the next iteration. When a prototype is used, minimal requirements effort is expended at project inception. Most requirements are defined after the initial prototype but before the development for the first real iteration begins.

## Manufacturing Needed

Some systems require a manufacturing phase after development. This is primarily a function of the media involved. Pure software systems require no manufacturing (other than the trivial creation of CD-ROMs), whereas systems that include physical components do. When manufacturing is required, care must be taken during requirements elicitation and specification to ensure manufacturability and testability.

## Research Needed

Some systems require research, invention, or innovation prior to starting the development activities. Usually, requirements are difficult to express when innovative research is needed. In such cases, a set of goals is stated (which are rarely termed requirements). Then the research is performed. Requirements efforts do not commence in earnest until after the research effort is complete.

## Management Demand for Sequentiality.

If management enforces the idea that no task may be started until the previous task is completed, then elicitation must be completed before triage begins, and triage must be completed before specification can begin. Only the most conservative of management organizations still adhere to this ancient custom.

## Iterative Nature of Requirements Process Itself

Hickey and Davis (2002) describe requirements as an iterative process where each iteration uncovers additional requirements, and changes the current situation. These changes to the situation, and the new requirements uncovered, drive the analysts to modify their approach for the next iteration. This is a more realistic view of the requirements process than attempting to do all elicitation on one phase.

## Software-Intensive Applications

Traditionally, software had been developed using large iterations, with all the planning up front, with the assumption of high sequentiality. This approach was termed the *waterfall* model. It is typically represented by a linear PERT chart, as shown in Figure 1.7. Figure 1.8 shows where the requirements activities are performed during the development.

More modern software development projects use the so-called iterative model of software development (also called incremental). There are two general ways to plan the requirements for each iteration: by fixed time and by logical functionality sets. In the former, the length of time for each iteration is set in advance, and then the requirements are managed to ensure that only those requirements that can be satisfied in that time frame are included. Iteration length varies typically from a few weeks to a few months. In the latter way, logical subsets of requirements are grouped together and each iteration is scheduled to be reasonable with respect to the functions it is satisfying. In either case, the iterative method is typically represented as shown in Figure 1.9. Figure 1.10 shows where the requirements activities are performed during the development.

A more recent approach to software development is generally called *agile*. The agile movement (Cockburn, 2002; Highsmith and Cockburn, 2001) proposes a significant decrease in the power of project management and general management, and instead pushes many responsibilities down to the individual contributors. Readers wishing to learn the details of agile development should refer to the sources cited in the previous sentence. Here we discuss the implications of agile methods on requirements management itself. Instead of attempting to elicit requirements at the beginning of the development process, agile devel-

FIGURE 1.7. A WATERFALL MODEL.

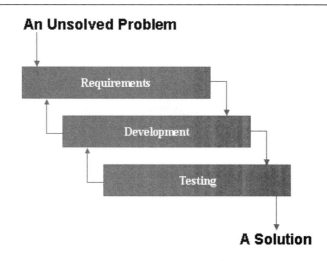

FIGURE 1.8. REQUIREMENTS ACTIVITIES WITHIN A WATERFALL MODEL.

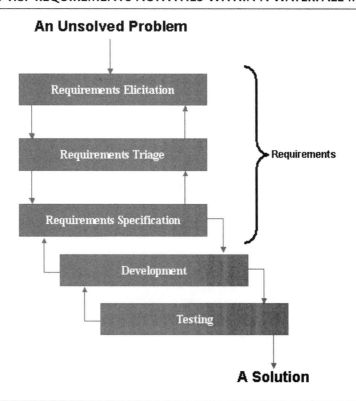

**FIGURE 1.9. AN ITERATIVE DEVELOPMENT MODEL.**

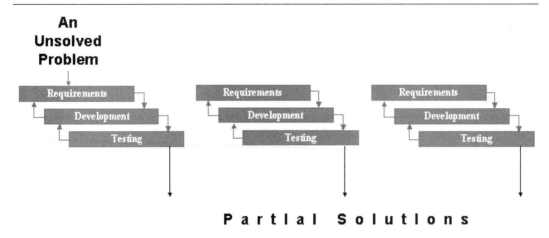

opment recommends that systems be built immediately. Agile developers construct iterations of the system in rapid succession, even as short as every day. A customer is required to be on-site with the development team at all times. Thus, requirements elicitation is performed constantly and is based primarily on the stimulation resulting from seeing system iterations. The omnipresent customer also has exclusive authority to select which requirements to include in each iteration. Thus, elicitation and triage are performed constantly, and specification is not performed *per se*.

Agile development is a reaction by software developers to what they perceive as too much control. The fact is that software development *is* difficult, and it requires a great deal of coordination. Agile development is likely to work well in situations where (a) the requirements are not changing, (b) there is only one customer (or there are more than one customer, but no conflicts exist among the stakeholders), (c) the problem is relatively simple, so that few misunderstandings concerning requirements are likely to arise.

## Maintenance Projects

Once a system is deployed, the life of the system, in the eyes of the user, has just begun. Now that the user has had an opportunity to put the product through its paces, there will likely be plenty of feedback regarding the software. This feedback falls into two general categories: (a) failures of the product to meet the intended requirements and (b) requests for new features. The demand for new features will accelerate in any system that is being used (Belady and Lehman, 1976). Rather than allowing the system to be under constant flux, system evolution should be managed as a series of well-planned releases. The length of time between subsequent releases is a function of (a) the rate of arrival of new requirements, (b)

FIGURE 1.10. REQUIREMENTS ACTIVITIES WITHIN AN ITERATIVE DEVELOPMENT MODEL.

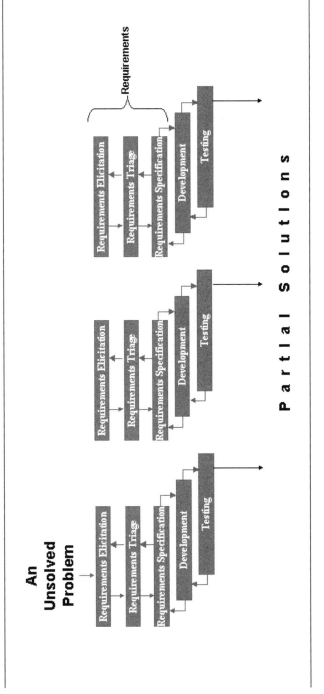

the overhead involved in producing and maintaining a release, and (c) the demand for early satisfaction. As each new requirement is discovered, it should be annotated just like the original requirements and documented in the same way that all previously approved requirements were. When the time arrives to initiate development of a new release, a triage meeting should be held. In principle, the management of post-deployment maintenance releases is no different than the management of predeployment iterations.

After a requirement is approved for a new release, multidirectional traces should be maintained between the change request, the new requirement, and all changes to the product and its documentation made in response to the change request. This enables the development team to (a) undo the changes if they prove erroneous and (b) reconstruct the history of changes made to the product.

Even with the best of processes in place, a product's entropy increases as it evolves (Lehman, 1978). The length of time that a system can survive is a function of the resiliency of the original architecture and the number of changes made over time. shows how the same system could last 7, 12, or 18 years before its entropy renders it no longer maintainable, based solely on the quality of the original architecture.

## System Procurement

Many projects are commissioned to solve a problem by procuring, or acquiring, an available system from a third party. In such cases, requirements should still be elicited as described earlier. However, rather than performing an explicit triage step, the team generally prioritizes the elicited requirements and performs a "best fit" analysis with the available solutions.

# Tool Issues

A requirements tool is a software application designed to assist the team in performing some combination of requirements elicitation, triage, and specification. Here is a list of the kinds of things such tools could do:

*Elicitation*

- Collect candidate requirements.
- Allow analysts to record lists of requirements as they are ascertained.
- Allow stakeholders to record their recommended requirements.
- Enforce discipline and/or protocol during elicitation sessions.
- Provide for anonymity during elicitation.
- Prompt for key missing information.

*Triage*

- Collect priorities and effort estimations.
- Allow analyst to record inclusion/exclusion of each requirement.

## FIGURE 1.11. LONGEVITY OF A PRODUCT IS A FUNCTION OF ORIGINAL ARCHITECTURE'S RESILIENCY.

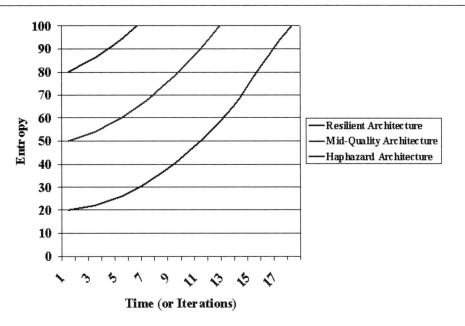

- Determine probability of completing a set of requirements within a given budget.
- Determine probability of completing a set of requirements within a given schedule.
- Allow analyst to refine requirements.

*Specification*

- Store requirements in a database.
- Determine ambiguities.
- Determine inconsistencies.
- Allow analyst to sort requirements based on multiple criteria.
- Allow analyst to cross-reference[2] requirements among themselves.
- Allow analyst to cross-reference[2] requirements to other products of the development effort (e.g., tests, designs).
- Provide the stakeholders with a simulation of the requirements (i.e., a prototype of the system).

---

[2] Also termed "traceability."

Requirements tools range from such basic tools as spreadsheets and word processors to extremely sophisticated tools such as special-purpose requirements-based simulation tools. In general, they fall into the following categories:

- *General-purpose tools that happen to be useful during requirements activities.* Word processors allow you to record requirements in natural language either in paragraph form or tabular form. Spreadsheets and databases provide the same capability but also give you the ability to easily define and record attributes such as effort, priority, and inclusion easily. Examples of these tools are Microsoft Word or any other word processor, Microsoft Excel or any other spreadsheet, and Microsoft Access or any other database.

  A majority of projects use these low-cost tools because they are already readily available on desktops with no additional cost. They also present no learning curve for the analysts, stakeholders, or project managers.

- *Meeting facilitation tools.* These tools are particularly helpful during elicitation. They enable stakeholders to record their suggested requirements easily, and even anonymously. They help to keep the discussion on-topic, can sort and filter the candidate requirements easily, and in some cases, can populate a requirements database tool. Two examples of such tools are Ventana's GroupSystems and Meetingworks' Connect.

  Facilitation tools have had surprisingly little impact on most companies. Analysts performing elicitation tend to either interview stakeholders or hold group sessions without tools.

- *Requirements database and traceability tools.* These tools include a database view that is already populated with common requirements attributes. They provide special sorting and filtering capabilities unique to requirements management. Many also provide a word-processed view, so you can update requirements in either the word-processed view or the database view and the other updates automatically. Furthermore, all of these tools make cross-referencing and establishing relationships among requirements easy. Some of these tools are integrated into a full development environment, thus facilitating referencing to and from requirements, designs, and tests. Examples of these tools include RequisitePro from IBM Rational Software, Caliber RM from Borland Software Corporation, and DOORS from Telelogic.

  Approximately 25 percent of all software development projects use requirements database and traceability tools. They significantly reduce the effort expended by analysts in recording and maintaining requirements, but have little impact directly on the stakeholders. One of their biggest advantages is to the project manager who can make intelligent and useful queries such as "Which requirements are high priority, included in the next release, and which are related to software components that Sally is working on."

- *Requirements risk analysis tools.* These tools help the project manager assess the likelihood that the selected requirements will be completed on schedule and within budget. Examples include OnYourMark Pro from Davis and the EstimatePro from Software Productivity Solutions, and part of Caliber RM from Borland Software Corporation.

  These tools have been in existence only since the late 1990s. Early adopters have started experimenting with them, but their adoption has been slow. The primary benefactor is the project manager and, indirectly, the company.

- *Requirements simulation tools.* These tools allow the requirements analyst to simulate the requirements after they have been written. In all cases, the requirements must first be written in a relatively formal notation. One example is Statemate Magnum from I-Logix.

  These tools have been in existence since the early 1970s. All of the vendors have had a hard time finding their niche. The primary benefactor of such tools appears to be the engineering analyst.

In summary, requirements tools can assist analysts in all aspects of requirements management. But no tool makes any aspect of requirements management easy. Elicitation still requires great listening skills. Triage still requires great diplomacy, and specification still requires incredible precision. The tools simply offload the more mundane aspects of the discipline.

## Trends in Requirements Management

### Research

The field of requirements research is one of the most active in universities. Recent research surveys (Finkelstein, 1994; Hsia et al., 1993; van Lamsweerde et al., 2000; Nuseibeh et al., 2000; and Potts, 1991) have defined the following trends:

- *Data and process modeling* is viewed as a critical activity in requirements. Much of the research since the 1970s has focused on the creation and analysis of modeling notations and techniques. Two somewhat contradictory trends occurring in this area include (1) the increasing emphasis on object-oriented modeling notations (e.g., UML) that focus on the system and (2) the recognition that modeling cannot focus on the system in isolation but must occur in an organizational context (Nuseibeh et al., 2000; Goguen and Jirotka, 1994; and Zave and Jackson, 1997). More recent emphasis has been on techniques to detect errors in models. See the special issue of the *Requirements Engineering Journal* guest edited by Easterbrook and Chechik (2002).
- *Increasing formality* to improve the quality and testability of requirements specifications has been a goal of requirements research (Hsia et al., 1993), especially for process control and life- and safety-critical systems (van Lamsweerde et al., 2000). For example, in the area of reactive systems for process control, specification notations and languages such as SCR Heninger, 1980), CORE (Faulk, 1992), and RSML (Leveson et al., 1994) have been developed to support automated consistency and completeness checking. Formal specification languages such as Z (Spivey, 1990) and others are designed to support requirements verification, visualization, and simulation.
- *Viewpoints* explicitly capture different perspectives or views of multiple stakeholders. Viewpoint integration can be used to check for consistency and aid in the resolution of conflicts among stakeholders (Easterbrook, 1994; Nuseibeh and Easterbrook, 1994). The earliest references to using viewpoints date back to 1981 (Orr, 1981).
- Since the beginning of requirements research, attempts have been made to *reduce ambiguity* in requirements. Obviously, the aforementioned activities of modeling and increasing

formality are aimed at this goal. Additional research has been done to either reduce or detect ambiguity in natural-language specifications. This includes work as early as 1981 (Casey and Taylor, 1981) and extends to the current day (Duran et al., 2002).

- *Goal-oriented requirements elicitation* takes an organizational approach to completeness and consistency checking of requirements by explicitly identifying and representing organizational goals for the system, and then checking the requirements against those goals (van Lamsweerde et al., 2000). Research in this area has resulted in a variety of methods and notations for representing, analyzing, and resolving conflicts between goals including KAOS (Dardenne et al., 1993; van Lamsweerde et al., 1998) and NFR (Mylopoulos, 1992).

- Behavioral requirements have always been the primary emphasis in requirements research. However, *nonbehavioral requirements* have also been addressed for many years and continues to be the focus of many research efforts. Some efforts have spanned the wide range of nonbehavioral requirements, for instance Chung et al. (1993), Chung (2000), Cysneiros and Leite (2002), Kirner and Davis (1995) Mostert and van Solms (1995), and Mylopoulos (1992), and others emphasize specific kinds of nonbehavioral requirements such as security (Shim and Shim, 1992), safety (Berry, 1998; Hansen et al., 1998), and performance (Nixon, 1993).

- *Scenarios* are concrete descriptions of the sequence of activities that users engage in when performing a specific task (Carroll, 1995). Studies have shown than scenarios are extremely useful for requirements elicitation when users are having difficulty specifying goals or using more abstract modeling techniques (Weidenhaupt, 1998; Jarke, 1999; van Lamsweerde, 2000). Scenarios have also proven useful in systems design and testing, for example, in user interface design (Carroll, 1995), and for generating test cases (Hsia, 1994). Other scenario uses are described in an *IEEE Transactions on Software Engineering* special issue on scenarios in (Jarke and Kurki-Suonio, 1998). Finally, scenarios are closely related to the Jacobson's use cases (Jacobson et al., 1992) in object-oriented analysis and the user stories, which are a key component of XP (Beck, 2000).

- With the wide variety of requirements techniques now in existence, some researchers are focusing on the *criteria for technique selection*. For example, Hickey and Davis (2003, 2003a) describe the best way to select the right elicitation techniques. Similar research still needs to be conducted for model selection.

- The field of software (design and code) reuse has settled into a status quo now; modern programming languages include large libraries of reusable entities whose use has become standard. However, *requirements reuse* has not yet reached this level of maturity. Perhaps this is because reusing requirements has little direct benefit to increasing quality or productivity. Instead, the real potential benefit of requirements reuse comes from the second-order effect of reusing design and code components associated with the reused requirements. See Castano and Antenellis (1993), Homod and Rine (1999), van Lamsweerde (1997), and Maiden and Sutcliffe (1996) for some of the latest ideas on requirements reuse.

## Practice

It is surprising how little of the current research in the requirements field is making its way to practice (Davis and Hickey, 2002). From the inception of software engineering as a discipline in the 1970s until the current day, (a) the standard for documenting requirements has been the word-processed SRS, (b) analysts in specialized applications have advocated the use of models, and (c) a counterculture has existed that is firmly convinced that writing requirements is primarily a waste of time.

In spite of the enormity of these invariants, a few changes have occurred. Two of these changes are in the evolution of the modeling notations themselves. The first is the introduction of new notations that provide unique perspectives of the system under specification. Classic among these are the introductions of statecharts by Harel (Harel, 1988; and Harel and Politi, 1998). Second is the tendency for the industry to move from sets of specialized notations (which in theory force analysts to become skilled in multiple languages) to all-encompassing notations (which in theory force analysts to become skilled in just one language, albeit enormous), and back to the specialized languages in a cycle. We expect this cycle to continue indefinitely into the future.

Another trend is in the isolation of optimal "starting points" for requirements activities. For many years, analysts have struggled with the question of where to start because of the sheer enormity of requirements. We have thus seen structured analysis (DeMarco, 1979) augmented by events as starting points (McMenamin and Palmer, 1984), and object-oriented analysis (Booch, 1994) augmented with use cases as starting points (Jacobson et al., 1992). This trend will continue. Unfortunately, every situation demands starting points that are a unique function of situational characteristics.

## Summary

Project management cannot succeed without careful attention to requirements management. Requirements management is responsible for determining the real needs of the customers, as well as clearly documenting the desired external behavior of the system being constructed by the project. If either of these goals is ignored, the project is guaranteed to result in failure.

## References

Beck, K. 2000. *Extreme programming explained.* Boston: Addison-Wesley.

Belady, L., and M. Lehman.1976. A model of large program development. *IBM Systems Journal* 15 (3, March): 225–252.

Berry, D. 1998. The safety requirements engineering dilemma. *Ninth International Workshop on Software Specification and Design.* 147–149. Los Alamitos, CA: IEEE Computer Society Press.

Boehm, B. 1982. *Software engineering economics.* Upper Saddle River, NJ: Prentice Hall.

Booch, G., 1994. *Object-oriented analysis and design.* Redwood City, CA: Benjamin/Cummings.

———. Personal conversation with two of the authors; September 17, 2002, Colorado Springs, Colorado.

Booch, G., et al. 1999. *The Unified Modeling Language user guide.* Reading, MA: Addison-Wesley.

Borland Software Corporation, Inc. 2003. See www.borland.com/products or www.starbase.com/products.

Carroll, J., ed. 1995. *Scenario-based design: Envisioning work and technology in system development.* New York: Wiley.

Casey, B., and B. Taylor. 1981. Writing requirements in English: A natural alternative. 95–101. *IEEE Software Engineering Standards Workshop*. Los Alamitos, CA: IEEE Computer Society Press.

Castano, S., and V. De Antonellis. 1993. Reuse of conceptual requirements specification. 121–124. *International Symposium on Requirements Engineering*, January. Los Alamitos, CA: IEEE Computer Society Press,

Christensen, M., and R. Thayer. 2001. *The project manager's guide to software engineering's best practices*. Los Alamitos, CA: IEEE Computer Society Press.

Chung, L. 1993. *Representing and using non-functional requirements: A process-oriented approach*. Department of Computer Science. PhD. thesis, University of Toronto.

Chung, L., et al. 2000. *Non-functional requirements in software engineering*. Norwell, MA: Kluwer.

Cleland, D., and L. Ireland. 2000. *Project manager's portable handbook*. New York: McGraw-Hill.

Cockburn, A. 2002. *Agile software development*. Boston: Addison-Wesley.

Cysneiros, M., and J. Leite, 2002. Non-functional requirements: From elicitation to modeling languages. 699–700. *Twenty-fourth International Conference on Software Engineering*. Los Alamitos, CA: IEEE Computer Society Press.

Dardenne, A., et al. 1993. Goal-directed requirements acquisition. *Science of Computer Programming* 20:3–50.

Davis, A., 1993. *Software requirements: Objects, functions, and states*. Upper Saddle River, NJ: Prentice Hall.

———. 1995. Software prototyping. *Advances in Computers 40*. 39–63. New York: Academic Press.

———. 2002. Requirements management. In *Encyclopedia of software engineering*. 2nd ed., ed. J. Marciniak. New York: Wiley-Interscience.

———. 2003. Secrets of requirements triage. *Computer* 36 (3, March): 42–49.

Davis, A., and A. Zweig. 2000. The missing piece of software development. *Journal of Systems and Software* 53 (3, September): 205–206.

Davis, A., et al. 1993. Identifying and measuring quality in software requirements specifications. 141–152. *IEEE-CS International Software Metrics Symposium*. Los Alamitos, CA: IEEE Computer Society Press.

Davis, A., and A. Hickey. 2002. Requirements researchers: Do we practice what we preach? *Requirements Engineering Journal* 7(2):107–111.

Dean, D., et al. (1997–1998. Enabling the effective involvement of multiple users: Methods and tools for collaborative software engineering. *Journal of Management Information Systems* 14 (3, Winter): 179–222.

DeMarco, T. 1979. *Structured analysis and system specification*. Upper Saddle River, NJ: Prentice Hall.

Duran, A., et al. 2002. Verifying software requirements with XSLT. *ACM Software Engineering Notes* 27: 39 ff.

Easterbrook, S. 1994. Resolving requirements conflicts with computer-supported negotiation. In *Requirements engineering: Social and technical Issues*, ed. M. Jirotka and J. Goguen. 41–65. London: Academic Press.

Easterbrook, S., and M. Chechik 2002. Guest editorial: Special issue on model checking in requirements engineering. *Requirements Engineering* 7(4):221–224.

Faulk, S. 1997. Software requirements: A tutorial. In *Software Requirements Engineering*, ed. R. Thayer and M. Dorfman. 128–149. Los Alamitos, CA: IEEE Computer Society.

Faulk, S., et al.1992. The CORE method for real-time requirements *IEEE Software* (September): 22–33.

Finkelstein, A.1994. Requirements engineering: A review and research agenda. 10–14. *First Asia-Pacific Software Engineering Conference*. December. Los Alamitos, CA: IEEE Computer Society.,

Gause, D., and J. Weinberg 1989. *Exploring requirements: Quality before design*. New York: Dorset House.

Goguen, J., and C. Linde 1993. Software requirements analysis and specification in Europe: An overview. 152–164. *First International Symposium on Requirements Engineering*. Los Alamitos, CA: IEEE Computer Society Press.

Goguen, J., and M. Jirotka, eds.1994. *Requirements engineering: Social and technical issues*. Boston: Academic Press.

Gottesdeiner, E. 2002. *Requirements by collaboration*. Reading, MA: Addison-Wesley.

Hansen, K., et al. 1998. From safety analysis to software requirements. *IEEE Transactions on Software Engineering* 24 (7, July): 573–584.

Harel, D.1988. On visual formalisms. *Communications of the ACM* 31 (5, May): 514–530.

Harel, D., and M. Politi 1998. *Modeling reactive systems with statecharts*. New York: McGraw Hill.

Heninger, K.1980. Specifying software requirements for complex systems: New techniques and their application. *IEEE Transactions on Software Engineering* 6(1):2–13.

Hickey, A., and A. Davis. 2002. The role of requirements elicitation techniques in achieving software quality. *International Workshop on Requirements Engineering: Foundations for Software Quality (REFSQ)*. Los Alamitos, CA: IEEE Computer Society Press.

———. 2003a. Requirements elicitation and requirements elicitation technique selection: A model of two knowledge-intensive software development processes. *Proceedings of the Thirty-Sixth Hawaii International Conference on System Sciences*. Los Alamitos, CA: IEEE Computer Society Press.

———. 2003b. Elicitation technique selection: How do the experts do it?" International Joint Conference on Requirements Engineering (RE03). September. Los Alamitos, CA: IEEE Computer Society Press.

Highsmith, J., and A. Cockburn. 2001. Agile software development: The business of innovation. *Computer* (September): 120–122.

Hofmann, H., and F. Lehner 2001. Requirements engineering as a success factor in software projects. *IEEE Software* 18 (4, July/August): 58–66.

Homod, S., and D. Rine. 1999. Building requirements repository using requirements transformation techniques to support requirements reuse. *World Multi-Conference on Systemics, Cybernetics and Informatics*, Volume 2.

Hsia, P., et al. 1993. Status report: Requirements engineering. *IEEE Software* 10 (6, November): 75–79.

Hsia, P., et al. 1994. Formal approach to scenario analysis. *IEEE Software* 11(2):33–41.

IEEE. 1983. *IEEE standard glossary of software engineering terminology*. IEEE Standard 729. New York: IEEE Press.

———. 1986. *IEEE standard for software verification and validation plans*. IEEE Standard 1012. New York: IEEE Press.

———.1993. *A guide to software requirements specifications*. Standard 830-1993. New York: IEEE Press.

I-Logix Corporation. www.ilogix.com/products/magnum/index.cfm.

Jacobson, I., et al. 1992. *Object-oriented software engineering*. Reading, MA: Addison-Wesley.

Jarke, M., and R. Kurki-Suonio. 1998. Guest editorial: Introduction to the special issue. *IEEE Transactions on Software Engineering* 24(12):1033–1035.

Jarke, M. 1999. Scenarios for modeling. *Communications of the ACM* 42(1): 47–48.

Kirner, T., and A. Davis. 1996. Nonfunctional requirements for real-time systems. *Advances in Computers*.

Knapp, M., and J. Hall. 1997. *Nonverbal communication in human interaction*. Austin, TX: Holt, Rinehart and Winston.

Kotonya, G., and I. Sommerville. 1997. Integrating safety analysis and requirements engineering.259–271. *Fourth Asia-Pacific Software Engineering Conference*. Los Alamitos, CA: IEEE Computer Society.

Kowal, J. 1992. *Behavior models*. Upper Saddle River, NJ: Prentice Hall.

Lam, W., et al. 1997. Ten steps towards systematic requirements reuse. 6–15. *IEEE International Symposium on Requirements Engineering*. January Los Alamitos, CA: IEEE Computer Society Press. Also appears in *Requirements Engineering Journal* 2(2):102–113.

van Lamsweerde, A. 2000. Requirements engineering in the year 00: A research perspective. *Proceedings of the 22nd International Conference on Software Engineering*. 5–19. New York: ACM Press.

van Lamsweerde, A., et al.1998. Managing conflicts in goal-driven requirements engineering. *IEEE Transactions on Software Engineering* 24 (11, November): 908–926.

Lauesen, S. 2002. *Software requirements: Styles and techniques.* London: Addison-Wesley.

Leffingwell, D., and D. Widrig. 2000. *Managing software requirements.* Reading, MA: Addison-Wesley.

Lehman, M. 1978. *InfoTech State of the Art Conference on Why Software Projects Fail.* Paper #11, April.

Leveson, N., et al. 1994. Requirements specification for process-control systems. *IEEE Transactions on Software Engineering* 20 (9, September): 684–706.

McMenamin, J., and J. Palmer. 1984. *Essential systems analysis.* Upper Saddle River, NJ: Prentice Hall.

Maiden, N., and A. Sutcliffe. 1996. Analogical retrieval in reuse-oriented requirement engineering. *Software Engineering Journal* 11(5):281–292.

Meetingworks, Inc. 2003. www.meetingworks.com.

Meredith, J., and S. Mantel. 2003. *Project management: A managerial approach.* 5th ed. New York: Wiley.

Microsoft, Inc. 2003. www.microsoft.com..

Mostert, D., and S. von Solms. 1995. A technique to include computer security, safety, and resilience requirements as part of the requirements specification. *Journal of Systems and Software* 31 (1, October): 45–53.

Mylopoulos, J., et al. 1992. Representing and using nonfunctional requirements: A process-oriented approach. *IEEE Transactions on Software Engineering* 18(6, June): 483–497.

Nixon, B. 1993. Dealing with performance requirements during the development of information systems. 42–49. *IEEE International Symposium on Requirements Engineering.* Los Alamitos, CA: IEEE Computer Society Press.

Nuseibeh, B., et al. 1994. A framework for expressing the relationships between multiple views in requirements specifications. *IEEE Transactions on Software Engineering* 20 (10, October): 760–773.

Nuseibeh, B., and S. Easterbrook. 2000. Requirements engineering: A roadmap. *Proceedings of the 22nd International Conference on Software Engineering.* 35–46. New York: ACM Press.

Opdahl, A. 1994. Requirements engineering for software performance, *International Workshop on Requirements Engineering: Foundations of Software Quality.* June.

Orr, K. 1981. *Structured requirements definition.* Topeka, Kansas: Ken Orr and Associates.

Project Management Institute. 2000. *A guide to the project management body of knowledge.* Newtown Square, PA: Project Management Institute.

Potts, C. 1991. Seven (plus or minus two) challenges for requirements research. *Sixth International Workshop on Software Specification and Design.* Los Alamitos, CA: IEEE Computer Society.

Rational Software Corporation, Inc. 2003. www.rational.com/products.

Robertson, J., and S. Robertson. 2000. *Mastering the requirements process.* Reading, MA: Addison-Wesley.

Reinertsen, D. 1997. *Managing the design factory.* New York: Free Press.

Shim, Y., H. Shim, et al. 1997. Specification and analysis of security requirements for distributed applications. 374–381. *Ninth IEEE International Conference on Software Engineering and Knowledge Engineering.* June. Skokie, IL: Knowledge Systems Institute.

Siddiqi, J., and C. Shekaran. 1996. Requirements engineering: The emerging wisdom. *IEEE Software* 13(2):15–19.

Software Productivity Centre, Inc. 2003. http://www.spc.ca/products/estimate.

Spivey, J. 1990. An introduction to Z and formal specifications. *Software Engineering Journal* 4:40–50.

The Standish Group. Undated. *The CHAOS Chronicles* www.standishgroup.com.

Swartout, W., and R. Balzer 1982. On the inevitable intertwining of specifications and design. *Communications of the ACM* 25 (7, July): 438–440.

Telelogic, Inc. 2003. www.telelogic.com/products.

Thayer, R., and M. Dorfman 1994. *Standards, guidelines, and examples on system and software requirements engineering.* Los Alamitos, CA: IEEE Computer Society Press.

Ventana, Inc. 2003. www.ventana.com.

Wallace, D. 1994. Verification and validation. In *Encyclopedia of Software Engineering*, ed., J. Marciniak. 1410–1433. New York: Wiley.

Weidenhaupt, K., et al. 1998. Scenarios in system development: Current practice. *IEEE Software* 15(2): 34–45.

Wieringa, R. 1996. *Requirements engineering*. Chichester, UK: Wiley.

Wiegers, K. 1999. *Software requirements*. Redmond, WA: Microsoft Press.

Wood, J., and D. Silver 1995. *Joint application development*. 2nd ed. New York: Wiley.

Young, R. 2001. *Effective requirements practices*. Boston: Addison-Wesley.

Zave, P., and M. Jackson. 1997. Four dark corners of requirements engineering. *ACM Transactions on Software Engineering and Methodology* 6 (1, January): 1–30.

---

# CHAPTER TWO

# DESIGN MANAGEMENT

---

Peter Harpum

Design is of primary importance in the project and is carried out throughout the project life cycle. Design begins with the business case formulation for a project—how the project can most effectively and efficiently deliver the benefits to the organization that are required of it. At the other end of the life cycle, when the project's deliverables are being decommissioned (whether it is a nuclear power station, a financial service product, or computer software), design work is required to ensure that the products that the project made are effectively removed from the environment.

Central to the notion of design is creativity—creation of the business case, outline design, in-service improvements, and work in all other stages of the project that have some element of design. Creativity, however, is notoriously hard to define, and in many people's opinion even harder to manage. Much of the difficulty in managing design is found at the psychological interface between what is seen as an "instrumentalist" project management paradigm—that is, a tool used to predict the future—and the creative flair, and at times genius, needed for great design work (Allinson, 1997).

This chapter describes the following:

- Design in the context of projects, discussing its importance and specific characteristics
- The strategic design management considerations, including the philosophical approaches that can be taken
- Control of the design process, in terms of scope, schedule, budget, and quality.

# Design in the Context of Projects

## The Importance of Design in the Project

There are many different views on the role and process of design management. At one end of the spectrum is the approach that design is *the* dominant business process. This is common in industries that depend on a continuous supply of new products for sustaining profitability, such as consumer goods and computer software. In such industries the design function is often represented at Board level. Project managers in these companies are relatively junior compared to those managing design, often reporting to the design manager of the new product (Topalian, 1980; Cooper, 1995). At the other end of the spectrum, project management is the dominant process, and design managers have only a coordination role between different design groups. This situation is more likely to be found in industries that have a history of being implementation-oriented—that is, they are focused on making the project's products (the project "deliverables"). Construction projects are more likely to follow this arrangement (though signature architects may dominate project managers).

These differences in perception of how design is managed within the corporate context reflect on the various models of design management. Where design dominates the organizational culture, the strategies and tactics of design management center on the relationships between business-as-usual and individual projects to ensure corporate value is created through design. When the *delivery* of projects to external buyers is the dominant paradigm design management models more commonly address information flow, interface control, and the logistics of producing error-free, high-quality design work, on time and on budget (Gray et al., 1994). These types of projects often "buy-in" design services from third-party design consultants, since the core capabilities of firms that deliver such projects are in the management of the implementation phases.

## Characteristics of Design in Projects

Project management has its roots very much in the management science and systems engineering fields. The earliest modern tools for managing projects evolved from the scientific school of management thinking during the early part of the twentieth century: the Gantt chart and the little known Harmonygraph. These were later to form the foundation of scheduling techniques such as program evaluation and review technique (PERT) and the critical path method (and more recently Gantt–chart-based software), developed primarily to help plan large systems engineering programs in the U.S. defense sector (Morris, 1994).

This meant that from the beginnings of modern project management, designers were being asked to work to explicit schedules—schedules they often perceived to be inflexible and that did not reflect the reality of their work, and designers believed design could not be scheduled. This view persists to this day. Many designers resent having time constraints imposed on them, based on what they consider to be a mechanistic, and hence unrealistic, management paradigm: project management.

The separation of design from "making" is still only recent. Up to the beginning of the twentieth century it was normal for craftspeople to design and make whatever it was they were producing. In this way of working, the delivery of the "product" was the responsibility of one person. This approach, where design and making are inseparable, also applied to most of the large engineering projects of the time (one thinks of Brunel, Telford, Stephenson, and other great engineers of the seventeenth and eighteenth centuries). Many of these engineers were better at design than at the management of those responsible for the building of their creations. In short, many of these people are viewed in retrospect as more artist than scientist. Around the end of the nineteenth century, when the work required to turn the design solution into a physical reality became increasingly sophisticated, these two fundamental aspects of projects began to be separated (Lawson, 1997). Since then, the project manager has been striving to *reintegrate* designing and making, made challenging by the intrinsically different mind-set required for design in comparison to implementation (Harpum and Gale, 1999).

This difference in mind-sets between the two groups can be shown as a steady change across the life cycle of the project (see Figure 2.1). The designer works on the left of the diagram, where there is a greater freedom (a larger "action space"). This means multiple perspectives can be generated in order to solve a problem—therefore creating many possible solutions. As the project life cycle moves inexorably through its early stages, the degree of freedom gets smaller as the final design solution becomes clear. The action space continues to become smaller as the stages in the life cycle move into implementation and completion; there is less and less ability for the solution to be changed (and it becomes increasingly expensive in time and money to make any changes).

### FIGURE 2.1. THE RELATIONSHIP BETWEEN THE PROJECT LIFE CYCLE AND THE DESIGNER'S ACTION SPACE.

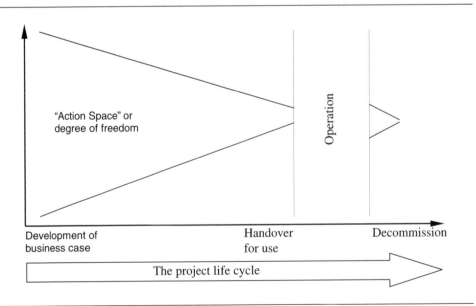

*Source:* Developed from an earlier version in Smith (2002).

Recent research shows that there can be significant advantages for projects where the power relationship between the project manager, the design manager, and the manager of the implementation phases of the project is more balanced than has been the case historically. This is being driven in part by increasing emphasis on the project front-end by project sponsors and other significant players in the project community, seeking to shorten delivery times and maximize value by reducing rework (and indeed aborted implementation work). It is also driven by the increasing projectization of many sectors of industry, including those where design management has traditionally played a strong role in product development—meaning in these industries that the project manager's authority has increased in relation to the design manager. An increasing awareness of the fundamental importance of the design phases of projects is also no doubt reshaping the relationship between creative designers and action-oriented implementation people.

Fundamentally, design is about creativity—creating solutions to problems. Therefore, managing design means managing creative people. This is inherently difficult. The large degree of freedom that design needs to be most effective, the requirement for room to create multiple solutions before any final one is chosen, tends to preclude artificial restriction. Creativity is about making connections between ideas that are often not obviously connected, and we are only just beginning to understand how this happens in our minds. Yet design within the project context (which actually means almost all design work carried out) must have some element of control placed on it. Nearly all projects have *some* time and cost constraints placed on them. And this is the conundrum at the heart of managing design. Design must be managed to ensure project success, but by its very nature, design rejects the concept of management (Allinson, 1997).

## How Designers Design

Effective management requires a comprehension of the activity that is to be managed. Managing design is no different. However, the traditional mechanistic, and predominantly implementation-oriented, paradigm of project management has tended not to acknowledge the creative (hence artistic) aspect of design. Therefore, it is important to have an understanding of the unpredictable nature of the creativity required to design to enable more effective management of this work. This means acknowledging that the design process is unpredictable.

A number of theories exist to describe how the creative process works. None has yet been able to make the process more predictable in terms of the time frame required to find a particular solution to a particular problem. However, the theories do help us understand the process through which creative thought moves. Perhaps the most common model of the creative process is the Assess-Synthesize-Evaluate model (Lawson, 1997), shown in Figure 2.2.

Each step is described as follows:

- *Assess.* A number of information inputs are considered by the creator in relation to the problem to be solved. Some of these inputs will be well known—for instance, known solutions to similar problems—others will be more fragmented in the thinker's mind, without any great clarity about their relevance to the problem.

## FIGURE 2.2. THE ASSESS-SYNTHESIS-EVALUATE MODEL OF CREATIVITY.

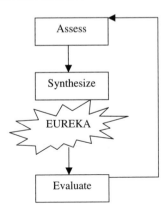

*Source:* After Lawson (1997). Copyright (1997), with permission from Elsevier.

- *Synthesize.* Conscious and subconscious mental activity in the thinker tries to make the linkages between the various information inputs that will provide solutions to the problem. This period can become intensely frustrating, as often the more the solutions are forced, the more difficult their crystallization becomes. The thinker may move away from actively trying to solve the problem, while the subconscious continues to ponder the solutions.
- *Eureka.* This is an intense, usually short period of time when what often seems to be a fully formulated solution comes into the thinker's mind.
- *Evaluate.* The thinker now evaluates the solution, usually finding that it is not quite as complete as at first seemed to be the case. The flaws in the solution are explored and the information is fed back into the assess stage of the process.

This creative process may cycle many times until a final solution is found; at other times a complete solution is found at the first attempt. There are many examples that demonstrate how difficult it is to know when, or how, a solution will be found (Csikszentmihalyi, 1996). Murray Gell-Mann, the Nobel-prize-winning physicist, articulates the reality of this unpredictability when recollecting a meeting of a group of physicists, biologists, painters, and poets and their discussion on creativity:

> First we had worked, for days or weeks or months, filling our minds with the difficulties of the problem in question and trying to overcome them. Second, there had come a time when further conscious thought was useless, even though we continued to carry the problem around with us. Third, suddenly, while we were cycling or shaving or cooking . . . the crucial idea had come. We had shaken loose of the rut we were in.'
>
> Gell-Mann, 1994

The unpredictability comes from the fact that it can be very difficult to forecast how long the synthesis phase of the process will take. Clearly some problems are harder to solve than others, and not all problems necessarily need great creative thought to produce acceptable solutions. The more difficult problems when designing in the project context are usually found early in the life cycle. It is normally the case that greater creativity is needed when developing concept designs than when working on the detailed solutions to the chosen concept. This accords with Figure 2.1, since in detail design, there is less freedom for the designer to work within; the solution has become constrained. (This is not always the case, of course. During detail design, a solution that was expected to be relatively easy may turn out to be very difficult indeed. This is where many projects encounter their first significant delays, as much more time is used up on the particularly difficult design task than was forecast.)

## Strategic Design Management Considerations

The first part of the chapter discussed the inherent difficulty of managing the unpredictable nature of creative design work. Project management is viewed as mechanistic and unsympathetic to design work. However, working within explicit processes can *facilitate* creativity, as less attention needs to be paid by the designer to ensure ad hoc processes are in place to meet the needs and constraints of the project (Luckman, 1984; Pugh, 1990). The next part of this chapter examines the strategic approaches and techniques associated with design management processes.

Aside from the creative aspect of design, there are also various ways of looking at the overall design process. The way design is approached affects the way that the design processes are organized. This in turn determines the effectiveness of the management control that can be brought to bear.

### Philosophical Approaches to Organizing Design Work

Organizing the way that the design work is carried out by teams, and the people in them, is not a trivial activity. From a high-level and generic point of view, the design process options are as follows (Simon,1981; Kappel and Rubenstein, 1999):

| Design Team Approach | *Designer's Personal Approach* | |
|---|---|---|
| | **Depth First** | Intuitive |
| Bottom-up | | |
| Top-down | | |
| Meet-in-the-middle | | |

- *Bottom-up.* Basic elements of the solution are created and then put together, changing them until an overall fit is achieved.
- *Top-down.* The desired end solution is conceptualized, and the designer then works backwards until all the basic elements have been completed.
- *Meet-in-the-middle.* As the name implies, top-down and bottom-up approaches are combined until the design is fully integrated.
- *Depth first.* The designer takes whichever possible solution is conceptualized first and attempts immediately to make it work.
- *Intuitive.* The designer considers several possible solutions but takes the one that intuitively seems to offer the best hope of working.

Hence, there are six possible approaches to the way in which the design work can be organized. The option chosen will have implications for the selection of team members. It is almost inevitable that the design process chosen will not perfectly match all the members of the design team. This does not spell disaster for the team, but it does mean that the manager of design ought to be aware of the possibility for mismatch, and motivate and lead individuals accordingly. It is worth pointing out here that even in groups that are inherently highly motivated, such as is found in "skunk works" environments, great care and effort is needed by the manager to get the best from each team member.

The models described previously are by necessity fairly abstract. Developing a design process for a particular project context requires "mapping" design activity in a more detailed way, showing generally what work must be done at consecutive points in the process. Such a map will include processes for (at the least) the following:

- Determining functions of the design solution (and often their physical structures)
- Elaborating specifications
- Searching for solution principles
- Developing layouts
- Optimizing design forms;
- Dividing design work into realizable modules.

## Systems Engineering

Some sectors of industry use systems engineering as a fundamental and core part of their design process. Indeed, in computer software and hardware design it is synonymous with the design process. Other industries, such as aerospace and electronics, are similar. However, formal systems engineering is little known in other industries. In some cases, this seems surprising, since the design solutions are quite similar in nature to computers, aircraft, and electronic circuits. For example, building design is clearly about creating a system with multiple subsystems (heating, ventilation, water, waste disposal, lighting, power), yet the discipline has made little contribution so far to building design (Groák, 1992).

There are two reasons for including a review of the subject in this chapter:

1. Significant parts of industry use formal systems engineering as a design approach.
2. Most design solutions (some would say all) are of the nature of a system, and knowledge of the formal approach may be beneficial to those not currently using it.

Many of the design solutions that are needed to satisfy project objectives can be classified as systems (indeed, in a purist sense, every solution is a system, or at the very least becomes part of a system). It is not easy to define a system in a readily understandable way, while at the same time being totally clear and unambiguous about what is meant. The term system (in the context of a design solution to a problem) implies that

- a number of elements must work together to deliver a consistent output;
- those elements are dependent on each other for their proper functioning.

Systems can be "open" or "closed." That is, they may be impacted by their external environment (open) or may be independent of their environment (closed). Open systems are usually part of a larger supra-system and also contain subsystems. Almost all systems that form the output from a project are open in some way or another, even if only because they are subject to climatic changes (an electrical circuit is affected by temperature, for instance). Many "softer" systems, such as financial products, customer service products, and the like, are by nature open to multiple environmental inputs.

The essence of the systems approach is well captured by Howard Eisner (1997) in his description of the key features and results of taking a systems approach:

1. Follow a systematic and repeatable process.
2. Emphasize interoperability and harmonious system operations.
3. Provide a cost-effective solution to the customer's problem.
4. Ensure the consideration of alternatives.
5. Use iterations as a means of refinement and convergence.
6. Satisfy all user and customer requirements.
7. Create a robust system.

Points 3, 4, 5, and 6 are well within the remit of much current design management. The other points, however, are not so obviously in the domain of much design work that is carried out. Taking a systems approach to design management is done in many technical industries, particularly defense contracting, aerospace, and computer hardware and software. The design solution to most project objectives in these industries is an engineering system. The approach has benefits, though, in many other less technically oriented sectors. As an example, a systems view of the design for a piece of clothing is not necessarily obvious. Careful consideration, however, shows that a clothes designer already works in a systems way—following much of the advice in the preceding list, although perhaps without being consciously aware of doing so. Specifically:

- Designing clothes follows a well-determined process.
- The sleeves, collar, cuffs, and panels of a blouse or shirt must obviously work together and be harmonious.
- The system needs to be robust; it must be easy to put on, cleanable, repairable, and work correctly with other clothes that will be worn with it.

There are two distinct aspects of a system. At a high level there is a system architecture, and below this there are the subsystems that together form the functioning system. Broadly speaking, the relationship between the development of the system and the design stages is shown in Table 2.1.

The architecture of the system defines the best combination of subsystems to meet the business and technical requirements, as well as the definition of the functions to be carried out by each subsystem. However, system architecting is more than providing the framework for subsystems to work within. It includes defining the approach that should be taken toward the creation of the functional subsystems, as well as identifying the most cost-effective arrangement of these subsystems. This means that a specification for each subsystem must be written, and the interfaces between the subsystems clearly delineated and documented (Eisner, 1997).

Developing the subsystems is very much the domain of designers expert in their particular field, and this applies whatever system is being delivered to meet project objectives (engineering, financial services, organizational change, etc.). See Figure 2.3.

There are two main advantages to creating a system architecture:

1. Thinking carefully about the system, as distinct from a collection of individual deliverables to be put together at the end of the project, can help enormously to improve the

## TABLE 2.1. THE RELATIONSHIP BETWEEN SYSTEM DEVELOPMENT AND LIFE CYCLE STAGES.

| Life Cycle Stages | System Development |
|---|---|
| Concept design | Decide what type of system is most likely to meet the business and technical needs, expressed by the statement of requirements (or design brief). |
| Feasibility studies | Assessing whether the type of system decided on can be created successfully, by measuring against carefully set criteria (see the description of generic design stages in the *Design Management Techniques* section later in the chapter). |
| Outline design | The system architecture is developed with reference to the requirements, by understanding the various functions required of the system and how these can be achieved (including deciding on any trade-offs needed between the various requirements). |
| Detail design | The way in which each subsystem of the final delivered system will provide the function required of it. |

FIGURE 2.3. ILLUSTRATIVE SYSTEM ARCHITECTURE FOR AN INSURANCE PRODUCT.

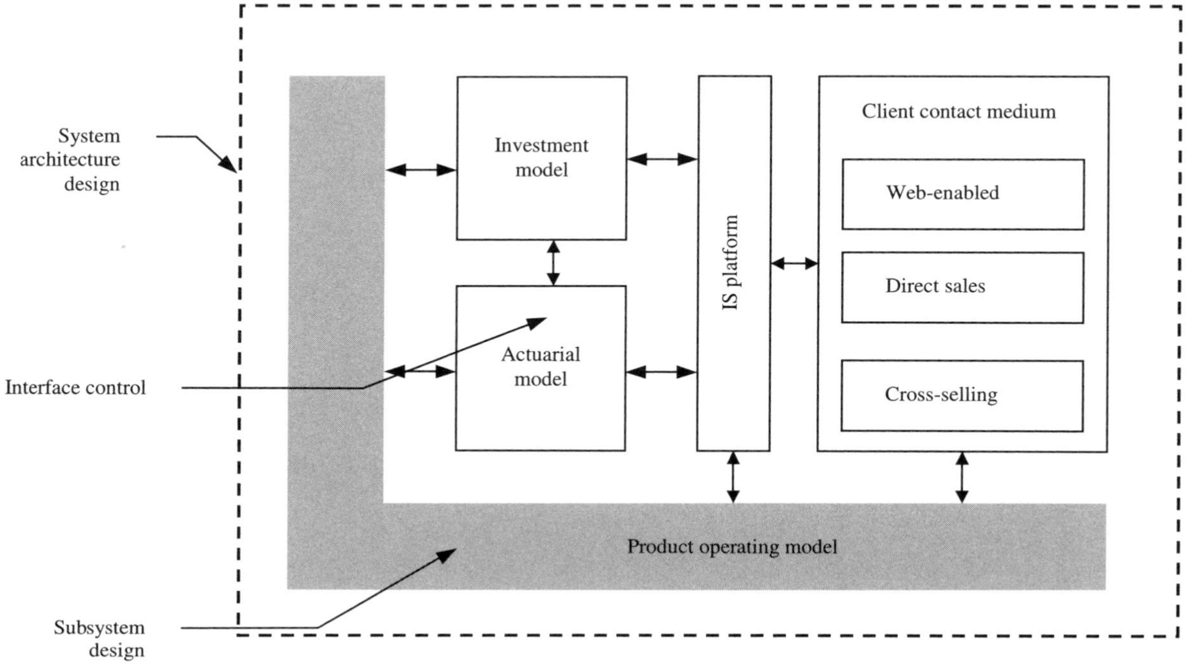

Client contact medium
- Web-enabled
- Direct sales
- Cross-selling

IS platform

Investment model

Actuarial model

Product operating model

System architecture design

Interface control

Subsystem design

way that the overall deliverable works—ensuring it provides a better solution to meet the project's objectives;
2. Designing the system architecture requires different skills than designing each functional subsystem, in whatever sector of industry the project exists; thinking systemically at an early stage can bring significant improvement in the overall solution that is decided on.

The need for effective interface definition and control becomes apparent as the subsystem design begins. Setting, and subsequently "freezing," the interface requirements between subsystems means that the designers of the subsystems can then work on designing their part of the overall solution without further reference to those working on adjoining systems. Each subsystem design only needs to satisfy the interface constraints. If these are met by the subsystem, the operation of the internal components in the subsystem is not relevant to other interfacing subsystems—hence, the term "black box." The need for information to constantly flow between the designers working on the separate systems is removed.

The work of defining interfaces is not trivial. The degree to which the overall design is broken down, and the size of the subsystems, is fundamental to effective system (and hence design) management. The crucial interface control issues are as follows:

- Level of disaggregation of the system to subsystems—which determines the number of interfaces.
- Amount of compromise that can be tolerated for each interface constraint (since subsystems frequently have conflicting interface constraint needs).
- Tolerance that the constraints should have: If the constraints are too tightly specified, optimization of subsystem design can be reduced dramatically; if too loosely specified, the overall design solution is likely to perform poorly.
- Need to freeze interfaces, and their constraints, at an appropriate time in the design project's life cycle. Freezing too soon will lead to suboptimization of the overall system, since not enough is known about the system's properties, whereas freezing too late will prevent the designers from making the technical (and quite likely commercial) decisions needed to deliver the subsystem on time.

A schedule of the interfaces showing freeze dates and required delivery dates for subsystem designs is a valuable design management tool.

## Life Cycle Management

Different approaches to the management of the project life cycle lead to different emphasis being put on design. There are two fundamentally different types of project life cycle. They are differentiated by what is considered to be the work of project management. The task-oriented life cycle includes the major activities that require to be managed, vis business case, feasibility studies, concept design, detail design, implementation, commissioning, handover, operations, and decommissioning. (And there are often others included such as procurement and testing.) These life cycles are usually drawn in a circular, or spiral, way. An example is shown in Figure 2.4.

## FIGURE 2.4. TASK-ORIENTED PROJECT LIFE CYCLE.

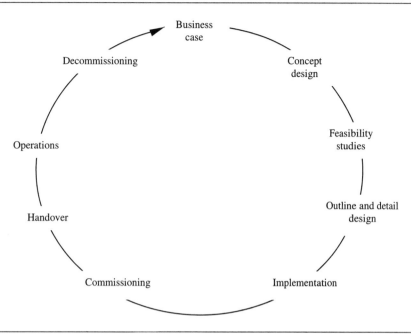

*Source:* After Wearne (1989).

In contrast to the task-oriented life cycle, the product-oriented life cycle de-emphasizes design (along with other processes). The life cycle only describes the management of a strictly limited set of "pure" project management tasks: start-up, plan, implement, closedown. All other tasks associated with the project's work packages, including design, are considered to be part of the *product* life cycle. The diagram typical of this type of life cycle is shown at Figure 2.5.

The danger is that the disassociation of design from project management implied by the product-oriented life cycle leads to insufficient attention being paid to the management of design (and indeed the management of other processes such as testing, handover, procurement, etc.). Design requires much attention. Decisions made at business case through to detail design fundamentally define the project's outputs. This means the cost of the project, the time likely to be needed to carry out the project, the type of resources needed, and the quality requirements of the products. If the conceptual design of the project's deliverable does not reflect the context of the project as a whole, and therefore the wrong design solution is chosen, the project has little chance of success.

## The Inputs to Design

The primary input at the highest level into the design process is the project objective. What change has the project been set up to create? This applies whether the project is internal

FIGURE 2.5. THE PRODUCT-ORIENTED PROJECT LIFE CYCLE.

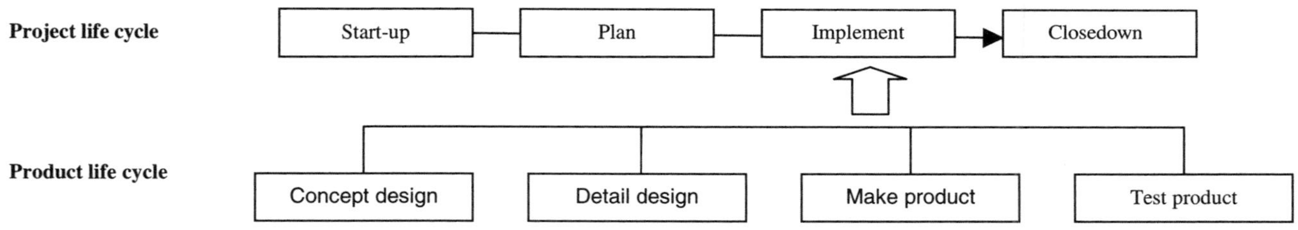

to the organization or is an external project, delivering change to a client's organization. From an understanding of the project objectives, the primary deliverables can be deduced. This sounds easy but in fact can be quite difficult. The process that links objectives to primary deliverables is requirements capture (requirements capture is discussed in detail in the chapter by Davis, Hickey, and Zweig). This means understanding what *both* the business and technical needs of the organization are to enable the project objectives to be satisfied. The requirements are a clear and concise statement of the problem that the design is to overcome, completely devoid of any suggestion of the solution.

It is clear that involving experienced designers in the capture of business requirements can significantly improve the understanding of the needs of the project. The reason for this is the designer brings knowledge of the ways in which similar business needs have been satisfied in the past.

It is more obvious that designers need to be involved in capturing technical requirements of the project, since they

- know when sufficient technical requirements have been captured to be able to proceed to concept design;
- bring knowledge of how similar technical needs have been met in the past; and
- will have a first-hand understanding of the requirements, enabling them to match them with solutions more quickly and easily in later design stages.

The documented output from requirements capture is usually called the "statement of requirements." In a number of industries where the idea of explicitly capturing business and technical requirements without an implied solution is relatively new, the input to the design process has usually been called the "design brief" (Barratt, 1999). This document is in many ways similar to the statement of requirements but is more often used when briefing professional design consultants with whom a contract will be placed to deliver design to a project. The brief is often more directive than a statement of requirements in that it specifies the expected design solution (for instance, that an office building is to be designed, normally with quite a lot of detail as to the expected final design[1]). In this sense it is often a contract document, and so has a different purpose to the statement of requirements.

# Design Management Techniques

## Stage Gate Control

The life cycle shown previously identified a number of design stages at the front end of the project. A more detailed explanation of these stages will help us to understand how the design solution is managed as it evolves through the life cycle. What matters is that the

---

[1] A statement of requirements might say that the business needs to increase the number of workers it employs—for which the solution *could* be more office space, or could be more home working, or hot desking using the existing office space.

evolving design is best controlled if the work is managed in discrete stages (British Standards Institute, 1996). The generic design stages can be described as follows:

**Concept**      A number of high-level design proposals are developed that will all lead to the project objectives being accomplished; each concept design must satisfy the business case developed in first stage of the project. The designs are at a low level of detail but are sufficiently well developed that the overall cost of the project can be estimated.

**Feasibility**  The feasibility of the various options are considered against a number of criteria, typically:

- Cost to make the project deliverables
- Amount of time that would be needed to complete the project
- Capability of the organization to make the deliverables
- Congruence with the technology strategies of the project participants
- Environmental impact the deliverables will have

(There are, of course, many more criteria that may be used to assess the design solutions proposed.)

**Outline**      The concept design, which may or may not have been further extended during the feasibility stages, is now developed to the outline level of detail. The major parts of the deliverables are defined in terms of form and function (and "delight" in most consumer-oriented industries). Outline design includes the following:

- Process design
- Space planning
- General arrangement drawings
- System architecture
- Design specification for major components/subsystems

**Detail**       The individual elements of the overall project deliverable are now broken down to a great level of detail. Each element of the design at this level will probably form a discrete work package in the implementation stage, as well as being a design work package in its own right. Individual components are designed, then integrated to form the work package.

The progression of the design work is controlled by "stage gates." These are shown in Figure 2.6.

A gated design process means that at certain points in the life cycle, the evolving design must pass through stage gates. Part of setting up the design management framework must include deciding which types of gates will be employed, and between which stages they are needed. The basic rules for passing through the gates are noted in the box in Figure 2.6. However, the specific rules that will be applied to the gates will differ according to industrial sector, and usually the criticality of the project to the organization.

Commonly there are three types of gates: "hard," "soft," or "fuzzy." A hard gate is where the design cannot be progressed to the next stage if the gate is not passed. The design process may not move into the following stage until sufficient rework has been done to allow the design to pass through the gate. Soft gates are ones in which the design is allowed to

### FIGURE 2.6.  THE GATED DESIGN PROCESS AND STAGE GATE RULES.

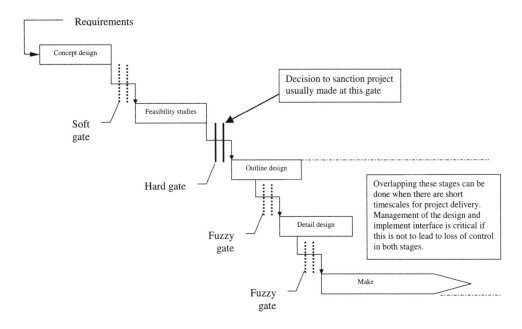

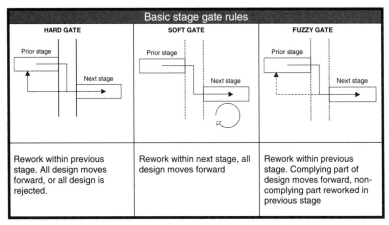

progress to the next stage, even if not being accepted as "compliant" (dependent on the gate's rules). However, a commitment must be made by the person responsible for the design to make changes to the design to ensure that it becomes compliant before the next gate. It is also possible to have "fuzzy" gates, which are essentially a combination of hard and soft gates. In a typical fuzzy gate process, parts of the design may be progressed to the next stage (those that comply with the rules), while the noncomplying parts must be reworked in the previous stage until they do comply. Fuzzy gates are typically used where stages are being overlapped to shorten the overall time to delivery for the project. This type of gate ensures proper attention is given to the rework needed, while not stopping the work in the next stage from progressing.

## Concurrent Engineering

Concurrent engineering is used to keep control of the design and implementation stages when they are overlapped to reduce overall project duration. It means that the early stages of the making of some or all of the deliverables begins before the final design of those deliverables has been decided. This has become common practice in industries supplying consumer goods where time-to-market is one of the dominant project success factors (Shtub et al., 1994). When you are overlapping project stages, it becomes crucial to take a holistic view of the overall project, including:

• Understanding and reviewing the strategic issues that drive the solution to the problem the project has been set up to solve
• Assessing the level of sophistication required in the project deliverable (which in most projects, in most industries, includes deciding on the level of technological innovation to be incorporated in the design solution);
• Assessing process capability to make the various possible solutions
• Deciding the appropriate level of compromise between core project control issues of schedule, budget, and quality and performance
• Determining the through-life costs of the deliverable to the owner—essentially, initial capital, operating, maintenance, and disposal costs (whether internal or external to the organization)
• Understanding the strategy for extending the value the deliverable could bring to the owner during its life

The processes that could help ensure that all these aspects are appropriately managed include value management, project strategy development, quality management, technology management, design for manufacturing/design for assembly, project control, testing, maintainability of the deliverable, product liability, uncertainty management, and others. Moreover, these different aspects of the management of the project design stages need to be managed simultaneously. Concurrent engineering is discussed in detail in the chapter by Thamhain.

## Design for Manufacturing

Generally speaking, the majority of a project's cost is incurred in the implementation stages of the life cycle. This applies whether the project deliverable is a physical object or artifact (typically in construction, mechanical and electrical engineering, electronics, computer hardware and software, and new product development) or nonphysical (such as a changed organization, a financial services product, or other service industry product). The "making" stages of projects typically account for between 75 and 90 percent of the total project expenditure, depending on industrial and technological factors affecting the project. Therefore, anything that reduces the cost of creating the project deliverables should be pursued. One of the biggest cost drivers in projects making physical deliverables is design work that does not take account of the most cost-effective processes for making the deliverables. In industries where there are long production runs for the product created during the project, or where the cost of production is very high (due, for instance, to stringent quality requirements), effective design can significantly improve production costs. Hence, it is clear that manufacturing specialists need to have significant input at the design stages.

The process of bringing in this expertise to design is called design for manufacturing (DFM). DFM aims to optimize the design at the earliest stages to take account of the processes that will be used to make the deliverables. This is not an easy or comfortable approach to design for many designers and manufacturing specialists alike. Figure 2.1 reminds us of the fundamental difference in mental models between designer and implementer. Getting these groups of people to work together effectively is a key task for the person managing the design work. It is important to recognize that for optimal effectiveness, DFM needs to be started at the earliest stages of design, when concepts are being generated for the various solutions to the design problem. There is little point in choosing a concept design to progress into detailed design work if the concept chosen cannot be supported by the existing capability of the organization to make the deliverable (or at least the high-value components of the deliverable). At the least, DFM allows a logical debate to take place about trading off the costs of new manufacturing capability against the attributes of the design that can create extra value in the final product.

The success of DFM can be ensured by recognizing, and acting on, the realization that differing cultures within design and manufacturing exist. The primary obstacle that this difference creates is that of effective communication. There are two key ways to improve communication between these two groups:

- *Plan for communication.* This means identifying where in the design project life cycle DFM will have most effect (invariably early on) and then ensuring appropriate DFM processes are created in time to be used most effectively. It also implies that DFM workshops and review meetings are built into the schedule.
- *Ensure common understanding.* It is far from obvious to designers that the manufacturing process capability required to actually make a design solution may not exist—particularly when an external client is doing the making. However, this lack of knowledge of manufacturing capability is also frequently found when the design will be made in-house. Equally, manufacturing specialists are rarely aware of the specific reasons why a particular feature of the design is necessary to create added value to the client.

The differences in awareness between designers and manufacturing specialists are to be expected. It is up to managers of design to manage the DFM process effectively for the greater good of the organization itself and, where applicable, the external client.

A related design management process is design for assembly (DFA). A major part of the "making" cost for a design solution is the time needed for the assembly of the various components forming the overall product. In such industries as aerospace, power engineering, electronics, and the manufacture of consumer goods, assembly time is heavily influenced by the ease of assembly of the product that will be sold. Consequently, the specialists in assembly processes must be brought into the design process in the same way as the manufacturing experts are involved in DFM. Unsurprisingly, the differences in culture between the designers and assembly specialists are just as evident in the DFA process as for DFM. Communication between the two groups is facilitated in the same way as for DFM: Plan to communicate, and create a situation where common understanding can be gained.

The management processes of DFA and DFM clearly interact with, and affect, the design solution chosen. It is quite possible that the design of a component that has been optimized for manufacturing is very difficult to assemble, adding time (and therefore expense) to the processes that will deliver the final product. Conversely, a design optimized for assembly may be expensive, or even impossible, to make using existing manufacturing process capability. It is incumbent on the design manager to ensure that the correct trade-offs are made between designing for maximum client value, low-cost manufacturing, and ease of assembly.

## Computer-Aided Design and Computer-Aided Manufacturing

Both computer-aided design (CAD) and computer-aided manufacturing (CAM) are part of the wider area of technology of computer-aided engineering (CAE). CAD is part of the fabric of much design work that is carried out, particularly for technically oriented projects (as opposed to business change and other "softer" projects). The initial manifestation of CAD in the mid-1970s was to replace the designer's drawing board, making the production, updating, storage, and transmission of technical design drawings more efficient. The rapid increases in computing power and associated increase in the sophistication of software means that the nature of design work in architecture, new product development, and all sectors of engineering has changed. Current CAD software packages are very powerful tools to help designers generate and test design ideas, working in three dimensions and allowing virtual models to be created. As such, the creative process in design has been affected by the ability to move through the assess-synthesize-evaluate cycle more quickly, and in more detail (particularly the evaluate stage). This means more options for the solution to the design problem can be generated before a concept design is chosen. The ease with which CAD systems share information is another factor that has impacted the way that design is managed. Specifications, drawings, and other design information is transmitted electronically between all groups involved in the design process, from the project owner or sponsor, via the project and design teams, to suppliers of equipment and end users.

Computer-aided manufacturing takes advantage of many aspects of CAD and integrates them with aspects of manufacturing that are computerized. CAM allows design information to be fed directly into such processes as material ordering, manufacturing scheduling, resource management, and testing and quality management. It is commonplace for design information to be fed directly into the manufacturing process and products made, tested, and quality checked without hardcopy information being generated, or indeed any solid "real" prototype being produced.

Essentially one must be aware of how CAD/CAM changes the way people work. The critical issue is in creating design organizations that can make maximum use of the technology available. Frequently this means dispersed "virtual" teams work together on the design processes. Document management is completely redefined with few hard copies of drawings made. Techniques for control and tracking of the design itself are different from those used previously. The fundamentals of design management are not changed by the technology, but the detailed way in which designers and design is managed must suit the tools used.

### Uncertainty (Risk) Management in the Design Process

Uncertainty in the design stages of a project should be actively managed. This is normally done by carrying out risk identification and assessment, and then implementing action plans to reduce the risks or minimize the effects of risks if they actually occur. It is becoming increasingly common to manage opportunities as well as risks, and there are often many opportunities to be found in the design stages. Some of the common risks and opportunities are shown in Table 2.2.

At the project level it can bring significant benefit if those involved in the design stages participate in the risk (and opportunity) management process. Often risks to work in the implementation stages of the project are not identified as having a possible effect on the design process. Design involvement in the overall risk management process can help to ensure these risks are picked up and mitigating actions incorporated into the design schedule and budget.

## Controlling the Design Process

The design work can be controlled as though it was a project in its own right—a project within a project. This approach is fiercely resisted by some designers, for the reasons given earlier.

There is merit, however, in using mechanistic control techniques, so long as

- the design manager (and the project manager as well) do not expect design work to be as predictably controlled as implementation work;
- the techniques are used sensitively—that is, there is explicit recognition of the inherent difficulties posed by controlling design in this way.

Planning for the project and planning for the design stages are inextricably linked. Many inputs to project planning will flow from the earlier stages of the design work. Likewise, these earlier stages will also define the plans for the remaining design stages.

### TABLE 2.2. COMMON RISKS AND OPPORTUNITIES IN THE DESIGN PROCESS.

**Risks**

| | |
|---|---|
| Technology—How well understood is the technology that the design solution is based on? | If the technology is mature, and there is great experience and knowledge in the design firm of working with the technology, there is probably little risk in this area. Conversely, if the technology is new, or the designers have little experience of working with it, then the risk of overrunning the time to produce the design deliverable is high. |
| Change—To requirements or brief | In some sectors, change to the requirements (and design brief) are quite likely as the market is very volatile: New product development is typical. Fast response and flexibility are needed to cope with this situation. |
| Process capability—For both design and making processes | Often the process capability to make the product that has been designed is unknown or untested (typically in precision engineering and similar sectors). Design-for-manufacturing and 'Designfor-assembly are therefore important techniques in this environment. |

**Opportunities**

| | |
|---|---|
| Step change in capability of product | If the right environment and context can be created for designers to work in, there is the possibility of designing a product with a step change improvement over existing products. (Skunk works design environments can help with this, isolating the design team and ensuring they are greatly motivated). |
| Early development of new products that use knowledge gained from the project | The insights and knowledge gained during the design stages of projects about new technology, the application of new manufacturing process capability, and also improved design processes themselves can all contribute to the fast development of additional products, whether they are for the use of internal or external clients. |
| Reduced overall project schedule and budget | There are often opportunities in the design stages to identify ways of delivering the design solution quicker, or of changing the design solution to make it faster to make or implement. |
| Increase value delivered by the project | The value management process begins at the earliest stages of the design phase of the project, and most of the outputs from the process will impact the design solution. Hence, it is vital that designers make a full contribution to all stages of the value process. |

Chapter 1 describes the project control process. The control diagram can be redrawn in the context of design, as shown in Figure 2.7.

The diagram shows that three fundamental control documents need to be generated: the scope of the design work, the schedule to carry out the design work, and the budget for this work. The specific nature of creating these documents for design is briefly described in the following.

## Scope of Design Work

The work required to carry out the outline and detailed design is defined by the solution chosen to deliver the project objective—the concept design. The exact nature of the work required to produce the design is dependent on the nature of the product's deliverables. Software projects involve writing code, creating system architectures, creating and documenting module interface requirements, and so forth. This is very different in *nature* to the work needed to create a new financial product (market analysis, actuarial calculations, investment risk strategies, etc.). What is fundamentally important is to work out what discrete deliverables are needed to make the final project product, then assign design work packages to each of these deliverables, and document this information in the outline design. This process can be quite complicated, although different industries have developed techniques to help this process.

## FIGURE 2.7. THE DESIGN PROCESS CONTROL.

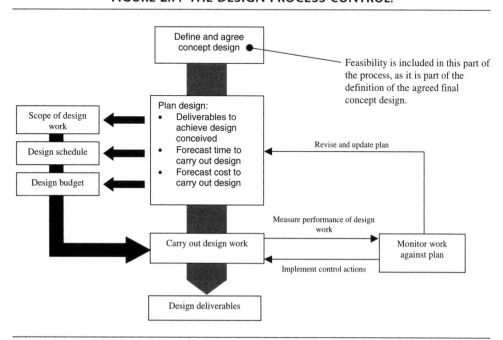

A work breakdown structure (WBS) for design can be created, showing how the individual elements of the design are related to each other. In many technical industries, the design WBS is in fact a description of the system architecture. As such, it will contain information on how each design element is configured—broadly speaking, a description of the interface between each part of the design and the other elements it is directly connected to. However the design work required to be carried out is captured, it forms the central part of the plan for how that work will be carried out. Changes to the scope should be managed by the project change control mechanism.

## Design Work Schedule

Making sure that the design stages are completed in a time frame that is to some extent or other predictable is the key challenge stemming from the creative nature of the work. Forecasting the time durations to complete each work package identified in the design scope is difficult. The durations for some work packages are more difficult to forecast than others—for instance, those that are innovative or in some other way new, or those where a great deal of iterative work is known to be required.

The process for scheduling design work must be based on knowledge of the individual deliverables (drawings, calculations, reports, specifications, and other documents) from the WBS. Reliance on previous experience and considered thought by experts in the field about the work required to be done leads to a set of forecast durations being established.

Scheduling the creation of the design deliverables means putting the work packages in a logical sequence and then calculating the total time required to complete all the work. The sequencing of the work is done by creating a dependency network (see the earlier chapter on project control). The most common tool then used to establish the time to carry out the work on the network's critical path is a Gantt chart—that is, linear scheduling. The difficulty with linear scheduling is that the iteration that is a fundamental aspect of the design process cannot be modeled effectively. Hence, it is not a very satisfactory way of scheduling design, leading to continually adjusting the schedule as the iterative cycles in the design work unfold.

The amount of iteration required between design work packages must be "built in" to the schedule in some way. Traditionally, this is done by adding time to the schedule where there is significant doubt about the likely duration of difficult design work. However, this rather defeats the purpose of creating a logically consistent schedule and can help lead to loss of control of the design stages of the project.

There is a method of scheduling that can overcome this difficulty by using a dependency structure matrix (DSM). The development of DSM originated with systems modeling (which is also where linear scheduling techniques were developed in the 1950s and 1960s). The essence of the technique is the creation of a matrix showing the activities within a system and their dependence on information from each other. The matrix can then be manipulated to show the most effective route through the activities based on information dependency (and that also identifies critical decisions in terms of their impact on other decisions). This means that iterative processes can be more clearly understood and management attention focused on critical information flows. When the project team is managing design schedules, critical design information flows can be spotted and where necessary educated guesses can be made at certain points to keep the overall information flow moving. The guesses are then validated when the true information becomes available and limited and more predictable amounts of rework can be carried out than would otherwise have been likely to be the case. Much work continues on making user-friendly stand-alone and Web-based software available that will carry out DSM scheduling (Austin et al., 2000).

One of the critical considerations when planning design work is to decide the extent of "front-end loading" that will be carried out during the project. *Front-end loading* is the practice of employing a significantly higher number of designers earlier in the design phases of the project than has normally been the case. This concentrates project resource in the project life cycle where there is the greatest opportunity to reduce the overall project time scale. The iterative cycles can be moved through rapidly and the final outline design solution articulated in a much shorter time. In essence, this means concentrating effort at the stage of the project where uncertainty can be removed most effectively—specifically, outline design and early detail design.

Ensuring that many experienced people work on these various options simultaneously helps to reduce the overall time taken at this stage and, with careful management, should lead to a more robust final design being arrived at. Detail design can then be started with less risk that the outline design will have to be revisited (which often means that the design process has to be stopped while the implications of technical risk in the outline design are reassessed). Loading of extra resource is also done at concept, feasibility, and detail stages, as shown in Figure 2.8. However, finding a large number of experienced designers is quite difficult, so there is likely to be a natural limit on how much front-end loading can be carried out.

## Design Work Budget

Forecasting the cost for carrying out design work is a straightforward process, since it is almost entirely the cost of designers' time, usually defined in terms of cost per hour. There is a very small cost element for fixed material costs. There are also overheads to consider (for equipment, offices, management, etc.). This means that the cost to produce the design is directly linked to the time taken to create the design, and, of course, the number of designers employed on the work. When the forecast has been developed, a cost breakdown structure can be built up to allocate budget to specific parts of the work breakdown structure.

## Quality

It is during the design phases of the project that much of the quality of the ultimate project deliverable is established or enabled. The quality process used in the design stages must also ensure high-quality design work per se. A number of aspects need to be covered:

- Accurately capturing the requirements of the client
- Putting in place a design process capable of developing an appropriate solution
- Ensuring that the solution developed satisfies the client's requirements.

## FIGURE 2.8. THE DESIGN RESOURCE PROFILE OR A PROJECT WITH FRONT-END LOADING.

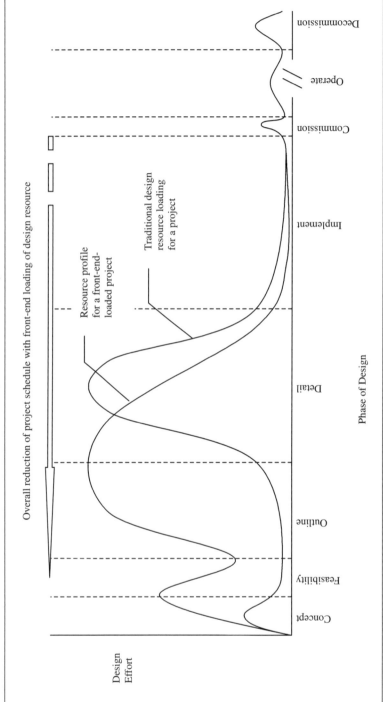

Overall reduction of project schedule with front-end loading of design resource

Resource profile for a front-end-loaded project

Traditional design resource loading for a project

Design Effort

Concept    Feasibility    Outline    Detail    Implement    Commission    Operate    Decommission

Phase of Design

*Source:* Harpum (2002).

Carrying out these activities effectively is dependent on an integrated process for achieving high-quality design. The most well established and comprehensive quality system for the design phases of a project is known as quality function deployment (QFD). QFD monitors the transformation of the client's requirements into the design solution, to ensure that quality is inherent in the solution (Hauser and Clausing, 1988). To do this, QFD integrates the work of people in the project's participant organizations in the following areas:

- Requirements capture (to understand client's business and technical requirements)
- Technology development (to understand what technology is available to be used)
- Implementation (typically DFM and DFA)
- Marketing (to understand the client's perceptions of the solution that satisfies the requirements)
- Management (to understand how the processes to ensure quality can be operationalized)

The primary set of considerations for the QFD team are as follows:

1. *Who* are the clients. In the broadest terms (i.e., the users of the project deliverable, the owner, other stakeholders).
2. *What* are the customer's business requirements. Which may or may not be explicitly stated in a design brief.
3. *How* will these requirements be satisfied. Including an evaluation at the highest level of abstraction, such as should the project actually build a road or a railway to meet the requirement to transport people from A to B?

Figure 2.9 shows how client requirements are matched to the individual elements (subsystems) of the design solution. The importance of the client (or user of the project's

### FIGURE 2.9. THE QUALITY FUNCTION DEPLOYMENT MATRIX.

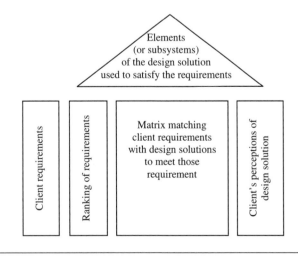

deliverables) to the design process is clear from the diagram. There is a clear and auditable trail from the collection of client requirements through to the client's perception of whether those requirements have been met in the solution proposed.

The client requirements are scored in order of their relative importance and ranked, after a weighting criteria is used. The "roof" of the matrix (QFD is also called the "house of quality") contains the elements of the design solution that will satisfy the requirements of the client. This part of the matrix represents the system that has been designed to meet the project objectives, whether that system is formally recognized as one (a system architecture with sub-systems) or not. The ability of the elements of the design solution to satisfy the requirements is then estimated using experience and judgment in the central matrix. The final aspect of the QFD matrix is to evaluate the client's acceptance of the design solution both at an overall and an elemental level of the matrix. It is important to understand that QFD does not generate the design solution; it enables the quality of the chosen solution to meet the client's requirements to be monitored with rigor and accuracy.

## Summary

This chapter set out to provide the context for design in projects, the strategic management considerations that arise, and how design can be controlled effectively. The way in which design is managed depends on the focus of value creation for the business sector. The relationship between design and the project, and hence between design management and project management, varies tremendously. In some sectors, project management is subservient to design management; in others, project management dominates design management.

The major challenge in managing design work in projects is ensuring the necessary level of integration is achieved with the "making" phases of the work. Because of the differing mental models of the people that work in these two fundamentally different stages of the project, this is not an easy task. Creativity *is* difficult to manage—not the least for the person who is doing the creating! However, there are a number of approaches to organizing design work at the personal and team level, both strategic and tactical, that can help to bring control to the process without threatening the freedom required to be creative.

## References

Allinson, K. 1997. *Getting there by design: An architects guide to design and project management.* Oxford, UK: Architectural Press.

Austin, S., A. Baldwin, B. Li, and P. Waskett. 2000. Application of the analytical design planning technique to construction project management. *Project Management Journal* 31(2):48–59.

Barratt, P. 1999. *Better construction briefing.* Oxford, UK: Blackwell Science.

British Standards Institute. 1996. *BS 7000: Design management systems: Part 4: Guide to managing design in construction.*

Cooper, R. 1995. *The design agenda: A guide to successful design management.* Chichester, UK: Wiley.

Csikszentmihalyi, M. 1996. *Creativity: Flow and the psychology of discovery and invention.* New York: HarperCollins.

Eisner, H. 1997. *Essentials of project and systems engineering management.* New York: Wiley.

Gell-Mann, M. 1994. *The quark and the jaguar.* London: Abacus.

Gray, C., W. Hughes, and J. Bennett. 1994. *The successful management of design.*, Reading, UK: Center for Strategic Studies in Construction, University of Reading.

Groák, S. 1992. *The idea of building: Thought and action in the design and production of buildings.* London: Spon.

Harpum, P. 2002. In *Engineering Project Management,* 2nd ed., ed. N. J. Smith. Oxford, UK: Blackwell Science, 238–263.

Harpum, P., and A. W. Gale. 1999. Achieving success by early project planning and start-up techniques. In *Managing business projects,* ed. K. A. Artto, K. Kähkönen, and K. Koskinen. Espoo, Finland: Project Management Association of Finland.

Hauser, R., and D. Clausing. 1988. The house of quality. *Harvard Business Review* 66 (May–June): 63–73.

Kappel, T. A., and A. H. Rubenstein. 1999. Creativity in design: The contribution of information technology. *IEEE Transactions on Engineering Management* 46 (2, May).

Lawson, B. 1997. *How designers think: The design process demystified.* 3rd ed. Oxford, UK: Elsevier Science.

Luckman, J. 1984. An approach to the management of design. In *Developments in Design Methodology,* ed. N. Cross. Chichester, UK: Wiley.

Morris, P. W. G. 1994. *The management of projects.* London: Thomas Telford.

Pugh, S. 1990. *Total design: Integrated methods for successful product engineering.* Wokingham, UK: Addison-Wesley.

Shtub, A., J. B. Bard, and S. Globerson. 1994. *Project management: Engineering, technology, and implementation.* Upper Saddle River: Prentice Hall.

Simon, H. A. 1981. *The science of the artificial.* Cambridge, MA: MIT Press.

Smith, N. J., ed. 2002, *Engineering project management.* Oxford, UK: Blackwell Science.

Topalian, A. 1980. *The management of design projects.* London: Associated Business Press.

Wearne, S. H., ed. 1989. *Control of engineering projects,* 2nd ed. London: Thomas Telford.

CHAPTER THREE

# CONCURRENT ENGINEERING FOR INTEGRATED PRODUCT DEVELOPMENT

Hans J. Thamhain

When Benjamin Franklin said "time is money," he must have anticipated our fiercely competitive business environment where virtually every organization is under pressure to do more things faster, better, and cheaper. Indeed, for many companies, speed has become one of the great equalizers to competitiveness and a key performance measure. New technologies, especially computers and communications, have removed many of the protective barriers to business, created enormous opportunities, and transformed our global economy into a hypercompetitive enterprise system. To survive and prosper, the new breed of business leaders must deal effectively with time-to-market pressures, innovation, cost, and risks in an increasingly fast-changing global business environment. Concurrent engineering has gradually become the norm for developing and introducing new products, systems, and services (Haque et al., 2003; Yam, 2003).

## The Need for Effective Management Processes

Whether we look at the implementation of a new product, process, or service or we want to build a new bridge or win a campaign, project management has traditionally provided the tools and techniques for executing specific missions, on time and in a resource-efficient manner. These tools and techniques have been around since the dawn of civilization, leading to impressive results from Noah's ark, ancient pyramids, and military campaigns to the Brooklyn Bridge and Ford's Model T automobile. While the first formal project management processes emerged during the Industrial Revolution of the eighteenth century, with focus on mass production, agriculture, construction, and military operations, the recognition of project management as a business discipline and profession did not occur until the 1950s with the emergence of formal organizational concepts such as the matrix, projectized organizations, life cycles, and phased approaches (Morris, 1997).

These concepts established the organizational framework for many of the project-oriented management systems in use today, providing a platform for delivering mission-specific results. Yet the dramatic changes in today's business environment often required the process of project management to be reengineered to deal effectively with the challenges (Denker et al., 2001; Nee and Ong, 2001; Rigby, 1995; Thamhain, 2001) and to balance efficiency, speed, and quality (Atuahene-Gima, 2003). As a result, many new project management tools and delivery systems evolved in recent years under the umbrella of integrated product development (IPD). These systems are, however, not just limited to product developments but can be found in a wide spectrum of modern projects, ranging from construction to research, foreign assistance programs, election campaigns, and IT systems installation (Koufteros et al., 2000; Nellore and Balachandra, 2001). The focus that all of these IPD applications have in common is the effective, integrated, and often concurrent multidisciplinary project team effort toward specific deliverables, the very essence of *concurrent engineering processes*.

## A Spectrum of Contemporary Project Management Systems

Driven by the need for effective multidisciplinary integration and the associated economic benefits, many contemporary project management systems evolved with a focus on cross-functional integration. Many of these contemporary systems evolved from the traditional, well-established *multiphased approaches to project management*. They often focus on specific project environments such as manufacturing, marketing, software development, or field services (Gerwin and Barrowman, 2002). Many mission-specific project management platforms emerged under the umbrella of today's integrated product development (IPD), including design for manufacture (DMF), just-in-time (JIT), continuous process improvement (CPI), integrated product and process development (IPPD), structured systems design (SSD), rolling wave (RW) concept, phased developments (PD), Stage-Gate processes, integrated phase reviews (IPR), and voice-of-the-customer (VOC), just to name a some of the more popular concepts. What all of these systems have in common is the emphasis on effective cross-functional integration and incremental, iterative implementation of project plans. This is precisely the focus of concurrent engineering (CE), perhaps one of the most widely used IPD concepts, today.

## Concurrent Engineering—A Unique Project Management Concept

*Concurrent engineering, CE,* is an extension of the multiphased approach to project management. At the heart of its concept is the concurrent execution of tasks segments, which creates overlap and interaction among the various project teams. It also increases the need for strong cross-functional integration and team involvement, which creates both managerial benefits and challenges (Wu, Fuh, and Nee, 2002). While concurrent engineering was originally seen as a method for primarily reducing project cycle time and accelerating product developments (Prased et al., 2003; Prased, 1998), today, the concept refers quite generally to the most resource- and time-efficient execution of multidisciplinary undertakings.

Moreover, the CE concept has been expanded from its original engineering focus to a wide range of projects, ranging from construction and field installations to medical procedures, theater productions, and financial services (Dimov and Setchi, 1999; Pilkinton and Dyerson, 2002; Skelton and Thamhain, 1993). The operational and strategic values of concurrent engineering are much broader than just a gain in lead time and resource effectiveness, but include a wide range of benefits to the enterprise, as summarized in Table 3.1. These benefits are primarily derived from effective cross-functional collaboration and full integration of the project management process with the total enterprise and its supply chain (Prasad et al., 1998, 2003). In this context, concurrent engineering provides a process template for effectively managing projects. Virtually any project can benefit from this approach as pointed out by the Society for Concurrent Product Development (www.soce.org).

As a working definition, the following statement brings the management philosophy of concurrent engineering into perspective:

> Concurrent engineering provides the managerial framework for effective, systematic, and concurrent integration of all functional disciplines necessary for producing the desirable

## TABLE 3.1. POTENTIAL BENEFITS OF CONCURRENT ENGINEERING.

- Better cross-functional *communication* and *integration*
- Decreased *time-to-market*
- Early detection of *design problems*, fewer *design errors*
- Emphasizes human side of *multidisciplinary teamwork*
- Encourages *power sharing, cooperation, trust, respect,* and *consensus building*
- Engages all stakeholders in *information sharing* and *decision making*
- Enhances ability to support *multisite manufacturing*
- Enhances ability for coping with *changing requirements, technology,* and *markets*
- Enhances ability for executing *complex projects* and *long-range* undertakings
- Enhances *supplier communication*
- Fewer *engineering changes*
- High-level of *organizational transparency*, R&D-to-marketing
- Higher *resource efficiency* and *personnel productivity*; more resource-effective project implementation
- Higher *project quality*, measured by customer satisfaction
- Minimizes "downstream" *uncertainty, risk,* and *complications*; makes the project *outcome more predictable*
- Minimizes design-build-rollout *reworks*
- Ongoing *recognition and visibility of team accomplishments*
- Promotes *total project life cycle thinking*
- Provides a *template or roadmap* for guiding multiphased projects from concept to final delivery
- Provides *systematic approach* to multiphased project execution
- Shorter *project life cycle* and execution time
- *Validation of work in progress* and deliverables

project deliverables, in the least amount of time and resource requirements, considering all elements of the product life cycle.

In essence, concurrent engineering is a systematic approach to integrated project execution that emphasizes parallel, integrated execution of project phases, replacing the traditional linear process of serial engineering and expensive design-build-rollout rework. The process also requires strong attention to the human side, focusing on multidisciplinary teamwork, power sharing, and team values of cooperation, trust, respect, and consensus building, engaging all stakeholders in the sharing of information and decision making. In addition, the process must start during the early project formation stages and continue over the project life cycle.

The concurrent engineering process is graphically shown in Figure 3.1, depicting a typical product development. In its basic form, the process provides a template or roadmap for guiding multiphased projects from concept to final delivery. One of the prime objectives for using concurrent engineering is to minimize "downstream" uncertainty, risk, and complications, and hence make the project outcome more predictable (Iansiti and MacCormack, 1997; Liker and Ward, 1998; Moffat, 1998; Noori, Munro, and Deszca, 1997; O'Connor, 1994; Sobek, 1998). However, concurrent execution and integration of activities does not just happen by drawing timelines in parallel but is the result of carefully defined cross-functional linkages and skillfully orchestrated teamwork. Moreover, concurrent phase exe-

## FIGURE 3.1. GRAPHICAL PRESENTATION OF CONCURRENT PROJECT PHASE EXECUTION.

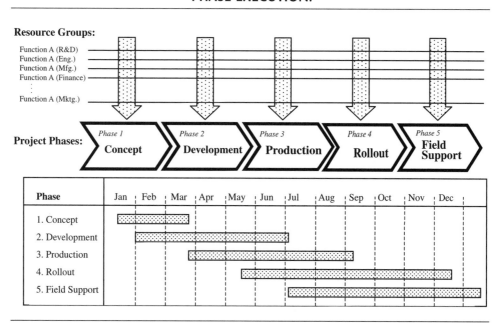

cution makes several assumptions regarding the organizational system and its people, as summarized in Table 3.2 and discussed in the next section.

For many managers and researchers the concept of concurrent engineering is synonymous with integrated product development because using concurrent engineering is to minimize "downstream" uncertainty, risk and complications, and hence make the project outcome more predictable (Iansiti and MacCormack, 1997; Liker, and Ward 1998; Moffat, 1998; Noori, Munro, and Deszca). For simplicity, concurrent engineering is often shown as a linear process, with overlapping activity phases, scheduled for concurrent execution, such as shown in Figure 3.1.

### TABLE 3.2. CRITERIA FOR SUCCESSFULLY MANAGING CONCURRENT ENGINEERING PROJECTS.

*Organizations and concurrent engineering teams must be able to (listed in approximate chronological order) do the following:*

- Allocate sufficient time and resources for up-front planning.
- Identify major task teams, their mission, and interfaces at the beginning of the project cycle.
- Work out the logistics and protocol for concurrent phase implementation.
- Lay out the master project plan (top level) covering the project life cycle.
- Establish consensus on project plan among project team members.
- Be willing to work with partial, incremental inputs, and evolving requirements throughout the team organization, throughout the project life cycle.
- Identify all project-internal and -external "customers" of its work, and establish effective communication linkages and ongoing working relations with these customers.
- Work flexibly with team members and customers, adjusting to evolving needs and requirements.
- Share information and partial results regularly during the project implementation.
- Identify the specific deliverables needed by other teams (and individuals) as inputs for their part of the project, including the timing for such deliverables.
- Establish effective cross-functional communication channels and specific methods for work transfer.
- Establish techniques and protocols for validating the work and its appropriateness to its "customers" on an ongoing basis.
- Work with partial results (deliverables) and incremental updates from upstream developments.
- Reiterate or modify tasks and deliverables to accommodate emerging needs of downstream task teams and to optimize the evolving project outcome.
- Prepare for its mission prior to receiving mission details (e.g., manufacturing is expected to work on pilot production setup prior to receiving full product specs or prototypes).
- Work as an integrated part of a unified and agreed-on project plan.
- Have tolerance for ambiguity and uncertainty.
- Establish reward systems that promote cross-functional cooperation, collaboration, and joint ownership of results.
- Have top management buy-in and support to concurrent engineering process.
- Have established a uniform project management system throughout the concurrent engineering team / organization.

## Criteria for Success

To make such concurrent project phasing possible, the organizational process must be designed to meet specific criteria that establish the conditions conducive to concurrent, incremental implementation of phased activities, such as summarized in Table 3.2. By its very definition, concurrent engineering is synonymous with cross-disciplinary cooperation, involving all project teams and support groups of the enterprise, internally and externally, throughout the project life cycle. The CE process relies on organizational linkages and integrators that help in identifying problems early, networking information, transferring technology, satisfying the needs of all stakeholders, and unifying the project team. It is important to include all project stakeholders in the project team and its management, not only enterprise-internal components, such as R&D, engineering, manufacturing, marketing, and administrative support functions but also external stakeholders, such as customers, suppliers, regulators, and other business partners.

Taken together, the core ingredient of successful concurrent engineering is the development and effective management of organizational interfaces. For most organizations, these challenges include strong human components that are more difficult to harness and to control than the operational processes of project implementation (Prased, 1998). They involve many complex and constantly changing variables that are hard to measure and even more difficult to manage, especially within self-directed team environments that are often required for realizing the concurrent engineering process (Banly and Nee 2000; Hall et al., 1996). While procedures provide (1) the baseline and infrastructure necessary to connect and integrate the various pieces of the multidisciplinary work process and (2) an important starting point for defining the communication channels, and are necessary for effectively linking the core team with all of its support functions, the resulting process is only as good as the team that implements it.

## Defining the Process

After reaching a principle agreement with major stakeholders, the concurrent engineering process should be defined, showing the major activity phases or stages of the project to be executed. Even more advantageous for future projects would be the ability to define phases that may be common to a *class of projects* that is being executed by the enterprise over time. To illustrate, let us use the example of a new product development, shown in Figure 3.1, which proceeds through five project phases: (1) concept development; (2) detailed development; (3) pilot production; (4) product rollout, launch, and marketing; and (5) field support. Each phase or stage is defined in terms of principle scope, objectives, activities, and deliverables, as well as functional responsibilities. Each project phase must also include cross-functional interface protocols, defining the specific collaborations and organizational linkages needed for the concurrent development. While the principle cross-functional interfaces can be summarized graphically, as shown in the upper part of Figure 3.1, more sophisticated group technology tools, such as the quality function deployment (QFD) matrix, shown in Figure 3.2, are usually needed for defining (1) the specific cross-functional requirements,

## FIGURE 3.2. QUALITY FUNCTION DEPLOYMENT (QFD) MATRIX, A TOOL FOR DEFINING INTERFACES.

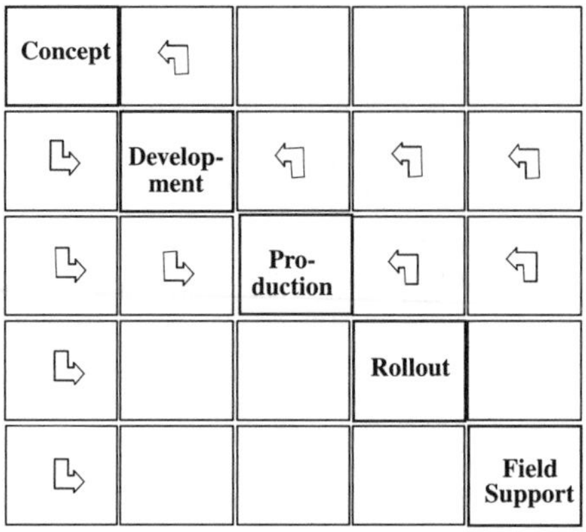

| Concept | ⤦ | | | |
|---|---|---|---|---|
| ⤵ | Develop-ment | ⤦ | ⤦ | ⤦ |
| ⤵ | ⤵ | Pro-duction | ⤦ | ⤦ |
| ⤵ | | | Rollout | |
| ⤵ | | | | Field Support |

(2) the methods of work transfer (often referred to as technology transfer), and (3) the stakeholder interactions necessary for capturing and effectively dealing with the changes that ripple through the product design process.

The best time for setting up these interface protocols is during the definition phase of a specific project when the team organization is most flexible regarding lines of responsibility and authority. To illustrate, Figure 3.2 shows the specific inputs and outputs required during the various phases of a product development process. Each arrow indicates that a specific input/output requirement exists for that particular interface. Most likely, some interface requirements exist for each project phase to each of the others. In our example, the total number of potential interfaces is defined by the 5 × 5 matrix, which equals 25 interfaces (this explains why the QFD Matrix is also referred to as the N-Squared Chart). The QFD Matrix is a useful tool for identifying specific interface personnel and input/output requirements. That is, for each interface, key personnel from both teams have to establish personal contacts and negotiate the specific type and timing of deliverables needed. In many cases, multiple interfaces exist simultaneously, necessitating complex multiteam agreements over project integration issues. An additional challenge is the incremental nature of deliverables resulting from the concurrent project execution. For downstream phases, such as production, to start their work concurrently with earlier project phases, such as product development, it is necessary for all interfaces to define and negotiate (1) what part of the phase deliverables can be transferred "early," (2) the exact schedule for the partial deliverable, and (3) the validation, iteration, and integration process for these partial deliverables.

Yet, another important condition for concurrent engineering to work is the ability of "downstream task leaders" to guide the "upstream" design process toward desired results, and to define the upstream gate criteria on which they depend as "customers." This interdisciplinary integration is often accomplished by participating in project and design reviews, by soliciting and providing feedback on work-in-progress, and by cross-functional involvement with interface definitions and technology transfer processes. Interface diagrams, such as the QFD Matrix shown in Figure 3.2, can help to define the cross-functional roadmap for establishing and sustaining the required linkages for each task group.

## Hidden Challenges and Benefits

In spite of all its potential benefits for more effective project implementation, including higher quality, speed, and resource effectiveness, project implementation, concurrent engineering holds many organizational challenges regarding its management. Some of the toughest challenges relate to the compatibility of concurrent engineering with the organizational culture and its values. Concurrent engineering requires a collaborative culture and a great deal of organizational power sharing, which is often not present in an enterprise to the degree required for concurrent engineering to succeed. Designing, customizing, and implementing a new project management system usually affects many organizational subsystems and processes, from innovation to decision making, and from cross-functional communications to the ability of dealing effectively with risk and organizational conflict. Hence, integrating concurrent engineering into a business process and its physical, informational, managerial, and psychological subsystems without compromising business performance is an important issue that must be dealt with during the implementation phase. Strong involvement of people from all levels throughout the organization is required for concurrent engineering to become institutionalized and to be used effectively by the people in the organization.

Why are companies doing it? Few companies go into a major reorganization of their business processes lightly. At best, introducing a new process is painful, disruptive, and costly. At worst, it can destroy established operational effectiveness and the ability to compete successfully in the marketplace. Obviously, companies that adopt concurrent engineering have powerful reasons for using this contemporary concept of project management. These companies are able to use concurrent engineering as an organizational platform to *increase project effectiveness, quality, and ultimately reduce recourse needs and cycle time.*

## Understanding the Organizational Components

The preceding benefits are not always obvious, looking at the basic concept of concurrent engineering, because they are often derivatives of more subtle organizational characteristics that unfold within a well-executed concurrent project management system. These characteristics need to be understood and skillfully exploited for project leaders and managers to gain the full benefits *of concurrent engineering.*

1. *Uniform Process Model.* The concurrent engineering concept provides a uniform process model or template for organizing and executing a predefined class of projects, such as specific new product developments.

   *Primary benefits:* Time and resource savings during the project/product planning and start-up phase.

*Secondary benefits:* Standardized process model breaks the project cycle into smaller, predefined modules or phases, resolving some of the project complexities, predefining potential risks and areas requiring managerial interactions and support. Standardized platform for project execution provides basis for continuous process improvement and organizational learning.

2. *Integrated Product Development (IPD).* Because of its focus on cross-functional cooperation, concurrent engineering promotes an integrated approach to product development and other project work.

*Primary benefits:* Promotes unified, collective understanding of project challenges and search for innovative solutions. Helps in team integration: identifying organizational interfaces, lowering risks and reducing cycle time.

*Secondary benefits:* Responsibilities for team and functional support personnel are more visible.

3. *Gate Functions.* The concurrent engineering platform is similar to other multiphased project management concepts, such as Stage-Gate, structured systems design, or rolling wave concepts, hence encouraging the integration of predefined gates, providing for performance reviews, sign-off criteria, checkpoints, and early warning systems.

*Primary benefits:* Ensures incremental guidance of the product/project execution and early problem detection, provides cross-functional accountability, helps in identifying risk and problem areas, minimizes rework, highlights organizational interfaces and responsibilities.

*Secondary benefits:* Stimulates cross-functional involvement and visibility; identifies internal customers, promotes full life cycle planning, focuses on win strategy.

4. *Standard Project Management Process.* The concurrent engineering concept is compatible with the standard project management process, its tools, techniques and standards. Predefined gates provide performance and sign-off criteria, checkpoints, and early warning systems, ensuring incremental guidance of the product development process and early problem detection.

*Primary benefits:* Provides cross-functional accountability, helps in identifying risk and problem areas, minimizes rework, highlights organizational interfaces and responsibilities.

*Secondary benefits:* Stimulates cross-functional involvement and visibility, identifies internal customers, promotes full life cycle planning, focuses on win-strategy.

5. *QFD Approach.* Using the quality function deployment (QFD) concept, built into the concurrent engineering process, helps to define cross-functional interfaces and provides pressures on both the performing and receiving organization toward closer cooperation and "upstream" guidance of the product development.

*Primary benefits:* Provides an input/output model for identifying work flow throughout the project/product development process, identifies organizational interfaces and responsibilities.

*Secondary benefits.* Stimulates cross-functional involvement and visibility; identifies internal customers, promotes full life cycle planning.

6. *Early testing.* Concurrent engineering encourages early testing of overall project or product functionality, features, and performance. These tests are driven by team members of both downstream and upstream project phases. Downstream members seek assurances for problem-free transfer of the work into their units, and upstream members seek smooth transfer and sign-off for their work completed.

   *Primary benefits:* Early problem detection and risk identification, opportunity to "fail early and cheap," less rework.

   *Secondary benefits:* Stimulates cross-functional involvement and cooperation, assists system integration.

7. *Total organizational involvement and transparency.* Because of its emphasis on mutual dependencies among the various phase teams, strong cross-functional involvement and teamwork is encouraged, enhancing the level of visibility and organizational transparency.

   *Primary benefits:* Total development cycle/system thinking, enhanced cross-functional innovation, effective teamwork, enhanced cross-functional communications and product integration, early warning system, improved problem detection and risk identification, enhanced flexibility toward changing requirements.

   *Secondary benefits:* Total team recognition; enhanced team spirit and motivation, conducive to self-direction and self-control.

Taken together, the top benefits of concurrent engineering refer to time, resource, and risk issues that ultimately translate into increased project performance: (1) reducing project start-up time, (2) reducing project cycle time, (3) detecting and resolving problems early, (4) promoting system integration, (5) promoting early concept testing, (6) minimizing rework, (7) handling more complex projects with higher levels of implementation uncertainty, (8) working more resource effective, and (9) gaining higher levels of customer satisfaction.

# Recommendations for Effective Management

A number of specific suggestions may help managers understand the complex interaction of organizational and behavioral variables involved in establishing a concurrent engineering process and managing projects effectively in such a system. The sequence of recommendations follows to some degree the chronology of concurrent engineering system design-implementation-management. Although each organization is unique with regard to its business, operation, culture, and management style, field studies show a general agreement on the type of factors that are critical to effectively organizing and managing projects in concurrent multiphase environments (Denker, 2001; Harkins, 1998; Nellore, 2001; Pilllai, 2002; Prasad, 1977; Thamhain and Wilemon, 1998).

## Phase I: Organizational System Design

***Take a Systems Approach.*** The concurrent engineering system must eventually function as a fully interconnected subsystem of the organization and should be designed as an integrated part of the total enterprise (Harque, Pawar, and Barson, 2003). Field studies emphasize consistently that management systems function suboptimal, at best, or fail because of a poor understanding of the interfaces that connect the new system with the total business process (Kerzner, 2001; Moffat, 1998). System thinking, as described by Senge (2001), Checkland (1999), and Emery and Trist (1965), provides a useful approach for front-end analysis and organization design.

**Build on Existing Management Systems.** Radically new methods are usually greeted with anxiety and suspicion. If possible, the introduction of a new organizational system, such as concurrent engineering, should be consistent with already established project management processes and practices within the organization. The more congruent the new operation is with the already existing practices, procedures, and distributed knowledge within the organization, the more cooperation management will find from their people toward implementing the new system. The highest level of acceptance and success is found in areas where new procedures and tools are added incrementally to already existing management systems. These situations should be identified and addressed first. Building upon an existing project management system also facilitates incremental enhancement, testing, and fine-tuning of the new concurrent engineering process. Particular attention should be paid to the cross-functional workability of the new process.

**Custom-Design.** Even for apparently simple situations, a new concurrent engineering process should be customized to fit the host organization, its culture, needs, norms, and processes (Hull, Collins, and Liker, 1996). For reasons discussed in the previous paragraph, the new system has a better chance for smooth implementation and for gaining organizational acceptance if the new process appears consistent with already established values, principles, and practices, rather than a new order to be imposed without reference to the existing organizational history, values or culture (cf. Swink, Sandvig, and Mabert, 1996; Kerzner, 2001).

## Phase II: System Implementation

**Define Implementation Plan.** Implementation of the new concurrent engineering system is by itself a complex, multidisciplinary project that requires a clear plan with specific milestones, resource allocations, responsibilities, and performance metrics. Further, implementation plans should be designed for measurability, early problem detection and resolution, and visibility of accomplishments, providing the basis for recognition, and rewards.

**Pretest the New Technique.** Preferably, any new management system should be pilot tested on small projects with an experienced project team. Asking a team to test, evaluate, and fine-tune a new concurrent engineering process is often seen as an honor and professional challenge. It also starts the implementation with a positive attitude, creating an environment of open communications, candor, and focus on actions toward success.

**Ensure Good Management Direction and Leadership.** Organizational change, such as the implementation of a concurrent engineering system, requires top-down leadership and support to succeed. Team members will be more likely to help implement the concurrent engineering system, and cooperate with the necessary organizational requirements, if management clearly articulates its criticality to business performance and the benefits to the organization and its members. People in the organization must perceive the objectives of the intervention to be attainable and have a clear sense of direction and purpose for reaching these goals. Senior management involvement and encouragement are often seen as an endorsement of the team's competence and recognition of their efforts and accomplishments (Thamhain and Wilemon, 1998). Throughout the implementation phase, senior management can influence the attitude and commitment of their people toward the new concept of concurrent engineering. Concern for project team members, assistance with the use of the tool, enthusiasm for the project and its administrative support systems, proper funding, help in attracting and retaining the right personnel, support from other resource groups—all will foster a climate of high involvement, motivation, open communications, and desire to make the new concurrent engineering system successful.

Involve people affected by the new system. The implementation of a new management system involves considerable organizational change with all the expected anxieties and challenges. Proper involvement of relevant organizational members is often critical to success (Barlett and Ghoshal, 1995; Nellore and Balachandra, 2001). Key project personnel and managers from all functions and levels of the organization should be involved in assessing the situation, evaluating the new tool, and customizing its application. While direct participation in decision making is the most effective way to obtain buy-in toward a new system (Pham, Dimov, and Setchi, 1999), it is not always possible, especially in large organizations. Critical factor analysis, focus groups, and process action teams are good vehicles for team involvement and collective decision making, leading to ownership, greater acceptance, and willingness to work toward successful implementation of the new management process (Thamhain, 2001).

***Anticipate Anxieties and Conflicts.*** A new management system, such as concurrent engineering, is often perceived as imposing new management controls, seen as disruptive to the work process and creating new rules and administrative requirements. People responses to such new systems range from personal discomfort with skill requirements to dysfunctional anxieties over the impact of tools on work processes and performance evaluations (Sundaramurthy and Lewis, 2003). Effective managers seem to know these challenges intuitively, anticipating the problems and attacking them aggressively as early as possible. Managers can help in developing guidelines for dealing with problems and establishing conflict resolution processes, such as information meetings, management briefings, and workshops, featuring the experiences of early adopters. They can also work with the system implementers to foster an environment of mutual trust and cooperation. Buy-in to the new process and its tools can be expected only if its use is relatively risk-free (Stum, 2001). Unnecessary references to performance appraisals, tight supervision, reduced personal freedom and autonomy, and overhead requirements should be avoided, and specific concerns dealt with promptly on a personal level.

***Detect Problems Early and Resolve.*** Cross-functional processes, such as concurrent engineering, are often highly disruptive to the core functions and business process of a company (Denker, Steward, and Browning, 2001; Haque, 2003). Problems, conflict, and anxieties over technical, personal, or organizational issues are very natural and can be even healthy in fine-tuning and validating the new system. In their early stages, these problems are easy to solve but usually hard to detect. Management must keep an eye on the organizational process and its people to detect and facilitate resolution of dysfunctional problems. Round-table discussions, open-door policies, focus groups, process action teams, and management by wandering around are good vehicles for team involvement leading to organizational transparency and a favorable ambience for collective problem identification, analysis, and resolution.

*Encourage Project Teams to Fine-Tune the Process.* Successful implementation of a concurrent engineering system often requires modifications of organizational processes, policies, and practices. In many of the most effective organizations, project teams have the power and are encouraged to make changes to existing organizational procedures, reporting relations, and decision and work processes. It is crucial, however, that these team initiatives are integrated with the overall business process and supported by management. True integration, acceptance by the people, and sustaining of the new organizational process will only occur through the collective understanding of all the issues and a positive feeling that the process is helpful to the work to be performed. To optimize the benefits of concurrent engineering, it must be perceived by all the parties as a win-win proposition. Providing people with an active role in the implementation and utilization process helps to build such a favorable image for participant buy-in and commitment. Focus teams, review panels, open discussion meetings, suggestion systems, pilot test groups, and management reviews are examples for providing such stakeholder involvement.

*Invest Time and Resources.* Management must invest time and resources for developing a new organizational system. An intricate system, such as concurrent engineering, cannot be effectively implemented just via management directives or procedures, but instead requires the broad involvement of all user groups, helping to define metrics and project controls. System designers and project leaders must work together with upper management toward implementation. This demonstrates management confidence, ownership, and commitment to the new management process. This will also help to integrate the new system with the overall business process. As part of the implementation plan, management must allow time for the people to familiarize themselves with the new vision and process. Training programs, pilot runs, internal consulting support, fully leveraged communication tools such as groupware, and best-practice reviews are examples of action tools that can help in both institutionalizing and fine-tuning the new management system. These tools also help in building the necessary user competencies, management skills, organization culture, and personal attitudes required for concurrent engineering to succeed.

## Phase III: Managing in Concurrent Engineering

*Plan the Project Effectively.* As for any other project management system, effective project planning and team involvement is crucial to success. This is especially important in the concurrent engineering environment where parallel task execution depends on continuous cross-functional cooperation for dealing with the incremental work flow and partial result transfers. Team involvement, early in the project life cycle, will also have a favorable impact on the team environment, building enthusiasm toward the assignment, team morale, and ultimately team effectiveness. Because project leaders have to integrate various tasks across many functional lines, proper planning requires the participation of all stakeholders, including support departments, subcontractors, and management. Modern project management techniques, such as phased project planning and Stage-Gate concepts, plus established standards such as PMBOK, provide the conceptional framework and tools for effective cross-functional planning and organizing the work toward effective execution.

**Define Work Process and Team Structure.** Successful project management in concurrent engineering requires an infrastructure conducive to cross-functional teamwork and technology transfer. This includes properly defined interfaces, task responsibilities, reporting relations, communication channels, and work transfer protocols. The tools for systematically describing the work process and team structure come from the conventional project management system; they include (1) a *project charter*, defining the mission and overall responsibilities of the project organization, including performance measures and key interfaces; (2) a *project organization chart*, defining the major reporting and authority relationships; (3) *responsibility matrix* or *task roster*; (4) *project interface chart*, such as the N-Squared Chart discussed earlier; and (5) *job descriptions*.

**Develop Organizational Interfaces.** Overall success of a concurrent engineering depends on effective cross-functional integration. Each task team should clearly understand its task inputs and outputs, interface personnel, and work transfer mechanism. Team-based reward systems can help to facilitate cooperation with cross-functional partners. Team members should be encouraged to check out early feasibility and system integration. QFD concepts, N-Square charting, and well-defined phase-gate criteria can be useful tools for developing cross-functional linkages and promoting interdisciplinary cooperation and alliances. It is critically important to include into these interfaces all of the support organizations, such as purchasing, product assurance, and legal services, as well as outside contractors and suppliers.

**Staff and Organize the Project Team.** Project staffing is a major activity, usually conducted during the project formation phase. Because of time pressures, staffing is often done hastily and prior to defining the basic work to be performed. The result is often team personnel that is suboptimally matched to the job requirements, resulting in conflict, low morale, suboptimum decision making and ultimately poor project performance. While this deficiency will cause problems for any project organization, it is especially unfavorable in a concurrent engineering project environment that relies on strong cross-functional teamwork and shared decision making, built on mutual trust, respect, and credibility. Team personnel with poorly matched skill sets to job requirements is seen as incompetent, affecting their trust, respect, and credibility and ultimately their "concurrent team performance." For best results, project leaders should *negotiate the work assignment* with their team members one-to-one, at the outset of the project. These negotiations should include the overall task, its scope, objectives, and performance measures. A thorough understanding of the task requirements develops often as the result of personal involvement in the front-end activities, such as requirements analysis, bid proposals, project planning, interface definition, or the concurrent engineering system development. This early involvement also has positive effects on the buy-in toward project objectives, plan acceptance, and the unification of the task team.

***Communicate Organizational Goals and Objectives.*** Management must communicate and update the organizational goals and project objectives. The relationship and contribution of individual work to overall business plans and their goals, as well as of individual project objectives and their importance to the organizational mission, must be clear to all team personnel. Senior management can help in unifying the team behind the project objectives by developing a "priority image" through their personal involvement, visible support, and emphasis of project goals and mission objectives.

***Build a High-Performance Image.*** Building a favorable image for an ongoing project, in terms of high-priority, interesting work; importance to the organization; high visibility; and potential for professional rewards are all crucial for attracting and holding high-quality people. Senior management can help develop a "priority image" and communicate the key parameters and management guidelines for specific projects (Pham, Dimov, and Setchi, 1999). Moreover, establishing and communicating clear and stable top-down objectives helps in building an image of high visibility, importance, priority, and interesting work. Such a pervasive process fosters a climate of active participation at all levels and helps attract and hold quality people, unifies the team, and minimizes dysfunctional conflict.

***Build Enthusiasm and Excitement.*** Whenever possible, managers should try to accommodate the professional interests and desires of their personnel. Interesting and challenging work is a perception that can be enhanced by the visibility of the work, management attention and support, priority image, and the overlap of personnel values and perceived benefits with organizational objectives (Thamhain, 2003). Making work more interesting leads to increased involvement, better communication, lower conflict, higher commitment, stronger work effort, and higher levels of creativity.

***Define Effective Communication Channels.*** Poor communication is a major barrier to teamwork and effective project performance, especially in concurrent engineering environments, which depend to a large degree on information sharing for their concurrent execution and decision making. Management can facilitate the free flow of information, both horizontally and vertically, by workspace design, regular meetings, reviews, and information sessions (Hauptman and Hirji, 1999). In addition, modern technology, such as voice mail, e-mail, electronic bulletin boards, and conferencing, can greatly enhance communications, especially in complex organizational settings.

***Create Proper Reward Systems.*** Personnel evaluation and reward systems should be designed to reflect the desired power equilibrium and authority/responsibility sharing needed for the concurrent engineering organization to function effectively. Creating a system and its metrics for reliably assessing performance in a concurrent engineering environment is a great challenge. However, several models, such as the Integrated Performance Index (Pillai, Joshi, and Rao, 2002), have been proposed and provide a potential starting point for customization. A QFD philosophy, where everyone recognizes the immediate "customer" for whom a task is performed, helps to focus efforts toward desired results and customer satisfaction. This customer orientation should exist, both downstream and upstream, for both company-internal and -external customers. These "customers" should score the performance of the deliverables they received and therefore have a major influence on the individual and team rewards.

***Ensure Senior Management Support.*** It is critically important that senior management provides the proper environment for a project team to function effectively (Prasad, 1998). At the onset of a new project, the responsible manager needs to negotiate the needed resources with the sponsor organization and obtain commitment from management that these resources will be available. An effective working relationship among resource managers, project leaders, and senior management critically affects the credibility, visibility, and priority of the engineering team and their work.

***Build Commitment.*** Managers should ensure team member commitment to their project plans, specific objectives, and results. If such commitments appear weak, managers should determine the reason for such lack of commitment of a team member and attempt to modify possible negative views. Anxieties and fear of the unknown are often a major reason for low commitment (Stum, 2001). Managers should investigate the potential for insecurities, determine the cause, and then work with the team members to reduce these negative perceptions. Conflict with other team members and lack of interest in the project may be other reasons for such lack of commitment.

***Manage Conflict and Problems.*** Conflict is inevitable in the concurrent engineering environment with its complex dynamics of power and resource sharing, and incremental decision making. Project managers should focus their efforts on problem avoidance. That is, managers and team leaders, through experience, should recognize potential problems and conflicts at their onset, and deal with them before they become big and their resolutions consume a large amount of time and effort (Haque, 2003).

***Conduct Team Building Sessions.*** A mixture of focus team sessions, brainstorming, experience exchanges, and social gatherings can be powerful tools for developing the concurrent work group into an effective, fully integrated, and unified project team (Thamhain and Wilemon, 1999). Such organized team-building efforts should be conducted throughout the project life cycle. Intensive team-building efforts may be especially needed during the formation stage of the concurrent engineering team. Although formally organized and managed, these team-building sessions are often conducted in a very informal and relaxed atmosphere to discuss critical questions such as (1) how are we working as a team? (2) what is our strength? (3) how can we improve? (4) what support do you need? (5) what challenges and problems are we likely to face? (6) what actions should we take? and (7) what process or procedural changes would be beneficial?

***Ensure Personal Drive and Involvement.*** Project managers and team leaders can influence the concurrent engineering environment by their own actions. Concern for their team members, the ability to integrate personal needs of their staff with the goals of the organization, and the ability to create personal enthusiasm for a particular project all can foster a climate of high motivation, work involvement, open communication, and ultimately high engineering performance.

***Provide Proper Direction and Leadership.*** Managers can influence the attitude and commitment of their people toward concurrent engineering as a project management tool by their own actions. Concern for the project team members, assistance with the use of the tool, and enthusiasm for the project and its administrative support systems can foster a climate of high motivation, involvement with the project and its management, open communications, and willingness to cooperate with the new requirements and to use them effectively.

***Foster a Culture of Continuous Support and Improvement.*** Successful project management focuses on people behavior and their roles within the project itself. Companies that effectively manage projects, and reap the benefits from concurrent engineering, have cultures and support systems that demand broad participation in their organization developments. Ensuring organizational members to be proactive and aggressive toward change is not an easy task, yet it must be facilitated systematically by management. Our continuously changing business environment requires that provisions are being made for updating and fine-tuning the established concurrent engineering process. Such updating must be done on an ongoing basis to ensure relevancy to today's project management challenges. It is important to establish support systems—such as discussion groups, action teams and suggestion systems—to capture and leverage the lessons learned and to identify problems as part of a continuous improvement process.

# Summary

In today's dynamic and hypercompetitive environment, proper implementation and use of concurrent engineering is critical for expedient and resource-effective project execution. The full range of benefits of concurrent engineering is in fact much broader than just a gain in lead time and resource effectiveness, but includes a wide spectrum of competitive advantages to the enterprise, ranging from increased quality of project deliverables to the ability of executing more complex projects and to higher levels of customer satisfaction. These benefits are primarily derived from effective cross-functional collaboration and full integration of the project management process with the total enterprise and its supply chain. However, these benefits do not occur automatically!

Designing, implementing, and managing in concurrent engineering requires more than just writing a new procedure, delivering a best-practice-workshop, or installing new information technology. It requires the ability to engage the organization in a systematic evaluation of specific competencies, such as for concurrent engineering, assessing opportunities for improvement, and designing a project management system that is fully integrated with the overall enterprise system and its strategy. Too many managers end up disappointed that the latest management technique did not produce the desired result. Regardless of its conceptual sophistication, concurrent engineering is just a framework for processing project data, aligning organizational strategy, structure, and people. To produce benefits for the firm, these tools must be fully customized to fit the business process and be congruent with the organizational system and its culture.

One of the most striking finding, from both the practice and research of concurrent engineering, is the strong influence of human factors on project performance. The organizational system and its underlying process of concurrent engineering is equally critical, but must be effectively integrated with the human side of the enterprise. Effective managers understand the complex interaction of organizational and behavioral variables. During the *design and implementation* of the concurrent engineering system, they can work with the various resource organization and senior management, creating a win-win situation between the people affected by the intervention and senior management. They can shake up conventional thinking and create a vision without upsetting established cultures and values. To be successful, both *implementing* concurrent engineering and *managing* projects through the system requires proactive participation and commitment of all stakeholders. It also requires congruency of the system with the overall business process and its management system.

Taken together, leaders must pay attention to the human side. To enhance cooperation among the stakeholders, managers must foster a work environment where people see the significance of the intervention for the enterprise and personal threats and work interferences are minimized.

One of the strongest catalysts, to both the *implementation of concurrent engineering* and the *management of projects*, is professional pride and excitement of the people, fueled by visibility and recognition of work accomplishments. Such a professionally stimulating environment seems to lower anxieties over organizational change, reduce communications barriers and conflict, and enhance the desire of personnel to cooperate and to succeed, a condition critically important for developing the necessary linkages for effective cross-functional project integration. Effective project leaders are social architects who understand the interaction of organizational and behavioral variables and can foster a climate of active participation and minimal dysfunctional conflict. They also build alliances with support organizations and upper management to ensure organizational visibility, priority, resource availability, and overall support for the project undertaking.

While no single set of broad guidelines exist that guarantees success for managing in concurrent engineering, project management is not a random process! A solid understanding of modern project management concepts, their tools, support systems, and organizational dynamics, is one of the threshold competencies for leveraging the concurrent engineering process. It can help managers in both, developing better project management systems and in leading projects most effectively through these systems.

# References

Atnahene-Gimo, K. 2003. The effects of centrifugal and centripetal forces on product development speed and quality. *Academy of Management Journal* 43(3, June):359–373.

Bauly, J. and A. Nee. 2000. New product development: implementing best practices, dissemination and human factors. *International Journal of Manufacturing Technology and Management* 2:(1/7):961–982.

Bishop, S. 1999. Cross-functional project teams in functionally alligned organizations. *Project Management Journal* 30(3, September):6–12.

Chambers, C. 1996. Transforming new product development. *Research Technology Management* 39(6, November/December):32–38.

Checkland, P. 1999. *Systems thinking, systems practice.* Hoboken, NJ: Wiley, 1999.

Cooper, R., and Kleinschmidt, E. 1993. Stage-Gate systems for new product success. *Marketing Management* 1(4):20–29.

Denker, S., D. Steward, and T. Browning. 2001. Planning concurrency and managing iteration in projects. *Project Management Journal* 32(3, September):31–38.

Emery, F. 1969. *Systems thinking.* Harmondsworth, UK: Penguin.

Emery, F., and E. Trist. 1965. The causal texture of organizational environments. *Human Relations* 18(1):21–32.

Gerwin, D., and N. Barrowman. 2002. An evaluation of research on integrated product development. *Management Science* 48(7, July):938–954.

———. 2002. An evaluation of research on integrated product development. *Management Science* 48(7, July):938–954.

Githens, G. 1998. Rolling wave project planning. *Proceedings of the 29th Annual Symposium of the Project Management Institute.* Long Beach, CA, October 9–15.

Goldenberg, J., R. Horowitz, and A. Levav. 2003. Finding your innovation sweet spot. *Harvard Business Review* 81(3, March):120–128.

Haddad, C. J. 1996. Operationalizing the concept of concurrent engineering. *IEEE Transactions on Engineering Management* 43(2, May):124–132.

Haque B., K. Pawar, and R. Barson. 2003. The application of business process modeling to organizational analysis of concurrent engineering environments. *Technovation* 23(2, February):147–162.

Harkins, J. 1998. Making management tools work. *Machine Design* 70:(12, July):210–211.

Hauptman, O. and K. Hirji. 1999. Managing integration and coordination in cross-functional teams. *R&D Management* 29(2, April):179–191.

Hull, F., P. Collins, and J. Liker. 1996. Composite form of organization as a strategy for concurrent engineering effectiveness. *IEEE Transactions on Engineering Management* 43:(2, May):133–143.

Kerzner, H. 2001. *The project management maturity model.* Hoboken, NJ: Wiley.

Koufteros, X., M. Vonderembse, and M. Doll. 2002. Integrated product development practices and competitive capabilities: The effects of uncertainty, equivocality, and platform strategy. *Journal of Operations Management* 20(4, August):331–355.

LaPlante, A., and A. Alter. 1994. Corning Inc.: The stage-gate innovation process. *Computerworld* 28(44, October):81–84.

Litsikas, M. 1997. Break old boundaries with concurrent engineering. *Quality* 36(4, April):54–56.

Moffat, L. 1998. Tools and teams: Competing models of integrating product development projects. *Journal of Engineering and Technology Management* 1(1, March):55–85.

Morris, P. W. G. 1997. *The management of projects.* London: Thomas Telford.

Nee, A. and S. Ong. 2001. Philosophies for integrated product development. *International Journal of Technology Management* 21(3):221–239.

Nellore, R., and R. Balachandra. 2001. Factors influencing success in integrated product development (IPD) projects. *IEEE Transactions on Engineering Management* 48(2, May):164–174.

Neves, T., G. L. Summe, and B. Uttal. 1990. Commercializing technology: what the best companies do. *Harvard Business Review* (May/June):154–163.

Noori, H., M. Hugh, and G. Deszca. 1997. Managing the P/SDI process: best-in-class principles and leading practices. *Journal of Technology Management* 13(3, 1997):245–268.

O'Connor, P. 1994. Implementing a stage-gate process: A multi-company perspective. *The Journal of Product Innovation Management* 11: (3, June): 183–200.

Paashuis, V., and D. Pham. 1998. *The organisation of integrated product development.* Berlin: Springer-Verlag.

Pham, D., S. Dimov, and R. Setchi. 1999. Concurrent engineering: A tool for collaborative working. *Human Systems Management* 18(3/4):213–224.

Pilkinton, A., and R. Dyerson. 2002. Extending simultaneous engineering: electric vehicle supply chain and new product development. *International Journal of Technology Management* 23(1,2,3,):74–88.

Pillai, A., A. Joshi, and K. Raoi. 2002. Performance measurement of R&D projects in a multi-project, concurrent engineering environment. *International Journal of Project Management* 20(2, February):165–172.

Prasad, B. 1976. *Concurrent engineering fundamentals: Integrated product and process organization.* Vol. 1. Englewood Cliffs, NJ: Prentice Hall.

———. 1977. *Concurrent engineering fundamentals: Integrated product development.* Vol. 2. Englewood Cliffs: Prentice Hall.

———. 1998. Decentralized cooperation: a distributed approach to team design in a concurrent engineering organization. *Team Performance Management* 4(4):138–146.

———. 2002. Toward life-cycle measures and metrics for concurrent product development. *International Journal of Computer Applications in Technology* 15(1/3):1–8.

———. 2003. Development of innovative products in a small and medium size enterprise. *International Journal of Computer Applications in Technology* 17(4):187–201.

Prasad, B., F. Wang, and J. Deng. 1998. A concurrent workflow management process for integrated product development. *Journal of Engineering Design* 9(2, June):121–136.

Rasiel, E. 1999. *The McKinsey way.* New York: McGraw-Hill.

Rigby, Darrel K. 1995. Managing the management tools. *Engineering Management Review (IEEE)* 23(1, Spring):88–92.

Senge, P. M. 1990. *The fifth discipline: The art and practice of the learning organization.* New York: Doubleday/Currency.

Senge, P. and G. Carstedt. 2001. Innovating our way to the next industrial revolution. *Sloan Management Review* 42(2):24–38.

Shabayek, A. 1999. New trends in technology management for the 21 century. *International Journal of Management* 16(1, March):71–76.

Skelton, T., H. Thamhain, J. Hans. 1993. Concurrent project management: A tool for technology transfer, R&D-to-market. *Project Management Journal* 24(4, December).

Sobek, D. K., K. Jeffrey, K. Liker, C. Allen, and A. Ward. 1998. Another look how Toyota integrates product development. *Harvard Business Review* (July–August):36–49.

Stum, D. 2001. Maslow revisited: Building the employee commitment pyramid. *Strategy and Leadership* 29(4, July/August):4–9.

Sundaramurthy, C. and M. Lewis. 2003. Control and collaboration: Paradoxes of governance. *Academy of Management Review* 28(3, July):397–415.

Swink, M., J. Sandvig, and V. Marbert. 1996. Customizing concurrent engineering processes: Five case studies. *Journal of Product Innovation Management* 13(3, May):229–245.

Thamhain, H. 2003. Managing innovative R&D teamsteams. *R&D Management* 33(3, June):297–311.

Thamhain, H. 1994. A manager's guide to effective concurrent project management. *Project Management Network* 8(11, November):6–10.

———. 1996. Best practices for controlling technology-based projects. *Project Management Journal* 27(4, December):37–48.

———. 2001a. Leading R&D projects without formal authority. In *Management of Technology: Selected Topics*, ed. T. Khalil. Oxford, UK: Elsevier Science.

———. 2001b. The changing role of project management. Chap. 5 in *Research in management consulting*, ed. Anthony Buono. Greenwich, CT: Information Age Publishing.

———. 2002. "Criteria for effective leadership in technology-oriented project teams," Chapter 16 in *The Frontiers of Project Management Research* (Slevin, Cleland and Pinto, eds.), Newtown Square, PA: Project Management Institute, pp. 259-270.

Thamhain, H. and D. Wilemon. 1998. Building effective teams for complex project environments. *Technology Management* 4: 203–212.

Wu, S., J. Fuh, and A. Nee. 2002. Concurrent process planning and scheduling in distributed virtual manufacturing. *IIE Transactions* 34(1, January):77–89.

Yam, R, W. Lo, H. Sun, and P. Tang. 2003. Enhancement of global competitiveness for Hong Kong/China manufacturing industries. *International Journal of Technology Management* 26(1):88–102.

CHAPTER FOUR

# PROCESS AND PRODUCT MODELING

Rachel Cooper, Ghassan Aouad, Angela Lee, and Song Wu

It is widely recognized that modeling processes and information is a complex task. This chapter looks at the various techniques that can be used for both process and product modeling. Beginning with a discussion of the importance of process management in managing projects, the chapter then defines a process, describes the various approaches to modeling processes, and illustrates the development of different process map generations. It also provides examples of processes in the management of new products in the manufacturing and construction industry sectors.

The second part of the chapter covers product modeling and object modeling paradigms. Included is the use of 3D and virtual reality to support visual modeling through work developed for the construction industry. The last part of the chapter discusses current research and how trends in the use of information communication technologies will influence development and management of both process and product modeling.

## The Importance of Managing Projects

Today, companies introduce new products in a variety of ways, ranging from chaotic to systematic. However, it is unwise to constantly rely on luck to salvage the organizational procedure of the work at hand (Peppard and Rowland, 1995). There are still companies that mistakenly believe that an idea or impetus will easily become a successful new product. Furthermore, once a superficially attractive idea has been articulated, many companies push ahead into development but may forget or overlook some important steps, and so will consequently slip from the desired schedule and will incur increased costs. Unstructured development, a chaotic or random approach, usually leads to problems, especially when change in a new product's specification occurs. It has been shown that without a formal structure in which to freeze a specification or evaluate changes, "creeping elegance often runs amok" (Elzinga et al., 1995). (See also the chapter by Cooper and Reichelt.)

In addition, those companies that have a process outlook—large and small, public and private, domestic and global—are now finding themselves in an era of inherent competition. Firms operate in dynamic environments and not stable ones, as both external competition and internal environments evolve over time. White (1996) proposes that in the near future around 75 percent of many organizations currently in business will no longer exist, either due to takeover or decline, whilst the others will emerge as international giants. Within such an environment, processes must also continuously change to enable firms to remain effective and profitable throughout changing conditions (Moran and Brightman, 1998). Those organizations undertaking improvements in productivity, quality, and operations need to reconsider their working practices (Elzinga et al., 1995). Katzenbach (1996) reports that organizational change is becoming everyone's problem and that customers require it, shareholder performance demands it, and continued growth depends upon it. Modeling and managing organizational processes are critical factors contributing to successful organizational change.

## Definition of a Process

Research has found that every successful organization needs a ". . .formal blueprint, roadmap, template or thought process" for driving a new project (Cooper, 1994). Table 4.1 illustrates the various approaches to defining a 'process'.

More simply stated, a *process* has an input and an output, with the process receiving and subsequently transforming the input into the desired output (see Figure 4.1). A process can be visible, and at the same time, it can be invisible. We all tend to do familiar things in the same way, in a manner we are used to, and do not reflect upon the fact that "now I am performing an activity" or "now I have completed this task." However, in order to model a task or a process, we need to describe the "what happens," thus providing a simplified description of real-world phenomena. Often, nouns, verbs, and adjectives are used to depict a process (Lundgren, 2002). The noun usually refers to a person, place, or object; a verb is a word or a phase that describes a course of events, conditions, or experiences; and the adjective specifies an attribute of the noun (see Figure 4.2). There is a flow relation between the noun, the verb, and the adjective—a car is painted and the result is a painted car.

## Approaches to Process Modeling

An understanding of processes can be reached in different ways. The project process is often depicted/modeled to enhance team coordination and communication through simple mech-

## TABLE 4.1. DEFINITION OF A PROCESS.

| Author | Definition |
|---|---|
| Davenport (1993) | A process is simply a "structured, measured set of activities designed to produce a specified output for a particular customer or market" and that they are "the structure by which an organisation follows that is necessary to produce value for its customers." |
| Cooper (1994) | Provides the thinking and action framework for transforming an idea into a product, and the processcan either be tangible or intangible, functionally based or organizationally based. |
| Oakland (1995) | "The transformation of a set of inputs, which can include actions, methods, and operations, into outputs that satisfy customer needs and expectations, in the form of products, information, services or—generally—results" |
| Zairi (1997) | "A process is an approach for converting inputs into outputs. It is the way in which all the resources of an organisation are used in a reliable, repeatable and consistent way to achieve its goals." |
| Bulletpoint (1996) | Suggests that regardless of the definition of the term "process," there are certain characteristics that this process should have:<br><br>• Predictable and definable inputs<br>• A linear, logical sequence of flow<br>• A set of clearly definable tasks or activities<br>• A predictable and desired outcome or result |

anisms such as flow and Gantt charts (a flowchart that encompasses time). To model more complex scenarios of real-world phenomena, techniques such as IDEF0 (Integrated Definition Language) and analytical reductionism/process decomposition are commonly used (Koskela, 1992).

## IDEF0

During the 1970s, the U.S. Air Force Program for Integrated Computer-Aided Manufacturing (ICAM) sought to increase manufacturing productivity through systematic application of computer technology. The ICAM program identified the need for better analysis and communication techniques for people involved in improving manufacturing productivity and thus developed a series of techniques known as the IDEF family (IDEF, 2002):

## FIGURE 4.1. A PROCESS.

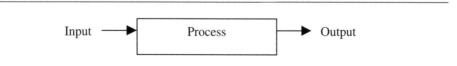

*Source:* Vonderembse and White (1996).

## FIGURE 4.2. DESCRIPTION OF A PROCESS.

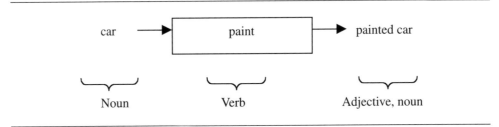

Source: Lundgren (2002).

1. IDEF0, used to produce a "function model"—a structured representation of the functions, activities, or processes within the modeled system or subject area.
2. IDEF1, used to produce an "information model," represents the structure and semantics of information within the modeled system or subject area.
3. IDEF2, used to produce a "dynamics model," represents the time-varying behavioral characteristics of the modeled system or subject area.

In 1983, the U.S. Air Force Integrated Information Support System program enhanced the IDEF1 information modeling technique to form IDEF1X (IDEF1 Extended), a semantic data modeling technique. Currently, IDEF0 and IDEF1X techniques are widely used in the government, industrial, and commercial sectors, supporting modeling efforts for a wide range of enterprises and application domains. For the purpose of this chapter, IDEF0 will be described as it most closely relates to the "functional" new product development process.

The Integrated Definition Language 0 for function modeling is an engineering technique for performing and managing needs analysis, benefits analysis, requirements definition, functional analysis, systems design, maintenance, and the baseline for continuous improvement (IDEF, 2002). IDEF0 models provide a "blueprint" of functions and their interfaces that must be captured and understood in order to make systems engineering decisions that are logical, integratable, and achievable, to provide an approach to:

- performing systems analysis and design at all levels, for systems composed of people, machines, materials, computers, and information of all varieties—the entire enterprise, a system, or a subject area;
- producing reference documentation concurrent with development to serve as a basis for integrating new systems or improving existing systems;
- communicating among analysts, designers, users, and managers;
- allowing team consensus to be achieved by shared understanding;
- managing large and complex projects using qualitative measures of progress; and
- providing a reference architecture for enterprise analysis, information engineering, and resource management.

The modeling language itself makes explicit the purpose of a particular activity and is composed of a series of boxes and arrows (see Figure 4.3). The boxes of the IDEF0 technique

## FIGURE 4.3. THE BASIC CONCEPT OF THE IDEF0 SYNTAX.

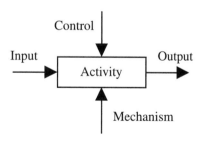

*Source:* IDEF (2002).

represent functions, defined as activities, processes, or transformations. Each box should consist of a name and number inside the box boundaries; the name is of an active verb or verb phrase that describes the function, and the number inside the lower right corner is to identify the subject box in the associated supporting text.

The arrows in the diagram represent data or objects related to the functions and do not represent flow or sequence as in the traditional process flowchart model. They convey data or objects related to functions to be performed. The functions receiving data or objects are constrained by the data or objects made available. Each side of the function box has a standard meaning in terms of box/arrow relationships. The side of the box with which an arrow interfaces reflects the arrow's role. Arrows entering the left side of the box are inputs; inputs are transformed or consumed by the function to produce outputs. Arrows entering the box on the top are controls; controls specify the conditions required for the function to produce correct outputs. Arrows leaving a box on the right side are outputs; the outputs are the data or objects produced by the function. Arrows connected to the bottom side of the box represent mechanisms; upward-pointing arrows identify some of the means that support the execution of the function.

The functions in an IDEF0 diagram can be broken down or decomposed into more detailed diagrams, until the subject is described at a level necessary to support the goals of a particular project (see Figure 4.4). The top-level diagram in the model provides the most general or abstract description of the subject represented by the model. This diagram is followed by a series of child diagrams providing more detail about the subject. Each subfunction is modeled; on a given diagram, some of the functions, none of the functions, or all of the functions may be decomposed individually by a box, with parent boxes detailed by child diagrams at the next lower level. All child diagrams must be within the scope of the top-level context diagram/parent box. In turn, each of these subfunctions may be decomposed, each creating another, lower-level child diagram.

## Analytical Reductionism/Process Decomposition

Analytical reductionism/process decomposition involves breaking the process down into levels of granularity, as demonstrated in Figure 4.5, with the lower-level subprocesses further

## FIGURE 4.4. IDEF DECOMPOSITION STRUCTURE.

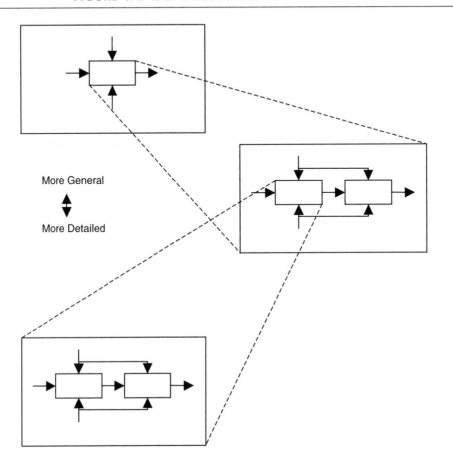

defining its corresponding upper-level process. It shares similarities with IDEF0, in that it breaks the parent process into subsequent more detailed child/subprocesses and then onto procedures and activities. The level at a process that differentiates from a procedure is, however, still a topic of discussion in the process management field.

A process (Koskela, 1992; Cooper, 1994; Vonderembse and White, 1996):

- converts inputs into outputs;
- creates a change of state by taking the inputs (e.g., material, information, people) and passing it through a sequence of stages during which the inputs are transformed or their status changed to emerge as an output with different characteristics. Hence, processes act

## FIGURE 4.5. PROCESS LEVELS.

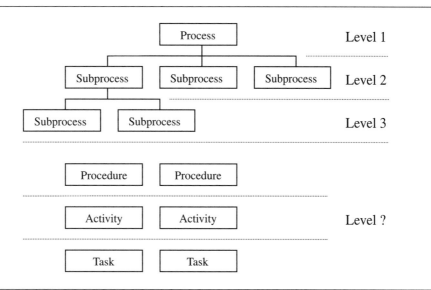

upon input and are dormant until the input is received. At each stage the transformation tasks may be procedural but may also be mechanical, chemical, and so on;

- clarifies the interfaces of fragmented management hierarchies;
- helps to increase visibility and understanding of the work to be done;
- defines the business/project activities across functional boundaries.

A procedure (Lee et al., 2001):

- is a sequence of steps. It includes the preparation, conduct, and completion of a task. Each step can be a sequence of activities, and each activity a sequence of actions. The sequence of steps is critical to whether a statement or document is a procedure of something else;
- is required when the task we have to perform is complex or is routine and is required to be performed consistently;
- defines the rules that should be followed by an individual or group to carry out a specific task; their definition is usually rigid, leaving no opportunity for individual initiative;
- supports the process.

## Process Management and New Product Development

According to Davenport (1994), the design of the project process should start with a high-level model that engages the management team. This is to avoid a too-detailed description of processes in the initial creative stage; a detailed model will only lower the motivation of the management team. The core process model should be used as a tool to communicate a shared project view on a high level of abstraction. The following section describes how this is applied to the NPD (new product development) process that is used in manufacturing.

The development of new products and services that can successfully compete in local, national, and global markets has become a key concern for a large majority of organizations (Cooper, 1992). The NPD process is fundamental for organizations to support this growth (HM Treasury, 1998). The process has received, and continues to receive, much attention by academia and practitioners to improve its effectiveness and efficiency, and its development has been examined with growing interest across various industrial sectors in lieu of the changing nature of the economic climate.

A new product is one that has not been previously manufactured by a company and is a necessary risk that companies must undertake (Crawford, 1992). Technological developments, shorter product life cycles, complexity of products, increasingly changing market demands, and globalized competition means that companies face a limited space in which they can succeed. NPD is a critical means by which the whole organization as a business as well as its employees can adapt, diversify, and in some cases, reinvent their firms to match evolving market and technical conditions. The fundamental aim of the NPD program is to get the right product to the market or customer as quickly as possible. The NPD process is composed of a number of activities (Crawford, 1944), initiated by the identification of the need or the adoption of an idea. A number of technical, financial, and business preliminary evaluations are then performed, followed by further detailed technical development follows. Finally, after a series of company and market tests, the finished product is launched onto the market (Crawford, 1994). Generically, these activities can be separated into three main broad categories (Cooper and Kleinschmidt, 1995):

- *Predevelopment activities.* Idea generating/establishing the need, followed by a number of preliminary market, technical, financial, and production assessments
- *Development activities.* The physical development of the product
- *Post-development activities.* The final launch of the product into the marketplace

From a historical point of view, NPD models can be classified into three main groups: sequential, overlapping, and Stage-Gate (Cooper and Kleinschmidt, 1995).

## NPD Process Models

In the 1960s, the NPD process was still in its first generation, following a simple linear sequential structural model whereby the development moved through different, almost mutually exclusive, phases in a logical step-by-step fashion (McGarth, 1996). These phases are shown in Figure 4.6.

The development proceeded to the next phase only after all the requirements of the preceding phase were satisfied, and in each succeeding phase, different intermediate results were created, with the outputs of one phase forming the major inputs to the next (Coughlan, 1991). In this sense, the major activities of the process were isolated from each other, creating an over-the-brick-wall effect, whereby each discrete activity played little or no regard to the

## FIGURE 4.6. TYPICAL SEQUENTIAL NPD PROCESS.

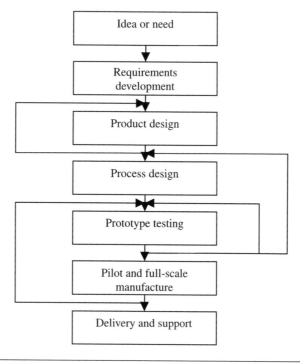

*Source:* Kagioglou (1999).

next activity (see Figure 4.7). This led to long lead times, late product launch, increased development costs, lack of effective information flow, and lack of flexibility for change in the process (Turino, 1990). However, this approach does offer high staff utilization in departments; it is favorable for breakthrough projects that require a revolutionary innovation or for very big projects where the shear size of personnel involved limits extensive communications between the members, and when product development is masterminded by a

## FIGURE 4.7. SEQUENTIAL OVER THE "BRICK WALL" APPROACH.

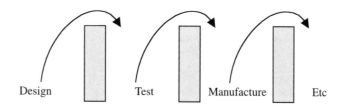

genius who makes the invention and hands down a well-defined set of product specifications (Takeuchi and Nonaka, 1986).

The need for change from the sequential to a more concurrent approach to NPD became increasingly apparent in the last two decades, where the manufacturing function had to be integrated into the design function so as to improve coordination and communication in the project. Thus, the NPD process steadily evolved into a more increasingly complicated second generation "coupling" model (Tidd, et al., 1997). Robert Cooper's NPD Stage-Gate model gained wide acceptance, as illustrated in Figure 4.8 (Cooper, 1990). It is presented as a series of stages and gates, which can vary from typically four to seven (Cooper, 1993). Each stage represented a number of activities that needed to be performed before progressing to a "decision" gate before the next stage; the stages represented multifunctional activities, involving a number of people from various departments relevant to the activities. These gates were clearly defined as "yes" or "no" decision points that provided organizations with the capability to measure and control the process and match subsequent funding to meeting the requirements at each gate (McGarth, 1996).

The Stage-Gate process was found to reduce development time and produce marketable results and optimized internal resources (Anderson, 1993). However, while enabling a higher degree of control and understanding of the progression of a project process, such gates required variable tasks to be checked off against predetermined lists. This often made the process both cumbersome and slow (Cooper and Kleinschmidt, 1992). Projects were forced to wait at each gate until all tasks were completed and not to stray from a process through which all projects had to progress. Any overlapping of activities was impossible (Devinney, 1995). Therefore, in order to overcome unnecessary delay and to enable smoother progression, the more recently developed third-generation "parallel" processes have sought to accommodate the need for certain tasks to overlap during a NPD program (Cooper, 1994). The main characteristic of the new process was the overlapping of the stages. Go/kill decisions were delayed to allow for flexibility and speed—the previous "hard" gates were replaced with "fuzzy" gates. (See also the chapter by Thamhain below on Concurrent Engineering.) In essence, these fuzzy gates allowed conditional-go decisions, enabling a degree that permitted the overlap of certain stages (see Figure 4.9). In addition, by being more outcome-focused, these processes have permitted organizations to build prioritization models that enabled projects to move through the process with more flexibility.

### FIGURE 4.8: SECOND-GENERATION NPD STAGE GATE PROCESS.

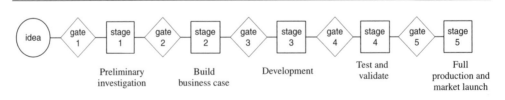

*Source:* Cooper (1990).

## FIGURE 4.9. THIRD-GENERATION FUZZY STAGE GATE OVERLAPPING NPD PROCESS.

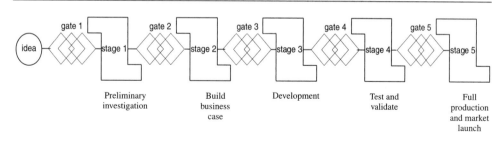

*Source:* Cooper (1994).

## Process Management in the Construction Industry

The evolution of the NPD process has often been cited as a learning point for other industries to improve their practice, in particular the construction industry (Howell, 1999). The UK construction industry is under increasing pressure to improve its practices (Hill, 1992; Howell, 1999). Indeed the construction industry has been criticized for its poor performance by several government and institutional reports, such as Emmerson (1962), Banwell (1964), Gyles (1992), Latham (1994), and more recently, Egan (1998). Most of these reports conclude, time and time again, that the fragmented nature of the industry, lack of coordination and communication between parties, the informal and unstructured learning process, adversarial contractual relationships, and lack of customer focus are what inhibits the industry's performance.

## The Traditional Design and Construction Process/RIBA Plan of Work

In 1959, the United Nations defined the building (project) process as ". . . the design, organisation and execution of building project' that has come to be recognized as . . . normal practice in any country or region . . . it is characterized by the fact that all operations follow a set pattern known to all participants in the building operation" (United Nations, 1959). However, this description is essentially untrue today. The nature of the design and construction process has grown in complexity since the 1950s, thus leading to an increased number of actors in the project.

The term largely associated with the "traditional building process" today usually refers to the practice where, upon perceiving a need for a new facility, a building client approaches an architect/engineer to initiate a process to design, procure, and construct a building to meet his or her specific needs. The process, in turn, almost invariably consists of the project being designed and built by two separate groups of disciplines who collectively form a temporary multiorganization for the duration of the project: the design group and the construction group (Mohsini and Davidson, 1992). The design group, typically, is coordinated by an architect/engineer. Depending upon the circumstances of the project at hand, it may also include other design professionals and specialists such as engineers, estimators, quantity surveyors, and so on. The principal function of this group is to prepare the design specifications of the work and other technical and contractual documents. The construction group, on the other hand, is usually coordinated by the main contractor and consists of a host of subcontractors and suppliers/manufacturers of building materials, components, hardware, and subsystems. This group is primarily responsible for the construction of the building project.

The two groups typically do not work coherently together (Kagioglou et al., 1998). The design activities in construction are usually isolated from the realities of the real issues facing production, as each function is expected to play a specific and limited role in any phase, thus contributing to the industry's problems, as highlighted by the many governmental and industrial reports (Emmerson, 1962; Banwell, 1964; Gyles, 1992; Latham, 1994; Egan, 1998). This factor has contributed to the problems of construction of poor supply chain coordination and fragmented project teams with adversarial relationships (Mohsini and Davidson, 1992). The Royal Institute of British Architects' (RIBA) Plan of Work (RIBA, 1997) fundamentally represents this practice. The model (see Figure 4.10) was originally published in 1963 as a standard method of operation for the construction of buildings, and it has become widely accepted as the operational model throughout the building industry (Kagioglou et al., 1998). However, it was designed from an architectural perspective, which has in some way restricted its applications to specific forms of UK construction contracts and is increasingly inappropriate for the newer types of contracts being used both in the United Kingdom and elsewhere, such as "partnering" frameworks.

## Generic Design and Construction Process Protocol (GDCPP)

The development and use of more generic and comprehensive process models for the new product development in construction has been a concern for researchers and the construction industry itself since the early 1990s. There is now wider use of such models in the industry. The Process Protocol is a generic model developed by the authors in conjunction with leaders in the construction industry and is an attempt to drive construction toward the third/new generation. This approach is one that, in light of increasing outsourcing and supply partnering in manufacturing, can be used to address process management in any extended enterprise.

The protocol is ". . . a common set of definitions, documentation and procedures that provides the basis to allow a wide range of organisations involved in a construction project to work together seamlessly' (Kagioglou et al., 1998b). It maps ". . . the entire project process from the client's recognition of a new or emerging need, through to operations and maintenance" (Cooper et al., 1998; Kagioglou et al., 1998c) by breaking down the design and construction process into four broad stages—Preproject, Preconstruction, Construction, and

## FIGURE 4.10. RIBA PLAN OF WORK.

| Predesign | A B | | | |
|---|---|---|---|---|
| Design | C D E | | | |
| Preparing to build | | F G H | | |
| Construction | | | J K L | |
| Post-construction | | | | M |

| | | | |
|---|---|---|---|
| Stage A: | Inception | Stage B: | Feasibility |
| Stage C: | Outline proposals | Stage D: | Scheme design |
| Stage E: | Detail design | Stage F: | Production info |
| Stage G: | Bills of quantities | Stage H: | Tender action |
| Stage J: | Project planning | Stage K: | Operations on site |
| Stage L: | Completion | Stage M: | Feedback |

## FIGURE 4.11. DETAIL OF THE PROCESS MAP.

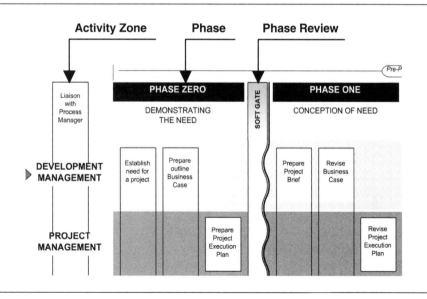

Post-construction stages—and ten phases (demonstrating the need; conception of need; outline feasibility; substantive feasibility study and outline financial authority; outline conceptual design; full conceptual design; coordinated design procurement and full financial authority; production information; construction; operation; and maintenance). These are represented by vertical columns (see Figure 4.11) separated by gates, soft and hard. As in Cooper's (1994) third-generation fuzzy Stage-Gate model, the soft gates allow flexibility of control, while the hard gates ensure that all work is progressing to program and are usually related to finance, and production signoff. The horizontal bands (see Figures 4.11 and 4.12) represent coordinated activities—namely, Development, Project, Resource, Design, Production, Facilities, Health and Safety and Legal, and Process Management—because in construction these are undertaken by numerous professional consultants and subcontractors as well as disciplines. In defining the activities rather than the disciplines, the Process Protocol emphasizes the need for team collaboration and coordination representing the fact that most construction projects are completed by a virtual enterprise of organization works.

The Process is based on six key principles (Sheath et al., 1996; Aouad et al., 1998; Kagioglou et al., 1998a; Cooper et al., 1998):

- *Whole project view.* The process has to cover the whole life of the project, from recognition of a need to the operation of the finished facility. This approach ensures that all the issues are considered from both a business and a technical point of view. The separation between the design and production functions, as described previously, has been pronounced as a key contributor to the inadequacies of construction (Harvey, 1971). The NPD process brought about the integration of multifunctions, thereby introducing those who do the building earlier into the design phase. Gunaskaran and Love (1998) argue that this will be invaluable. These specialist organizations have specific knowledge concerning the capability of the life cycle of materials, the overall performance of a product, and the programming of site operations.
- *Progressive design fixity.* Drawing from the "Stage-Gate" approach in manufacturing processes, the protocol adapts a phase review process that applies a consistent planning and review procedure throughout the project. The benefit of this approach is fundamentally the progressive fixing and approval of design information throughout the process. This is particularly useful in bringing the risk and cost of late changes to the attention of clients who are not familiar with the impact of their design changes on a project cost.
- *A consistent process.* The generic properties of the Process Protocol will allow a consistent application of the phase review process. This, together with the adoption of standard approach to performance measurement, evaluation, and control, will help facilitate the process of continual improvement in design and construction. Using the same basic generic process uniformly yields the most productive results (Kuczmarski, 1992). Everyone involved in the process develops a comfortable and consistent level of working; they see why category analysis works, and they understand the purpose of strategic roles better. The most important underlying factor to any development process is making it understandable and actionable by all people concerned.

# FIGURE 4.12. THE PROCESS PROTOCOL HIGH-LEVEL MAP.

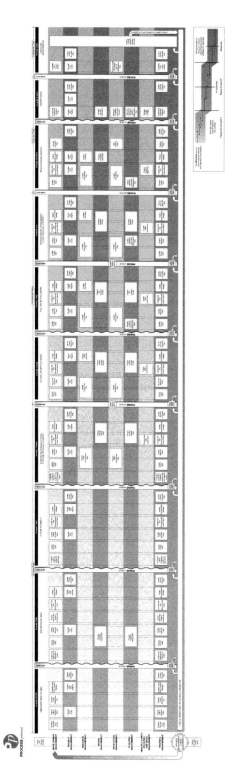

*Source:* www.processprotocol.com.

- *Stakeholder involvement/teamwork.* Project success relies upon the right people having the right information at the right time. The proactive resourcing of phases through the adoption of a "stakeholder" view will help to ensure that appropriate participants from each of the functions are consulted earlier in the process than is traditionally the case. Again, research suggests that full team participation improves the process by bearing all project input simultaneously, hence avoiding or reducing further revisions as to reduce time and money (Jassawalla and Sashittal, 1998).
- *Coordination.* Researchers have long argued that the employees are the critical building block of an organization (Crawford, 1977b). Successful teams bring together diverse information on every aspect that impacts customer satisfaction, and they overcome the shortcomings of hierarchical structures and generate quality decisions (Hoffman, 1979). Therefore the process map emphasizes the need to coordinate across activities the key actors in the process
- *Feedback.* Because of the nature of the supply chain in construction, rarely is knowledge or lessons learned in construction systematically incorporated back into the process. According to Li and Love (1998), construction problem-solving reliant on tacit knowledge has traditionally moved with individuals from project to project; cumulative project knowledge is not collected. Therefore, real benefits in cost, schedule, quality, and safety for future projects can only emerge if construction knowledge can be effectively harnessed in planning and executing future work is to be incorporated into the process (Kartam, 1996; Kumaraswamy and Chan, 1998). The Process Protocol recommends the use of a legacy archive; a central repository or information spine (Hinks et al., 1997) that can take the form of an electronic information management system. There have traditionally been such systems available to manufacturing industry but only recently have they been introduced to aid the collection and coordination of project knowledge and information (usually Web-enabled) to connect disparate suppliers and subcontractors.

These principles based on 20 years of research in manufacturing and construction.

## Product Modeling

As discussed in the previous section, process modeling involves the modeling of processes in a project and can often include the data and material that flows between them. Conversely, product modeling is used to model the elements specific to a product and the related process relationships; visual models are commonly produced through the mapping of conceptual data and process models and describe the information infrastructure of the product under development. The rapid prototyping of buildings/products using 3D/virtual-reality (VR) technologies enable developers and clients to quickly assess and evaluate their requirements before committing fully to the project. This section of the chapter considers aspects of the way information is used in product modeling and, by example, the use of IT specifically to model the construction product is detailed.

The UK construction industry has currently not fully adopted and envisaged the benefits of product modeling, unlike other industries such as aerospace and manufacturing. This is largely attributed to its deployment on an ad hoc basis without context or framework, leading to the development of unreliable information models that become unusable over time. Thus, efforts and resources of product modeling are wasted. In addition, the construction industry is divided for historical rather than logical reasons. These divisions tend to reflect the traditional roles performed by the disciplines (as discussed previously), and not the information required to complete the project. This leads to problems associated with information and project team integration.

It has recently been cited by a number of leading researchers (Lee et al., 2003; Dawood et al., 2003; Fischer, 2000; Graphisoft, 2003; Rischmoller et al., 2000) that object technology, coupled with client server applications and the Web environment, will provide the best way forward to enable project collaboration and information sharing, thus evoking a central project-based information database (building information/product model) and exchange between professionals. Graphical schema languages such as Entity Relationship Diagrams, NIAM, and IDEF1X were commonly used to undertake information modeling within the construction industry (Bjork and Wix, 1991; Rasdorf and Abudayyeh, 1992) until the early 1990s. Now UML (Unified Modeling Language) has become more popular because of its wide use in the software industry. However, the use of such modeling techniques is not advocated as appropriate for the industry, as they imply a separation between the data and the processes performed on the data. To overcome this problem, object-oriented models can be developed to describe the static information as well as the behavior of objects. This has proven to be more advantageous, as the resulting information model is richer and more natural, thus more usable for construction and other industries. This, it is anticipated, will enable effective coordination and communication of information among all project team members.

## Object Modeling

Unlike traditional data modeling techniques, the object-oriented paradigm models can be viewed as a collection of objects "talking" to each other via messages. The behavior of one object may result in changes in another object. This is done through message sending. For example, if the object "column" has been moved, it should send a message to the object "beam" (to which it is related) to tell it to resize itself reflecting the "object" change. This way of modeling is very powerful and is peculiar to the object-oriented world. In such a world, objects can be composed of other objects. These objects can be images, speech, music, or possibly a video. The object-oriented paradigm also supports the notions of encapsulation, abstraction, inheritance, and polymorphism (Martin and Odell, 1992) that were considered as critical in handling the complex task of information modeling. Encapsulation permits objects to have properties (data) and actions (operation). For instance, an object "beam" can have properties such as "length," "width." and so on and behaviors or actions such as "move beam," "calculate load on beam," and so on. Abstraction allows the analyst to abstract information according to requirements. For instance, the information about a beam can be abstracted in terms of properties, shape, materials, and so forth. Inheritance allows information in the parent object (beam) to be inherited by the child object (cantilever beam). Polymorphism allows objects to have one operation that can have different implementations. For instance, an operation such as "calculate area" can be attached to an object called "beam." However, the implementations of this operation differ according to whether the beam is a "rectangular beam," a "T beam," and so on.

Another major benefit of object orientation is the support of the notion of reusability. With such a notion, integrated databases can be developed from reusable object-oriented components that can be assembled as required. This is very similar to the way a building designer uses reusable plans that can be configured to his or her requirements. The object-oriented paradigm also supports the notion of "perspectives." This notion allows the construction professional to view the information from their own perspective. For instance, the architect is interested in features such as color, aesthetics, and texture, whereas the construction planner is interested in features such as time and resources. To illustrate this point, take the concept of a wall. This can be viewed from different perspectives. An architectural wall has attributes such as dimensions, color, and texture. A construction planner wall has attributes such as dimensions time and cost. It is therefore logical to store common information such as length and width in "wall" that can be inherited by the architectural wall, and so on through inheritance.

Object modeling is aimed at the identification of concepts/data—relationships between the concepts, attributes, and operations that are to be supported by the database. This task should be done independently of any implementation platform. Figure 4.13 shows an illustration of an object model incorporating objects, relationships, attributes, and operations.

## Activity/Process Modeling

As described earlier in the chapter, the activities performed within a construction project can be modeled using techniques such as activity hierarchy, data flows, IDEF0 techniques, and flowcharts. These techniques describe the information flows between processes. This is useful in understanding how information is communicated between processes. Figure 4.14 shows different representations of different process, data, and matrix models.

### Product Data Technologies

In the context of this chapter, product data technology (PDT) refers to techniques of data modeling, data exchange, and data management, which are aimed at the integration of product information through standard data models. Historically, the initial requirement for a standardized data model came from the need for different versions of CAD application to share their graphic files. IGES (the Initial Graphics Exchange Specification) was devel-

## FIGURE 4.13. AN ILLUSTRATION OF AN OBJECT MODEL.

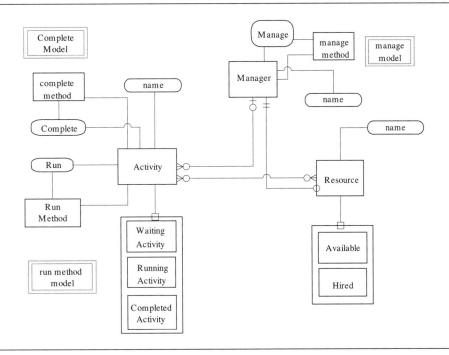

oped as a protocol for this purpose. However, graphical and geometrical data is only part of the information required in a building project. IGES is not able to support the exchange of other type of data such as construction, thermal, light, and so on. Therefore, a new project, PDES (Product Data Exchange Specification), was proposed in the United States in the early 1980s to overcome these limitations. In the same period, similar efforts were made in other countries, for example, the SET (Standard d'Echange et de Transfert) in France and the VDAFS (Verband der Deutschen Autombilindustrie Flaechen Scnittstelle) in Germany. In 1983, all these initiatives were coordinated into a major international program under the umbrella of the International Standard Organisation, Standard for Exchange of Product data (STEP). Thus, this became a comprehensive ISO standard (ISO 10303) that describes how to represent and exchange digital product information. In the construction industry, IFC (Industry Foundation Classes) was developed as a standard for exchange building product data, which is compliant with STEP. IFCs are an interoperable data standard that are linked to any proprietary software application.

## A Methodology for Modeling Information

Figure 4.15 illustrates how product models can be produced starting at a strategic/contextual level. This type of approach to developing product models helps in deriving a framework that ensures that all models are developed within a context. Activity hierarchy techniques

## FIGURE 4.14. MODELING TECHNIQUES.

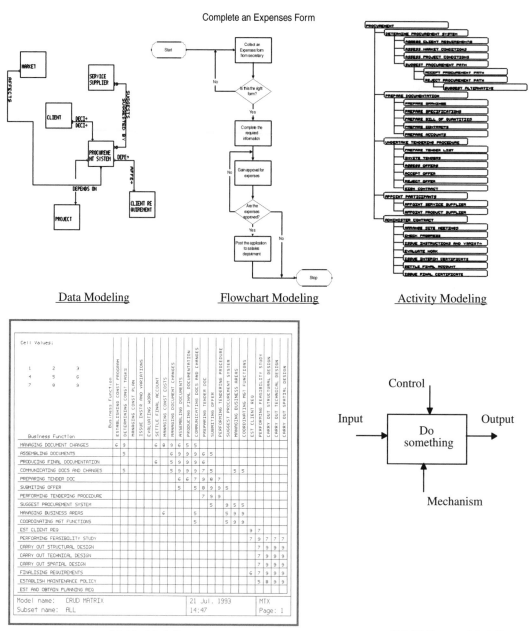

Data Modeling     Flowchart Modeling     Activity Modeling

Matrix Modeling     IDEF0/Process Modeling

## FIGURE 4.15. METHODOLOGY FOR MODELING INFORMATION.

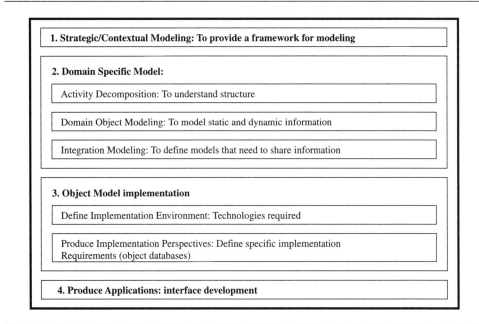

1. Strategic/Contextual Modeling: To provide a framework for modeling

2. Domain Specific Model:

Activity Decomposition: To understand structure

Domain Object Modeling: To model static and dynamic information

Integration Modeling: To define models that need to share information

3. Object Model implementation

Define Implementation Environment: Technologies required

Produce Implementation Perspectives: Define specific implementation Requirements (object databases)

4. Produce Applications: interface development

and matrix modeling could be used to define a contextual model (Graphisoft, 2003). The next step is to model information for specific domains using object-oriented techniques. Object as well as process models can be developed at this stage. The models should comply with standards such as the IFCs (Industry Foundation Classes). Following this, an environment for implementation must be defined; this includes technologies such as databases, interfaces, and so on. The last step is to implement the models in databases and define the interface requirement for developing and linking into applications. The interfaces to VR/3D applications can be developed at this stage.

The work on information modeling and integration in construction was initiated more than 30 years ago. However, fruitful practical results have failed to emerge because of the complex structure of the construction industry and because information technology has not been exploited properly. This chapter emphasized the importance of establishing a framework into which models from different domains and of varying abstraction levels can fit. The inclusion of object types viewed from different perspectives but shared across different domains and abstraction levels is seen as a major step forward in integrating information throughout the construction industry. Such structuring is considered essential for the development of accurate, understandable models. "nD modeling" has recently emerged in construction as the next step forward—the integration of process and product modeling (Lee et al., 2003).

## nD modeling

The building design process is complex, encapsulating a number and variety of factors in order to satisfy the client's requirements. In fact, it is rarely the case that there is one homogeneous "customer," but a number and variety of stakeholders who will be the end users of the building. These stakeholders are increasingly demanding the inclusion of design features such as maintainability, environmental friendliness, accessibility, crime deterrence, acoustical soundness, and energy performance. Each of these design parameters that the stakeholders seek to consider will have a host of social, economic, and legislative constraints that may be in conflict with one another. Furthermore, as each of these factors vary—in the amount and type of demands they make—they will have a direct impact on the time and cost of the construction project. Perspectives of design are usually balanced between aesthetics, ecology, and economics—a three-dimensional view of design that acknowledges its social, environmental, and economic roles is now necessary. The criteria for successful design therefore will include a measure of the extent to which all these factors can be coordinated and mutually satisfied to meet the expectations of all the parties involved (Lee et al., 2002a).

The volume of information required to interplay these scenarios to enable the client to visualize design changes and to assist with decision making—changing the design, planning schedules, and cost estimates—can be laborious, time-consuming, and costly (Lee et al., 2002b). There is now a need to allow users to create, share, contemplate and apply knowledge from multiple perspectives of user requirements. Conceptually, this will involve taking three-dimensional modeling in the built environment to an *n* number of dimensions, and thus integrates the process and information flow within a construction environment. Indeed market, regulatory, social, economic, and environmental factors are becoming so complex in the development of product in both construction and manufacturing that nD modeling is becoming a necessity.

This chapter has only been able to touch on the main approaches to process and product modeling in two industries; however, it does illustrate how critical both are to effective and efficient futures.

## The Future

Imagine a system that:

> given an idea, illustrates alternatives, illustrates constraints, and enables the understanding of both quantitative (time, cost, legislation) and qualitative (aesthetics, usability) dimensions. It enables all the stakeholders to participate and allows users to virtually experience the product concept. The system will determine the build specification, the manufacturing resources, and the production processes, and it will provide the drawings and the tool sets. It is a system where we can use our knowledge, in conjunction with the other stakeholders, to achieve the best solution, at the right cost, in a faster time, and in a sustainable manner.

This is the Holy Grail for process and product modeling. The need will not go away. Indeed, as companies, systems products, and markets get even more complex, we need the models to guide and help us make decisions. However, organizational behavior illustrates that we are not automatons; we will never work to a detailed and prescribed process and procedure, when situations demand innovation, creativity, and constant change to enable us to compete. Yet we do need systems to help us work through a complex world for the benefit of its inhabitants. The challenge is to understand what systems are the most appropriate, how we can best introduce them into organizations, and the impact that they will have on our work behavior and the future of the organizations who use them.

# References

Alshawi, M. 1996. *SPACE: Integrated environment.* Internal paper. University of Salford, July 1996.

Ammermann, E., R. Junge, P. Katranuschkov, and R. J. Scherer. 1994. *Concept of an object-oriented product model for building design;* Technische Universität, Dresden.

Anderson, E. J. 1994. *Management of manufacturing, models and analysis.* Wokingham, UK: Addison-Wesley.

Aouad G., M. Betts, P. Brandon, F. Brown, T. Child, G. Cooper, S. Ford, J. Kirkham, R. Oxman, M. Sarshar, and B. Young B. 1994. *Integrated databases for design and construction: Final report.* Internal report. University of Salford, July 1994.

Aouad, G. 1999. *Trends in information visualisation in construction.* IV 99; London, 1999.

Aouad, G., P. Brandon, F. Brown, T. Child, T., G. Cooper, S. Ford, J. Kirkham, R. Oxman, and B. Young. 1995. The Conceptual modelling of construction management information. *Automation in Construction* 3:267–282.

Aouad, G., J. Hinks, R. Cooper, D. Sheath, M. Kagioglou, and M. Sexton. 1998a. An information technology IT map for a generic design and construction process protocol. *Journal of Construction Procurement,* 4 (1, November): 132–151.

Aouad, G., M. Kagioglou, and R. Cooper. 1999. IT in construction: A driver or and enabler? *Journal of Logistics and Information Management* 12:130–137.

Aouad, G., J. Kirkham, P. Brandon, F. Brown, G. Cooper, S. Ford, R. Oxman, M. Sarshar, and B. Young. 1993. Information modelling in the construction industry: The information engineering approach. *Construction Management and Economics.* 11(5):384–397.

Arditi, D., and H. M. Gunaydin. 1998. Factors that affect process quality in the life cycle of building projects. *Journal of Construction Engineering and Management* 124(3):194–203.

ATLAS 1992. *Architecture, methodology and tools for computer integrated large scale engineering: ESPRIT Project 7280.* Technical Annex Part 1: General Project Overview.

Augenbroe, G 1993. *COMBINE.* Final report. Delft University.

Banwell, H. 1964. *Report of the Committee on the Placing and Management of Contracts for Building and Civil Engineering Work.* HMSO, London.

Bjork, B. C., 1989. Basic structure of a proposed building product model. *Computer Aided Design,* 21 (2, March): 71–78.

———. 1991. A unified approach for modelling construction information. *Building and Environment* 27(2): 173–194.

Bjork, B. C. and J. Wix. 1991. *An Introduction to STEP.* VTT and Wix McLelland Ltd., Bracknell, England.

Brandon, P., and M. Betts. 1995. *Integrated construction information.* London: Spon.

Bulletpoint. 1996. *Creating a change culture: Not about structures, but winning hearts and minds.* Wesley, New York.

Burbidge, J. L. 1996. *Period batch control.* Oxford, UK: Oxford University Press.

Cooper, R. G. and E. J. Kleinschmidt. 1987a. New products: What separates winners from losers? *Product Innovation Management Journal* 4:169–184.

———. 1987b. Success factors in product innovation. *Industrial Marketing Management Journal* 7:9–21.

———. 1995. Benchmarking the firm's critical success factors in new product development. *Journal of Product Innovation Management* 12:374–391.

Cooper, R. G. 1984. The performance impact of product innovation strategies. *European Journal of Marketing* 18(5):223–229.

———. 1990. Stage-Gate system: A new tool for managing new products. *Business Horizons* (May–June): 44–54.

———. 1993. *Winning at new products: Accelerating the process from idea to launch.* Reading, MA: Addison-Wesley.

———. 1994. Third-generation new product processes. *Journal of Product Innovation Management* 10(6–14).

———. 1999. From experience: The invisible success factors in product innovation. *Journal of Production Innovation Management* 16:115–33.

Cooper, R., M. Kagioglou, G. Aouad, J. Hinks, M. Sexton, and D. Sheath. 1998. Development of a generic design and construction process. *European Conference on Product Data Technology*, BRE, 205–214.

Coughlan, P. D. 1991. Differentiation and integration: The challenge of new product development. *Proceedings of the 5th Annual Conference of the British Academy of Management.* June 28.

Crawford, C. M. 1977a. *New products management.* Homewood, IL: Irwin.

———. 1977b. Product development: Today's most common mistakes. *University of Michigan Business Review* 6:7–8.

———. 1992. The hidden costs of accelerated product development. *Journal of Product Innovation Management* 9(3):161–176.

Crawford, K. M. and J. F. Fox. 1990. Designing performance measurement systems for just-in-time operations. *International Journal of Production Research* 28(11):2,025–2,036.

Davenport, T. H. 1993. *Process innovation: Reengineering work through information technology.* Boston: Harvard Business School Press.

Dawood, N., E. Sriprasert, and Z. Mallasi. 2003. Product and process integration for 4D visualisation at construction site level: A uniclass-driven approach. In *Developing a vision of nD-enabled construction.* A. Lee. et al.. Construct IT report, Salford, 64–68.

Devinny, T. M. 1995. Significant issues for the future of product innovation. *Journal of Product Innovation Management* 12:70–75.

Egan, J. 1998. *Rethinking construction.* Report from the Construction Task Force, Department of the Environment, Transport and Regions, UK.

Elzinga, D. J., T. Horak, L. Chung-Yee, and C. Bruner. 1995. Business process management: survey and methodology. *IEEE Transactions on Engineering Management* 24(2):119–128.

Emmerson, H. 1962. *Studies of problems before the construction industries.* HMSO, London.

Fenves, S. J. 1990. Integrated software environment for building design and construction. *Computer-aided design* 22 (1, January/February).

Fischer, M. 1997. 4D Modelling. *Proceedings of Global Construction IT.*

———. 2000. Benefits of 4D models for facility owners and AEC service providers. *Construction Congress VI.* ASCE. Orlando, FL. February, 990–995.

Froese, T. and B. Paulson. 1994. OPIS: An object model-based project information system. *Microcomputers in Civil Engineering* 9:13–28.

Graphisoft 2003. *The Graphisoft virtual building: Bringing the information model from concept into reality.* Graphisoft white paper.

Griffin, A. 1997. PDMA research on new product development practices: updating trends and benchmarking best practices. *Journal of Product Innovation Management* 14:429–458.

Gunasekaran, A. and P. E. D. Love. 1998. Concurrent engineering: A multidisciplinary approach for construction. *Logistics Information Management* 11(5):295–300.

Gyles, R. 1992. *Royal commission into productivity in the New South Wales building industry.* R. Gyles QC, Government Printer, London.

Harvey, J. P. 1971. *The master builders: Architecture in the Middles Ages.* London: Thames and Hudson.

Hill, T. J. 1992. Incorporating manufacturing perspectives in corporate strategy. In *Manufacturing Strategy.* C. A. Voss. Oxford, UK: Chapman & Hall.

Hinks, J., G. Aouad, R. Cooper, D. Sheath, M. Kagioglou, and M. Sexton. 1997. IT and the design and construction process: A conceptual model of co-maturation. *The International Journal of Construction* (July): 56–62.

HM Treasury 1998. *Innovating for the future.* Department of Trade and Industry. HMSO, London.

Hoffman, L. R. 1979. *The group problem solving process: Studies of a valance model.* New York: Praeger.

Howard, H. C. 1991. Linking design data with knowledge-based construction.

Howell, D. 1999. Builders get the manufacturers in. *Professional Engineer* (May): 24–25.

IDEF 2002. www.idef.com.

Jassawalla, A. R. and H. C. Sashittal. 1998. An examination of collaboration in high-technology new product development processes. *Journal of Product Innovation Management* 15:237–254.

Kagioglou, M. 1999. *Adapting manufacturing project processes into construction: A methodology.* Unpublished PhD thesis. Salford, UK: University of Salford.

Kagioglou, M., R. Cooper, G. Aouad, J. Hinks, M. Sexton, and D. Sheath. 1998a. *Final report: Generic design and construction process protocol.* Salford, UK: The University of Salford.

———. 1998b. *A generic guide to the design and construction process protocol.* Salford, UK: The University of Salford.

———. 1998c. Cross-industry learning: The development of a generic design and construction process based on the Stage-Gate new product development process found in the manufacturing Industry. *Proceedings of the Engineering Design Conference.* Brunel, UK.

Kartam, N. 1994. ISICAD: Interactive system for integrating CAD and computer-based construction systems. *Microcomputers in Civil Engineering* 9:41–51.

———. 1996. Making effective use of construction lessons learned in project life cycle. *Journal of Construction Engineering and Management* (March): 14–21.

Katzenbach, J. 1996. *Real change leaders.* London: Nicholas Brealey.

Khurana, A. and S. R. Rosenthal. 1998. Towards holistic "front ends" in new product development. *Journal of Product Innovation and Management* 15:57–74.

Koskela, L. 1992. *Application of the new production philosophy to construction.* Technical report no. 72. Center for Integrated Facility Engineering, Stanford University.

Kuczmarski, T. D. 1992. *Managing new products: The power of innovation.* Upper Saddle River, NJ: Prentice Hall.

Kumaraswamy, M. M. and D. W. M. Chan. 1998. Contributors to construction delays. *Construction Management and Economics Journal* 16(1):17–29.

Latham, M. 1994. *Constructing the team: Final report of the government/industry review of procurement and contractual arrangements in the UK construction industry.* London: The Stationery Office.

Lee, A., M. Betts, G. Aouad, R. Cooper, S. Wu, and J. Underwood. 2002b. Developing a vision for an nD modelling tool. Key note speech. *Proceedings of CIB W78 Conference—Distributing Knowledge in Building (CIB w78),* 141–148. Denmark.

Lee, A., A. J. Marshall-Ponting, G. Aouad, S. Wu, I. Koh, C. Fu, R. Cooper, M. Betts, M. Kagioglou, and M. Fischer. 2003. *Developing a vision of nD-Enabled construction.* Construct IT report. Salford, UK.

Lee, A., S. Wu, G. Aouad, and C. Fu. 2002a. Towards nD Modelling. Submitted to the *European Conference on Information and Communication Technology Advances and Innovation in the Knowledge Society.* E-sm@art, Salford, UK.

Li, H. and P. E. D. Love. 1998. Developing a theory of construction problem solving. *Construction Management and Economics* 16:721–727.

Lundgren 2002. Process. Unpublished proposal.

Martin, J. and J. Odell. 1992. *Object oriented analysis and design.* Upper Saddle River, NJ: Prentice Hall.

McGarth, M. E. 1996. *Setting the pace in product development.* Boston: Butterworth-Heinemann.

MOB 1994. *Rapport final. Modeles objet batiment, appel d'offres du plan construction et architecture.* Programme Communication/Construction.

Mohsini, R. A. and C. H. Davidson. 1992. Detriments of performance in the traditional building process. *Journal of Construction Management and Economics* 10:343–359.

Moran, J. W. and B. K. Brightman. 1998. Effective management of healthcare change. *The TQM Magazine* 10(1):27–29.

Oakland, J. S. 1995. *Total quality management: The route to improving performance.* 2nd ed. Boston: Butterworth-Heinemann.

Peppard, J. and P. Rowland. 1995. *The essence of business process re-engineering.* Upper Saddle River, NJ: Prentice Hall.

Plossl, K. R. 1987. *Engineering for the control of manufacturing.* Upper Saddle River, NJ: Prentice-Hall.

Powell, J. 1995. Virtual reality and rapid prototyping for Engineering. *Proceedings of the Information Technology Awareness Workshop.* University of Salford, Salford, UK.

Rasdorf, N. J. and O. Abudayyeh. 1992. NIAM conceptual database design in construction management. *Journal of Computing in Civil Engineering,* 6(1):41–62

Rezgui, Y. A, G. Brown, R. Cooper, A. Aouad, J. Kirkham, and P, Brandon. 1996. An integrated framework for evolving construction models. *The International Journal of Construction IT* 4(1):47–60.

RIBA. 1997. *RIBA plan of work for the design team operation.* 4th ed. London: Royal Institute of British Architects Publications.

Riedel, J. C. K. H., and K. S. Pawar. 1997. The consideration of production aspects during product design stages. *Integrated Manufacturing Systems* 8(4):208–214.

Rischmoller, L., and R. Matamala. 2003. Reflections about nD Modelling and Computer Advanced Visualisation Tools (CAVT). In *Developing a vision of nD-enabled construction.* A. Lee, et al. Construct IT report. Salford, UK, 92–94.

Rischmoller, L., M. Fisher, R. Fox, and L. Alarcon, L. 2000. 4D planning and scheduling (4D-PS): Grounding construction IT research in industry practice. Proceedings of CIB W78 Conference on Construction Information Technology: Taking the construction industry into the 21st century. Iceland, June.

Schonberger, R. J. 1982. *Japanese manufacturing techniques: Nine hidden lessons in simplicity.* New York: Free Press.

Sheath, D. M., H. Woolley, R. Cooper, J. Hinks, and G. Aouad. 1996. A process for change: The development of a generic design and construction process protocol for the UK construction industry. *Proceedings of the CIT Conference.* Institute of Civil Engineers. Sydney, Australia, April.

Sower, V. E., J. Motwani, and M. J. Savoie. 1997. Classics in production and operations management. *International Journal of Operations and Production Management* 17(1):15–28.

Takeuchi, H. and I. Nonaka. 1986. *The new product development game.* Cambridge, MA: Harvard Business Press.

Tidd, J., J. Bessant, and K. Pavitt. 1997. *Managing innovation.* Chichester, UK: Wiley.

United Nations 1959. *Government policies and the cost of building.* Geneva: ECE.

Vonderembse, M. A. and G. P. White. 1996. *Operations management: Concepts, methods and strategies.* New York: West Publishing.

Watson, A. and A. Crowley. 1995. CIMSteel integration standard. In .., *Product and process modelling in the building industry*, ed. R. J. Scherer. 491–493. Rotterdam: A. A. Balkema.

White, A. 1996. *Continuous quality improvement: A hands-on guide to setting up and sustaining a cost effective quality programme.* Gloucester: Judy Piakus.

Zairi, M. 1997. Business process management: A boundary-less approach to modern competitiveness. *Business Process Management* 3(1):64–80.

CHAPTER FIVE

# MANAGING CONFIGURATIONS AND DATA FOR EFFECTIVE PROJECT MANAGEMENT

Callum Kidd, Thomas F. Burgess

Configuration management (CM) has a severe image problem in many modern organizations: It is too often viewed as nothing more than glorified change control or version management—a costly exercise in form filling, with little or no technical content. As a value-added business activity, configuration management is, almost invariably, rated as less significant than, for example, quality management or project management (Kidd, 2001). The irony is that neither of these activities is possible without an effective configuration management process. Quality management, for example, requires us to know when configurations meet stated requirements. But how can we be sure that we are measuring against the most current list of requirements? Can we be sure that the reasons for making any changes were identified and impacts assessed prior to a decision being made? What effect will those changes have on the project schedule, and on total cost? Answering these questions is the business of configuration management. Effective configuration management is an essential part of an overall project management activity. To treat it as anything less is a recipe for disaster.

## What Is Configuration Management?

Configuration management is a technique used by many companies to support the control of the design, manufacture, and support of a product. ISO10007 (ISO 1997) defines configuration management as:

> a management discipline that applies technical and administrative direction to the
> development, production and support life cycle of a configuration item. This discipline is

applicable to hardware, software, processed materials, services, and related technical documentation.

It is important to understand at this point that the term configuration is a generic name for anything that has a defined structure or is composed of some predetermined pattern. Software, hardware, buildings, process plant, assets, and even the human body comes under the broad definition of a configuration. From a management perspective, it is often better to use the generic name of configuration, as it often avoids the software/hardware bias that causes confusion within the organization. Managing the definition of that pattern or structure from concept through to disposal is commonly termed configuration management.

According to Daniels (1985):

Very simply Configuration Management is a management tool that defines the product, then controls the changes to that definition.

In essence, the key to configuration management is founded on good business sense and straightforward practice in handling documentation. However, regardless of the routines practiced by some adept companies, in general, many companies have been comparatively poor in their control of the depth and uniformity of the relevant documentation. These deficiencies came in to focus in the United States during the late 1950s in the arms race to produce reliable, working defense materiel. As with any substantial program that is faced with tight deadlines and severe competition, the magnitude of change that was generated by the various collaborators to ensure compatibility among elements was enormous. The emphasis on hitting deadlines meant that when the various parties took stock after a successful missile flight was finally made, the realization dawned that adequate technical documentation to complete an identical missile was not available. Records of part identification, build statements, changes applied, changes implemented, and technical publications reflecting the build standard were missing. Such situations generated the impetus to systematically deal with product specifications and their modifications throughout the development and build life cycle.

The impetus to improve management in this key area was pushed forward by the customers, who generally were governments or their armed forces, and the developers, who often were large companies involved in defense work. To establish better control, the involved parties drew up configuration management standards that decreed how the projects were to be managed. In standards such as the EIA649 standard (EIA 1998), configuration management covers the full product life cycle from "concept through to de-commission."

The majority of case studies and written examples of configuration management come from highly technical and complex environments. Perhaps the following household example will demonstrate the application of CM in an environment that will be familiar to the majority of us. Consider a washing machine in your home. Bought in 1998, it has provided some years of trouble-free, reliable service. The 12-month warranty passed some years ago, but to date, there have been no problems with the appliance. Over the past few days, you notice that a patch of water has appeared in front of the machine. It appears during the wash cycle and looks to be getting worse. You phone the service engineer, and he tells you that it is most likely the seal on the pump. He asks the make and model, then arranges a visit. He arrives with the new seal and detaches the pump assembly. He checks the product identification number (PIN) on the side of the unit, then checks his catalogue. The seal he has brought does not look like the one on this pump. But how can that be? Surely each model will have common components? Not necessarily so. In Figure 5.1, you can see a simplified product breakdown structure (PBS) for the washer. The model, XYZ, consists of a number of assemblies common across the full model range. One of those assemblies, though, had a seal problem that was not identified until late 1998. Depending on when your machine was manufactured, it may contain the old seal on the pump assembly. But how do you know? Simple. The PIN gives detailed information on date of manufacture and batch number. The PIN, not the model number, will tell you which seal is on your pump unit. A further check in the catalogue will determine if the old seal and the new one brought by the service engineer are interchangeable. If not, it will provide another alternate part. This is simple configuration management in action—the same principle keeps aircraft in the air, cars on the road, and software working.

## The Configuration Management Process

Configuration management is probably best seen as a process for managing the following:

- The composition of a product
- The documentation and other data and products defining the product that supports it

The process may be related to a single product or to an associated collection of products, often referred to as systems and subsystems.

## FIGURE 5.1. PRODUCT BREAKDOWN STRUCTURE OF A WASHING MACHINE.

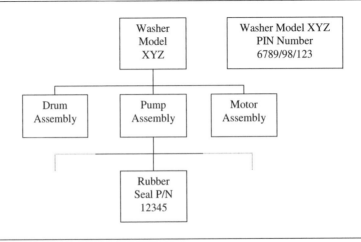

Efforts to develop a global consensus standard of best CM practice have resulted in the publication of ANSI/EIA 649, the most widely used CM practice model to date (Kidd, 2001). Configuration Management is traditionally defined in terms of the four interrelated activities:

- Configuration identification
- Configuration change management
- Configuration status accounting
- Configuration verification and audit

This structure (see Figure 5.2) is followed in ANSI/EIA 649, where the four activities sit beneath the overall planning activity. Each of these four areas is dealt with next.

## Configuration Identification

*Configuration identification* is the key element of the CM process. According to ANSI/EIA 649, configuration identification is the basis from which the configuration of products are defined and verified, products and documents are labeled, changes are managed, and accountability is maintained. Typical activities include the following:

- Define product structure and select elements to be managed
- Assign unique identifiers
- Define product attributes, interfaces, and details in product information

Configuration identification can be problematic because of the way in which we dissociate the development of the product and system structuring from change management. The difficulty of identifying configurations is further exacerbated by the fact that there may be several structures, or views, of each configuration, depending on which phase of the life cycle is under consideration.

## FIGURE 5.2. GENERIC CONFIGURATION MANAGEMENT ACTIVITIES.

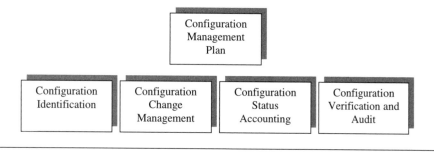

## Configuration Change Management

*Configuration change management* is a process for managing product changes and variances. According to ANSI/EIA 649, the purpose and benefits of the change management process include the following:

- Enable decisions to be based on knowledge of complete change impact
- Limit changes to those that are necessary or offer significant benefit
- Facilitate evaluation of cost savings and trade-offs
- Ensure customer interests are considered
- Provide orderly communication of change information
- Preserve control at product interfaces
- Maintain and control a current baseline
- Maintain consistency between product and information
- Document and limit variances
- Facilitate continued supportability of the product after change.

Change management is the most commonly recognized aspect of configuration management. It is also, unfortunately, the principal source of the reputation of CM being cumbersome and overly bureaucratic. There needs to be a documented process for change, through which all changes must progress. The processing of all changes through a single change board activity is where most organizations see unnecessary bureaucracy in the configuration management process. For this reason, it is important that clear rules exist whereby change classifications can help streamline the approval/implementation process, and changes that are considered minor, or low impact changes, can be directed to those empowered to do so.

## Configuration Status Accounting

*Configuration status accounting* allows the organization to view the current configuration at any stage of the life cycle. It is the means by which a company ensures that its product data and documentation are consistent. At certain points in the life cycle of a project, the configuration status information may need to be reported directly to the customer. Arguably, a more important reason for performing configuration status accounting is to report on the effectiveness of the configuration management process. This should start early in the life cycle of the project by defining target goals that are measurable. The reports from configuration status accounting should then be used throughout the project to identify areas for process improvement.

ANSI/EIA 649 states the typical configuration status accounting activities as the following:

- Identify and customize information requirements
- Provide availability and retrievability of data consistent with needs of various users
- Capture and reporting of information concerning
  - Product status
  - Configuration documentation

- Current baselines
- Historic baselines
- Change requests
- Change proposals
- Variances

## Configuration Verification and Audits

*Configuration verification and audits* are performed on two levels. First, configuration management is responsible for the functional and physical audits of the product. This determines if the product meets the requirements defined by the customer in terms of form, fit, and function. Second, the process itself is subject to audit. Few organizations have applied effective metrics to assess the CM process. Cost of change, cycle time of change, and defect analysis are all ways of assessing the effectiveness of the CM process.

ANSI/EIA 649 defines the purpose and benefits of the verification activity as including the following:

- Ensure that the product design provides the agreed to performance capabilities
- Validate the integrity of the configuration information and data
- Verify the consistency between the product and its configuration information
- Provide confidence in the establishment of baselines
- Ensure a known configuration is the basis for operation and maintenance and life cycle supportability documentation

# Configuration and Data Management

The relationship between physical documentation and digital data has been one of great debate in configuration management circles for many years. Historically, many found it hard to manage both with the same process, and as such there has been a rise in the development of both the hardware configuration management and software configuration management practices, with the latter being the life cycle management of both digital data and software. Essentially, the process was the same; what differed in the majority of cases were terminology, perception, and practice. The technological advances in digital product modeling and a growing interest in the management of product data meant that the divide between managing physical and digital representations was fast becoming an issue that needed a resolution.

To establish a common platform for the practice of managing configurations, it may be helpful at this point to understand the nature of managing data, information, and knowledge, as distinct from a physical, tangible product. The terms "data," "information," and "knowledge" are frequently used or referred to in much of the literature relevant to data management or information management (Checkland and Howell, 1998). Data, according to Tricker (1990) is an entity, and is used to refer to things that are known. Taggart and Silbey (1986) consider data to be groups or strings of characters recognized and understood by people. Data can be either "hard"—that is precise, verifiable, and often quantitative—

or "soft"—that is judgmental and often qualitative. Data has a cost; it can be sold, lost, or stolen and is considered to be an entity that is precise and verifiable and forms the foundation (building blocks) for information.

If data are the building blocks for information, it can be considered that information is formed from individual pieces of data knitted together in a cohesive manner. Taggart and Silbey (1986) provide the view that information is data that has usefulness, value, or meaning. Tricker (1990) states that information is a function resulting from the availability of data, the user of that data, and the situation in which it is used. Information, therefore, can be considered data that has meaning and usefulness and occurs as a result of a process and is understood to be the legacy of human endeavor.

Knowledge on the other hand can be made up of a number of factors, including experiences, education, and acquired information. Davenport et al. (1998) consider that knowledge is information combined with experience, context, interpretation, and reflection. Tricker puts it another way and suggests that it is the aggregate of data held together with understanding. In other words, it is the sum of what is known. Earl (1996) suggests that data is gathered from events and that this data, through manipulation, interpretation, and presentation, produces information. By testing and validation, the information leads to the acquisition of knowledge. From these interpretations of data, information, and knowledge, it can be considered that data is a verifiable and precise entity recognized by people. As a class, data should be easier to manage and control than the other, fuzzier categories.

A level of confusion stems from the different use of the term "data responsibility," particularly with regard to "data owner" and "data custodian." In a conventional representation of an information chain, the author, as originator or creator of the material, and therefore the owner of the information, sits at the top of the list (Basch, 1995). In other words, someone or some organization has to create the data initially and therefore has the authority over its attributes and use. As Van Alstyne et al. (1995) points out, ownership is a critical factor in the successful operations of information systems.

Within organizations, it is generally accepted that, legally, employees do not own the data they create on behalf of that organization. However, the creators of the data or the business function they work within would normally have the responsibility on behalf of the organization for ensuring that "their" data is not abused or misused; in effect, such employees are data owners in the nonlegal sense. Employees, other than the data owners, will also probably use the data; these can be designated data custodians. In some writing on data management, the distinction between custodianship and ownership is not drawn and the term data owner is used loosely to cover both categories of data responsibility.

One of the problems faced by large organizations is that potentially there are many data owners (and custodians) scattered throughout the organization (Brathwaite, 1983) who may adopt piecemeal approaches to data responsibility. Levitin and Redman (1998) point out that data is rarely managed well in organizations. Goldstein (1985) outlines the traditional solution to the problem of ensuring a consistent approach to data responsibility within organizations; he suggests that organizations should introduce a staff function, which he terms "information resource management" (IRM). Goldstein's reasoning behind this is that information is a basic organizational resource in the same way that people and money are. As such, information like these other resources should have a professional, high-level management group responsible for its effective use throughout the organization. The implication of this suggestion is that as data is the foundation for information, then if the organizational information has an owner, so has the data making up the information.

However, data responsibility issues are not simply contained within the organization's boundaries; ownership of data can be, and often is, protected via such as patents and copyright. In his paper "Ownership of Data," Cameron (1995) looked at some of the legal issues surrounding data where the oft-asked question is, "Is data a property or not?" Cameron puts forward the view that

> to be treated as proprietary data, the data must have been created by the owner, been created for the owner, or been purchased from its creator.
>
> Cameron, 1995; p. 47

In many commercial situations, the item being purchased is a product and not necessarily the underlying or ancillary data. In such circumstances, the ownership of the data is not usually transferred—as is the case with many software licenses where the purchaser has a license to use the software but does not own the underlying code (data). However, in circumstances where a customer purchases a "project" rather than a mass-produced product, as is more like the situation in the aerospace industry, for example, then the issue of transferring data ownership does becomes more of an issue.

## Life Cycle Management and Configuration Management

A lot of configuration management's work comes, as in the previous washing machine example, from component parts of a product entering into the project, or program, at different stages of its life cycle (see Figure 5.3 for a typical life cycle with stages). As the life cycle matures from concept through to disposal, the amount of information that comes

### FIGURE 5.3. LIFE CYCLE MANAGEMENT PHASES.

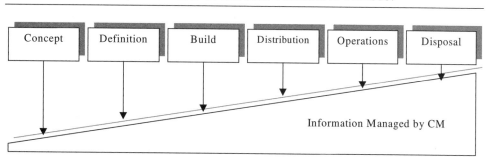

| Concept | Definition | Build | Distribution | Operations | Disposal |

Information Managed by CM

under configuration control becomes greater and greater. Indeed, it is true to say that the life cycle can be defined in terms of information, in that it begins with the first release of information and ends with retiring the last definitive information sets. The product itself may not appear for some time into that cycle and may be withdrawn long before the information is retired. In practice, "life cycle phases" are developed to suit the product type, the development method, the company doing the developing, and the industry; and they operate quite successfully despite such modifications. In the aerospace industry, life cycles can exceed 50 years, and such longevity does pose problems (Osborne, 2001). It is not uncommon for other industries with long product/program life cycles, such as process and nuclear power, similarly to emphasize configuration management. Indeed, the regulatory bodies of FDA (Food and Drug Administration), NIRMA (Nuclear Information and Records Management Agency) as well as CAA and FAA (aerospace regulatory bodies) make strict demands of the application of configuration management activities, although not always referring to them as "configuration management." Configuration management is increasingly accepted as a major factor in the design, development, and production of products and is becoming a major requirement in the in-service life of such essential products.

Within the field of configuration management, it can be argued that there are no new concepts as such; the elements of configuration management are the same as 20 years ago. The important issue concerns how the configuration management processes are implemented. One could argue on two fronts: (a) in the past, the implementation of the configuration management processes has not been as effective as it might be, and (b) the organizational context in which configuration management processes are implemented has changed and therefore the nature of the implementations need to change to reflect this fact.

The first level of argument includes the assertion that configuration management "paper" processes have been simply automated in a piecemeal fashion in nonintegrated software tools. Organizational philosophies in the past often consisted of managing the product within a "functional" environment. Thus, "islands of information" (Morton, 1994) were created, as piecemeal implementations were made on functionally based computer systems that rarely interacted sufficiently well with each other. However, configuration management activities cut across functional domains within the organization and thereby often create major problems in terms of contiguous data management across both the organization and throughout the product life cycle. Integrating the systems to ensure that data is available to the users at the required time and in the correct format can pose major challenges for the organization. It is also worthwhile recognizing that users are not necessarily passive recipients— quite often they are active in creating and modifying the data. This then raises issues about responsibility for where configuration management activities lie.

The second level of argument has a number of major strands that link to the prior point about functional organization and the lack of integration, and to the whole discussion about data, information, and knowledge.

- First, the availability of improved IT has enabled the implementation of a more integrated configuration management process (Osborne, 2001).
- Second, the adoption of teamworking by organizations with the consequent breaking down of functional barriers means that integrated approaches are required more than

ever before. As an illustration, the advent of Total Quality Management (TQM) (Oakland, 1992) has surfaced issues about the integration of quality responsibility with teams rather than with traditional functions.

- Third, the move toward increased levels of participation within industry—for instance, the "extended enterprise" (Schonsleben and Buchel, 1998) also unleashes an increasing pressure for integration of the configuration management processes, but this time the focus is *between* rather than *within* the organizations.
- A fourth strand relates to the relaxation by customers of a demand for adherence to their own specific configuration management standards, permitting organizations to put in place generic solutions to the CM "problem." This relaxation has originated, in part, from the dwindling of the public sector and the dominance of a business (private sector) ethos in many developed economies.

All these combine to create an organizational environment that could be characterized by an increased awareness of the role that configuration management could play and have heightened pressures to substantially alter configuration management practices. In particular, they accentuate the need for more integrative and effective CM, facilitated particularly through the use of IT. However, in practice, reports on configuration management practices do not feature prominently in the general CM literature, nor in the more specialized domain of software configuration management (Davies and Nielsen, 1992). In short, a knowledge gap exists.

The *Oxford English Dictionary* defines a configuration as "an arrangement of things." It follows, therefore, that managing configurations is concerned with managing arrangements or patterns. Questions arise as to where in the life cycle we manage those patterns and where exactly we stop managing the arrangement. To answer those questions, we need to consider the life cycle of the configuration itself.

Today, organizations in most business sectors place great emphasis on managing the product life cycle. However, this is only a small part of a much bigger picture. The information life cycle is much broader in scope and operation, and the product life cycle, system life cycle, project life cycle, and asset life cycle all lie within its phases. In many manufacturing industries, such as aerospace and automotive, there is a considerable period where the product exists in a purely digital form. Digital mock-ups and 3D models are representations of a product in an information form. The need to manage these representations with a common set of configuration management processes is still regarded as a major challenge.

At the other end of the configuration life cycle, when do we stop managing the pattern or arrangement? Prior to BOT/PFI (see the chapters by Ive and by Turner for a discussion of this type of project), the majority of projects closed out at the delivery phase, when we handed over to the client. The supportability of products has now become of strategic importance to many companies. In the aerospace community, there is considerable financial return for the supportability of in-service aircraft. Benefits are only realized when we ensure that we are certain of the current configuration status of each of those products at the time of service. The concept of the information life cycle has taken on a new level of importance. The flow from concept through to end of life must be well managed and maintained. The failure to do so will result in a catastrophic impact on the bottom line.

From a different viewpoint, in the majority of cases the product life cycle finishes with the disposal phase. In many cases it may be important to maintain the information post-disposal. There may be a legislative requirement, as in the medical device and pharmaceutical industries. Alternatively, it may be to assist in future development projects through support for modularization. Although many products are considered unique, the need to limit development costs means considerable benefits can be gained from cataloguing modules for future design projects.

In short, therefore, the configuration, or information, life cycle is the dominant life cycle when we are assessing the strategic impact of the full life of products and systems. It begins with the release of the first definitive set of information and ends with the retirement of the last. For many, the beginning will be the opening of the bid phase to the end of the contract; for the owner, however, it will be early in the development cycle and go through to the disposal phase. The salient point is that it is information, not the physical product itself, that has to be managed.

## Organization of Configuration Management

The existence of different organizational structures leads to a variety of views as to who owns the configuration management process; for example Sage (1995) describes CM as being owned by system engineering. However, other views exist where PM, quality management, engineering management, and logistics management all have a stake in the ownership of the enterprise configuration management process.

Integrated product development (IPD) practices are a recent, and significant, advent in organizational structures (see the chapters by Archer and Ghasemzadeh, and by Milosevic). The IPD philosophy is implemented through integrated product teams (IPTs) and encompasses concurrent engineering where the effective configuration of the system's life cycle takes on special significance, since simultaneous development activities need to be carefully coordinated and managed early on (see the chapter by Thamhain). Integrated product teams cause problems in identifying responsibilities for the different elements of the CM process. A typical response is that the development of the CM process and techniques within a company become the responsibility of a core CM discipline, whereas the day-to-day operational tasks become the responsibility of the IPT leaders. To discharge the CM responsibilities in such circumstances over the project life cycle requires a flexible, responsive structure.

While the above holds true for major manufacturing organizations, the software community has developed a very similar pattern for configuration management application. The majority of the changes to code and structure are carried out by the developers themselves. Organizationally, the planning of the CM activity is managed by a centralized CM function. It is fair to say at this point, however, that in the IT community, configuration management is a far more automated activity, with tools facilitating the change and versioning process.

# Changing Nature of Configuration Management in the Aerospace Industry

Aerospace is perceived as an industry that has a well-established and documented use of configuration management practice. (See the chapter by Roulston.) Further analysis, however, shows that the depth of such practice and innovation in application of the CM process is varied across the sector (see the research study that follows). Some of the world's highest-profile collaborative development projects such as Airbus and Eurofighter have suffered major cost overruns and schedule problems, due in part to CM not being coordinated at the outset. A European Commission Framework 4 project (AdCoMS; Project No. 22167) sought to establish a common CM platform for all partners in collaborative development programs, based on commercial best practice, rather than existing standards. Sadly, after a 2000 completion, little has been utilized by the consortium partners who developed it.

Interestingly, the Perry Initiative in the U.S. defense industry encouraged the use of commercial standards to replace those of the U.S. Department of Defense (DoD), where appropriate (Ciufo, 2002). Many years after this, it is still not uncommon to see traditional defense industry practices in configuration management being adopted. For many, the comfort of using an overly defined and regimented process became a barrier to change. Not surprisingly, therefore, the perception of value and organizational recognition of configuration management was, in many cases, poor.

Recent research has looked at the changing nature of CM application in the aerospace industry in Europe. Historically, aerospace has been a key player in the development and innovation of the configuration management process (Kidd, 2001), and the research identified the changing nature of its application within the rapidly changing environment of the aerospace industry. In the study, organizations were categorized as Tier 1, developing and manufacturing at a high level, and Tier 2, suppliers to Tier 1 organizations.

A population of 210 organizations were surveyed, with the nominated configuration managers being asked a total of 50 questions. A follow-up interview was undertaken with a cross section of the organizations in the initial investigation. A summary of the results was as follows:

1. *How do aerospace industry players define configuration management?* The use of international CM standards was evident across the whole of the product life cycle and drew on a good mix of standards. Customer needs and standards are perceived as important factors in defining the design of the CM process, while IT is identified as a key mechanism to support this. Overall, the responses indicated that the use of the CM plan was a major activity within the companies, with 78 percent of companies having a CM plan and 50 percent referring to it frequently. However, given the key role of the plan in CM thinking, even higher levels of reliance on a CM plan were to be expected, and therefore there is some evidence that companies are treating CM as a compliance issue rather than wholeheartedly believing in it. Sixty percent of respondents indicated that CM activities were generally not the responsibility of the quality function but were the province of a separate CM function, which had reporting links to other areas of the orga-

nization such as project management. Leaving aside the responsibility for CM activities, individual functional departments typically carry out the activities needed for CM within their own domain. The interviews highlighted strong views for a more active role for CM personnel. Views were expressed that companies should move to an organization form where a CM discipline managed the CM activities and requirements across the product life cycle rather than the functional fragmentation indicated previously. At present, little evidence is apparent of career progression, education, and training; the latter was particularly lacking in second-tier companies.

2.  *How do the companies value configuration management?* The responding companies demonstrated their low reliance (25 percent) on metrics to measure the performance of the CM activities and the low incidence of risk assessment. Again, this suggests CM is seen as a passive compliance activity rather than an area that, if managed properly, could deliver benefit. The CM processes within these companies were claimed to be flexible and supportive of customer and project requirements. External auditing of the CM process was undertaken against required standards by both tiers; however, first-tier companies were open to more external scrutiny.

3.  *How do the companies carry out the configuration management process?* Only just over half of the first-tier companies who responded to the questionnaire claimed to have an "end-to-end" process; however, there was a clear spread of activities across the differing company functions. (An end-to-end process is taken to cover the whole of the product life cycle rather than supporting limited parts such as the design process.) The second-tier companies were in a different position, because their activities often represented only part of the CM life cycle. The 80 percent view from both tiers indicated that there was a conceptual process for CM that was documented in line with the appropriate standards. The companies indicated that their CM procedures were developed in-house and supported the individual function's requirements. Companies indicated the uniqueness of their processes (76 percent) despite the common principles that underpin CM. The interviews probed further the view that an end-to-end process was employed and that this comprehensive scheme interacted with many other processes externally to the company. With the extra depth of information, it soon became clear that the end-to-end process is an intention rather than what is actually happening in companies. A life cycle process for configuration management is mainly a vision that most of the companies wish to attain but at present do not have.

4.  *Is configuration management recognized within the organization?* The key message here is the lack of recognition of the CM function, which stood at only 31 percent of companies overall but was particularly poor in second-tier companies (21 percent). Responsibility for CM is not vested in a specific senior manager; indeed, even at lower levels in the organization (20 percent), there is a clear absence of a designated CM manager. Fragmentation of CM activities is evident across individual functions. There is a lack of clear career progression and a lack of education and training provision. In the interviews, organizational structure and career recognition linked to education and training strategies was seen as a major requirement for the developing CM world and were viewed as much more important than apparent in the results of the questionnaire.

5. *Is configuration management covered by IT means?* Eighty percent of the questionnaire responses indicated that IT within the organizations was not fully covering the requirements for CM. The majority of the respondents indicated that the use of both IT systems and paper were the means to record and report the CM requirements. In the interviews, the development of the CM process was seen to be standards-driven, with best practice and experience adding to this development. Hence, the general feeling was that IT had not been a driving influence for CM. However, this appeared to be changing, and respondents saw IT as a driving force for process development, given the advances in the technology employed within the companies. This increased influence of technology was changing the manner in which processes were developed and deployed within the organization, and therefore this was having a major impact on configuration management in terms of process, data management, and status accounting.

6. *Is configuration management a stand-alone process, or is it covered by other, separate processes?* Questionnaire respondents were near unanimous in indicating that the CM process was not stand-alone and instead connected to many other wider spread processes, including processes external to the company. It was evident from the interview responses that the CM process cuts across all the different company functions and links to many required activities, particularly in first-tier companies. Thus, the CM process is not viewed as a single process but the interaction of many processes. Within all the companies who responded to the questionnaire, there appeared to be a good understanding of the requirements for CM and a good knowledge of the standards that were used. But the variation in the manner that the process was employed suggested that there was no single process that could have been developed that would fit the needs for all of them. There are many aspects that influence the requirements for the process for CM; one of these is the way that product development is organized. This could be a single integrated product development team, a separate function, an individual company, or a mixture of them all. Therefore, the process that needed to be employed was seen to differ significantly according to the requirement of the organization.

7. *Does configuration management add any value to the business?* This question was mainly addressed by looking to see views on the level of knowledge that CM personnel had and how CM data contributed to business activities. The positive responses indicated CM did add value to the business, with the CM personnel being seen as making a valuable contribution. CM activities were not seen as restricted to those individual functions with clear CM responsibilities within the organization, and CM personnel fulfilled a valuable role in advising on the requirements for projects that need to be undertaken. In total, 87 percent of those surveyed felt that CM added value to both program and the business as a whole. It can be surmised that that the remaining 13 percent felt that their efforts were either unrewarded or the CM process they were working with was inadequate for the purpose.

In summary, the preceding study of CM in the aerospace industry provided some interesting perceptions of the value of configuration management in the development, build, and maintenance of highly complex products. Seen as an industry that relies heavily on such practices to ensure integrity and reliability, it would appear that CM still carries a perception of being cumbersome and administrative to many. Part of the reason for this may be to do with the regulatory nature of CM in the defense sector. Many of the standards used in this sector were indeed user-unfriendly and relied on the use of prescribed documentation and process. However, this is changing, with the encouragement of companies to use commercial standards where appropriate and to innovate their own processes to include best practice. The focus on managing life cycles in the aerospace industry has breathed new life into configuration management. Many now see it is a part of their everyday work and not just the job of the configuration manager (Kidd, 2001). It could well be argued that CM is at a point today where quality management was in the late 1970s: transitioning from a control process to an enterprise-wide activity. Clearly, whether CM ultimately follows the same trajectory taken by QM over the last 30 years to reach such prominence depends upon the actions of all in organizations and not just CM professionals. Fostering this change argues for configuration management to be treated in a similar way to quality—that is, where everybody in the organization is exhorted to think about quality and be responsible for quality. Of course, this comparison with quality also points to the potential downside that people see CM as a bureaucratic impediment to be dealt with simply on a compliance basis.

## Summary

Configuration management is not just about managing products. It is about managing everything that defines the product or system across the full life cycle. When do we start doing CM? When we issue the first definitive information, not when we have a "configuration" to manage—by then it is too late. When do we stop doing CM? When we no longer have a need for the information, and we retire the last definitive information set. We live in a world characterized by rapid innovation. This also means rapid change, and we must develop better methods of incorporating change into products, systems, and services. As more and more organizations seek to exploit the benefits of life cycle support and service agreements, the role of configuration management becomes pivotal in maximizing value. If we do it badly, then the costs of maintaining poorly defined products will heavily impact the bottom line. For those who do it well, the benefits will set them apart from the competition.

For those of us working in project management, the role of CM should now be clear. How beneficial would it be to have the right information, in the right format, in the right place and at the right time? Would this assist in the decision-making process of managing projects? The clear answer is yes.

## Acknowledgment

We would like to acknowledge the research work carried out by Dave McKee and Colin Hillman from BAE SYSTEMS, and Kevin Byrne from CSC, while working on their master's theses with the CM Research Group at Leeds University. The results of this research contributed to this chapter.

# References

Anon. 1998. *EIA–649 National Consensus Standard for Configuration Management*, Electronics Industries Alliance.

Anon. 1997. *ISO 10007 Guidelines for Configuration Management*, ISO Geneva.

Basch, R. 1995. *Electronic information delivery*. Aldershot, UK: Gower.

Brathwaite, K. S. 1983. Resolution of conflicts in data ownership and sharing in a corporate environment. *Database* 15(1):37–42.

Cameron, D. M. 1995. Ownership of data: The evolution of "virtual" property, data as property. Presented at Toronto, Ontario, Canada, January.

Checkland, P., and S. Howell. 1998. *Information, systems and information systems: Making sense of the field*. London: Wiley.

Ciufo, C. A., 2002. Editorial. *COTS Journal* (Spring): 78.

Daniels, M. A. 1985. *Principles of configuration management*. Advanced Applications Consultants.

Davenport, T. H., D. W. De Long, and M. C. Beers. 1998. Successful knowledge management projects. *Sloan Management Review* (Winter).

Davies, L., and S. Nielsen. 1992. An ethnographic study of configuration management and documentation practices. *IFIP Transactions A—Computer Science and Technology* 8:179–192.

Earl, M. J. 1996. *Information management: The organizational dimension*, London: Oxford University Press.

Goldstein, R. C. 1985. *Database: Technology and management*. London: Wiley.

Kidd, C. R. 2001. The case for configuration management. *IEE Review* (September).

Levitin, A.V., and T. C. Redman. 1998. Data as a resource: properties, implications and prescriptions. *Sloan Management Review* 40(1):89–98.

Oakland, J. S. 1992. *Total Quality Management*. Oxford, UK: Butterworth-Heinemann.

Osborne, J. 2001. Avoiding potholes on the data highway. *Professional Engineering* (July): 39–40.

Sage, A. P. 1995. *System management for IT and software engineering*. London: Wiley.

Schonsleben, P., and A. Buchel, eds. 1998. *Organizing the extended enterprise*. London: Chapman and Hall.

Scott, M. A. 1994. *Information technology and the corporation of the 1990s*. New York: Oxford University Press.

Taggart, W. M., and V. Silbey. 1986. *Information systems: People and computers in organisations*, New York: Allyn and Bacon Inc.

Tricker, R. I. 1990. The management of organizational knowledge, Paper presented at the 1990 Conference on Systems Management, Hong Kong.

Van Alstyne, M., E. Brynjolfsson, and S. Madnick. 1995. Why not one big database? Principles of data ownership. *Decision Support Systems* 15:267–284.

# CHAPTER SIX

# SAFETY, HEALTH, AND ENVIRONMENT

## Alistair Gibb

Many readers may be wondering why safety, health, and environment (SHE) are included in a book about project management. Sadly, this view is not unusual. Even in "developed" countries, there is still a paucity of consideration of SHE issues for projects in all industrial sectors. This chapter introduces the reader to some of the key issues as they affect the overall management of a project. All tasks in all industrial and commercial sectors involve SHE risks; however, the intrinsic nature of most projects is such that steady state has not been achieved and the project conditions and environs are continually changing. This is particularly true for construction projects. Therefore, to provide a focus for this chapter, SHE issues have been considered mainly from a construction project perspective, although reference is made to other project scenarios where appropriate. The key principles apply to both large and small projects, although the implementation of them may vary (CII, 2001).

In the European Union, "construction" has been defined as all works associated with the project, including demolition and decommissioning. "Health" covers occupational health issues of construction workers, which are often overlooked in efforts to address the more immediate challenges of "safety." Safety and health implications of the completed buildings or facilities are also important but are outside the scope of this chapter except for maintenance aspects. "Environment" has become a much-used term, covering a broad spectrum of issues of the sustainability on the built environment. The sustainability of a project covers issues from construction and throughout the life cycle of the completed facility. Sustainability itself is a broad subject typically considered as relating to three main areas: environmental (planet), social (people), and economic (prosperity).

Once again, to maintain focus, this chapter concentrates on construction site aspects of the environment. Health and safety are typically covered together in much legislation and many publications. While environmental issues are different, there is often an overlap with health and safety in terms of management strategies and techniques. SHE is considered an integrated management task in many large, global organizations, although those responsible for it are often biased by background and training at least toward one particular aspect, often safety. It therefore cannot be taken for granted that all three aspects will be given the appropriate emphasis.

The causes of accidents, ill health, and environmental disasters are multifactorial and should not be considered simplistically (Hide et al., 2002; Reason, 1990; and others); however, it is accepted that effective project management will have a positive affect on SHE risks. The saying "if you can't manage health and safety, you can't manage" is supported by most writers on the subject. Griffith and Howard (2001) stress that the "management of health and safety is without doubt the most important function of construction management." Notwithstanding, SHE is still absent from many general management texts.

The chapter explains the importance of SHE, introduces SHE objectives and strategy, and highlights design and procurement activities and an action plan for construction. It briefly introduces life cycle issues and the measurement of success. This structure has been taken from the European Construction Institute's SHE manual (ECI, 1995). The ECI also has guidance documents dealing specifically with health and the construction environment (Gibb et al., 1999 and 2000), and readers may consult these publications for a more complete coverage of the subject.

## Why Are Safety, Health, and Environment (SHE) Essential Project Management Considerations?

### Moral Responsibility for SHE Management

International comparison of SHE performance is impossible, and it is decidedly unwise to even attempt it. Griffith and Howarth (2001) argue that there will be "considerable differences in, for example, economic climate, market forces, political environment, construction methods and availability of resources." Nevertheless, through my involvement with the international research network, Conseil Internationale de Batiment, it is obvious that the statistics throughout the world are unacceptably high. It really is not acceptable in the twenty-first century that someone working in construction cannot expect to complete a career in the industry without sustaining some form of injury or occupational disease. Furthermore, the issue of the environment has passed from a pressure group topic into the mind-set of the average person in the street—although they might not understand all the complexities, they believe that companies should take a responsible attitude toward caring for the environment.

### Legal Responsibility for SHE Management

In an international publication like this it is inappropriate to describe the legal arrangements of one particular country. Nevertheless, throughout the world, enshrined in the law of most countries is a duty of care to others, and in particular an employer's duty of care to those employed to work on their behalf.

Since the early 1990s European states have had the further legal requirement to ensure that designers overtly consider the health and safety of construction workers (EC, 1992). This same directive requires effective health and safety management systems to be used. A similar requirement for environmental management is enshrined in the ISO 14001 standards (ISO, 1996) and is expanded by Griffith (1994). A good summary of environmental law in the United Kingdom is provided by Stubbs (1998). In general, many other countries, including the United States, have not brought together the legal requirement for safety, occupational health, and environmental protection.

## Financial Necessity for SHE Management

"Humanitarian factors alone are more than enough to justify the effort required to eliminate worker injury. . . . . however, the significant cost of worker injury cries out for exposure to those who worry about the cost of safety programs. . . . Eliminating injury makes good business sense." (Nelson, Shell Oil Company, 1993). In the United Kingdom, the Egan report, "Rethinking Construction" (DTI, 1998), stated that "accidents can account for 3 to 6 per cent of total project cost." Nelson (1993) estimates that "the total cost of injury for the $450 billion U.S. construction industry ranges from $7 billion to $17 billion annually." The hidden costs can be many times more than the visible costs.

"In the Piper Alpha explosion (North Sea oil rig disaster), 167 lives were lost and £746 million (US$ 1243 million) was paid out by insurers, but estimates put the total loss, including business interruption, investigation costs, hiring and training replacement personnel and the like at over £2 billion (US$ 3.3 billion)" (Clarke, 1999).

The cost of accidents or environmental incidents include the following:

- *Management and organization.* Resources, administration, and accident investigation
- *Damage to reputation.* Adverse publicity and impact on industrial relations; impact on future tenders; liability; and compensation;
- *Loss of productivity.* On the day of the incident and for some time thereafter
- *Litigation and legal fees.*
- *Fines from statutory authorities and similar bodies.*
- *Delays to the project.* While the situation is normalized
- *Sick pay to injured personnel.*
- *Damage to property and materials.*
- *Increased insurance premiums.* Some countries make a direct correlation between SHE performance and insurance rates. In the United States this is called the EMR (Experience Modification Rating) and can have a significant financial effect
- *Medical costs.* Liability for these will vary between countries, but the costs can be substantial irrespective of who has to pay them.

Hinze has studied cost aspects of construction health and safety in the United States for many years (e.g., Hinze and Appelgate, 1991; Hinze, 1991 and 1996; CII, 1993a), and he argues strongly for serious consideration of the real costs of accidents and incidents, including the very substantial hidden costs. Those looking for a fuller discussion of financial issues can review the proceedings of the Conseil Internationale de Batiment W99 conference dedicated to the subject (Casals, 2001).

Any cost exercise should include the costs associated with setting up an effective SHE management system and procedures, where all parties recognize the implicit and explicit costs. If possible, SHE activities will be included in the contract agreements between all parties. Many of the explicit costs can then be linked to specific project activities such as scaffolding, or asbestos removal, but they can also be in the form of a SHE specification, priced as part of the contract.

The link between health and safety performance and project profitability has been debated for many years. In 2001, the UK construction and development organization Taylor Woodrow compared health and safety audit scores with the profitability of each project. Figure 6.1 shows the results with the main vertical scale being the audit score and the shades of the columns representing varying degrees of profitability (more than 2 percent above the expected return; within ±2 percent of expected return; more than 2 percent below expected return). The graph demonstrates that many of the poorer-performing projects from a health and safety viewpoint were also performing badly financially (shown as

## FIGURE 6.1. COMPARISON OF HEALTH AND SAFETY AUDIT SCORES WITH FINANCIAL PERFORMANCE.

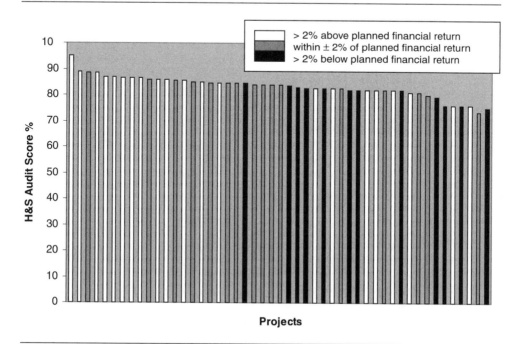

black columns that are grouped toward the lower end of the safety performance spectrum). Clearly the sample would need to be enlarged to argue this point more strongly; however, that there is a clear indication of a link did not go unnoticed within Taylor Woodrow's senior management team. This aspect is discussed further by Kunju-Ahmad and Gibb (2003).

## Cultural Challenge for SHE Management

There is a real cultural challenge for SHE management, with an acceptance among many involved in construction that the industry is inevitably dirty, unhealthy, and unsafe. Construction is often seen as a "macho" industry where brute force and a bravado attitude pervade. Those complying with the need to wear personal protective equipment, working with care, and putting safety first are too frequently taunted by their coworkers or branded as "difficult" by management. Furthermore, there are some national cultures that place less worth on the well-being of certain individuals, for example, those who work with their hands. We are part of a global human culture that tends to ignore the waste it produces and care for the environment only if it does not affect our everyday lives. These cultural misconceptions should be challenged and a positive safety culture cultivated at a project as well as a community level. Much has been written about achieving such a culture and the precise methods promoted vary with the application, but the principal remains where everyone looks out for the safety and health of themselves and of others along with having consideration for the environment.

# SHE Policy, Objectives, and Strategy

## SHE Policy and Objectives

The project's SHE objectives and strategy will be based upon a sound SHE policy for the stakeholders' organization. The policy will be a public-domain document, emanating from the executive board, which states the organization's corporate SHE philosophy in the context of its overall business activities. HSE (1997) stresses that "effective health and safety policies contribute to business performance by

- supporting human resource development;
- minimizing the financial losses which can arise from avoidable unplanned events;
- recognizing that accidents, ill-health and incidents result from failings in management control and are not necessarily the fault of individual employees;
- recognizing that the development of a culture supportive of health and safety is necessary to achieve adequate control over risks;
- ensuring a systematic approach to the identification of risks and the allocation of resources to control them; and
- supporting quality initiatives aimed at continuous improvement."

ECI (1995) advise that a "SHE policy should be clear, concise and motivating. The content should clearly express

- what the company intends to PREVENT (using words such as prevent, limit, protect, eliminate);
- what the company intends to IMPROVE (using words such as create, develop, carry out, replace); and
- what the company intends to COMPLY with (using words such as comply, demand, require)."

For instance, for the environment, the policy may aim to pursue progressive reduction of emissions, effluents, and discharges of waste materials that are known to have a negative impact on the environment with the ultimate aim of eliminating the negative impacts.

Typical strategic SHE objectives may include the early identification of major hazards, the examination of the impact on construction of SHE considerations during design, the development of a SHE framework for construction and the project life cycle, the development of a SHE plan by the principal contractor before site work begins, and compliance with this plan thereafter. SHE objectives must be achievable and therefore be in-line with other project management objectives such as time, cost, and quality. It is important to note, however, that it may be necessary to amend the time and cost parameters so that the SHE objectives can be achieved. This is another reason why SHE should be considered along with other project-wide issues as part of an overall project strategy rather than as a stand-alone issue. Many large organizations now incorporate SHE management holistically, within an overall quality management system. However, there is still some debate on this approach (e.g., Smallwood, 2001; Griffith and Howarth, 2001; CIRIA, 2000, Gibb and Ayode, 1996; Rwelamila and Smallwood, 1996).

## Project SHE Concept, Initial Risk Assessment and SHE Plan

The overall policy and objectives will be worked through at project level. Griffith and Howarth (2001) state that "project health and safety planning and management should be considered in two parts. The first part focuses on the client's project evaluation and design processes with the objective of producing a 'pre-tender' health and safety plan. The second part focuses on the site production processes with the objective for the appointed principal contractor to produce a construction phase health and safety plan." They go on to say that "it is the essential part of planning within each part which forms the basis for a systematic management approach, within which risk assessment is an important theme."

At this early stage, the emphasis will be on major hazards, with the output being an initial risk assessment and a preliminary SHE plan. The risk assessment process is described in more detail in the next section, although at this phase the exercise will be done at a fairly high level. Typical risks to be considered at the concept stage include those shown in Table 6.1 (ECI, 1995).

## TABLE 6.1. TYPICAL SHE RISKS TO BE CONSIDERED AT THE PROJECT CONCEPT STAGE.

| Safety Risks | Health Risks | Environment Risks |
|---|---|---|
| Climate | Infections | Emission |
| Natural hazards | Hygiene | Effluents |
| Transport | Worker accommodation | Wastes |
| Security factors | Medical facilities | Noise and vibration |
| Unskilled labor | Potable water | Light |
| Major risk factors, e.g., | Chemicals | Damage to surroundings |
| heavy lifts; excavations; | | Contaminated ground |
| demolition | | Heat |
| Concurrent operations | | Electricity |
| | | Pressurized systems |

*Source:* After ECI (1995).

Many companies in the engineering construction sector (petrochemical/power generation construction) use the HAZCON procedure. This is a two-part, formal procedure for early identification and assessment of SHE hazards in construction to enable all reasonably practicable steps to be taken to reduce or eliminate the risk. HAZCON 1 identifies major hazards to owner personnel, contractors, visitors, or the general public, along with actions and recommendations for hazard elimination or reduction. Risks may occur within the site or beyond its boundaries. HAZCON 1 uses checklists to aid the evaluation, and it is done as early as possible in the project, at least before the project scope and site details are finalized. HAZCON 2 is done later in the process, to provide a detailed assessment of construction hazards based on the completion of a significant level of engineering definition, at least including plans and elevations together with a draft overall construction method statement, contract plan, project schedule, and site layout drawings. It should also include a review of HAZCON 1 results to see whether the development of the scope has added or removed any major construction hazards. The HAZCON procedure and checklists are explained further in ECI's SHE manual (ECI, 1995). The follow-on procedure, HAZOP, relates to operating aspects of the constructed facility.

The SHE plan will include strategies for design, procurement, construction, commissioning, maintenance, decommissioning, and demolition. The SHE plan will also cover the following issues at a strategic level (ECI, 1995):

- "SHE management and leadership
  - including organization; communications and meeting schedule.
- SHE organization and rules
  - including policy statement; legislation; standards; procedures; basic rules; health; medical and welfare program; auditing; environmental; and sub-contractor strategy.
- SHE risk assessment and management
  - including, hazard identification; risk assessment; SHE performance and measurement; and emergency response procedure.

- SHE training
  - including employee orientation program; promotion and awareness; training program; and involvement of professionals.
- Personal protective equipment (PPE)
  - including risk assessment; PPE requirements and use.
- Incident/accident/injuries records and data
  - including reporting procedures
- Equipment control and maintenance
  - including SHE equipment and inspection; hygiene and housekeeping'.

# Design and Preconstruction activities

## Risk Assessment and Risk Avoidance

Risk assessment is an essential part of all business processes and again also necessary for SHE issues. In Europe, risk assessment and management is mandatory during both the design and construction phases. Designers are required to identify hazards and their associated risks and then to eliminate, reduce, or control the risks they have created. The designer's role in generating risk and identifying solutions has not yet been fully acknowledged outside Europe, and in many states risk assessment and control is left to the construction team. The design team will review the hazards identified at concept stage (through HAZCON 1 or similar) and develop the risk assessment in more detail, checking that no new hazards have become apparent.

It is important to understand two key terms: hazard and risk. According to the United Kingdom's Management of Health and Safety at Work regulations (1999), a hazard is "something with the potential to cause harm" and risk expresses "the likelihood that the harm from the hazard is realized."

Beilby and Dean (2001) identify "five steps to risk assessments:

- Step 1: Look for the hazards.
- Step 2: Decide who might be harmed and how.
- Step 3: Evaluate the risks and decide whether the existing precautions are adequate or whether more should be done.
- Step 4: Record your findings.
- Step 5: Review your assessment and revise it if necessary."

Most risk assessment methods follow a similar format. I favor a simple three-point scale where both hazard severity and likelihood are given a score of 1, 2, or 3. The risk is then the product of the hazard severity and the likelihood of occurrence. Often a risk matrix such as that shown in Figure 6.2 is used. More complicated systems are available, but they do not necessarily produce more accurate results.

As an example of this process, for health and safety (HSE, 1997) the following levels would apply:

## FIGURE 6.2. RISK ASSESSMENT MATRIX.

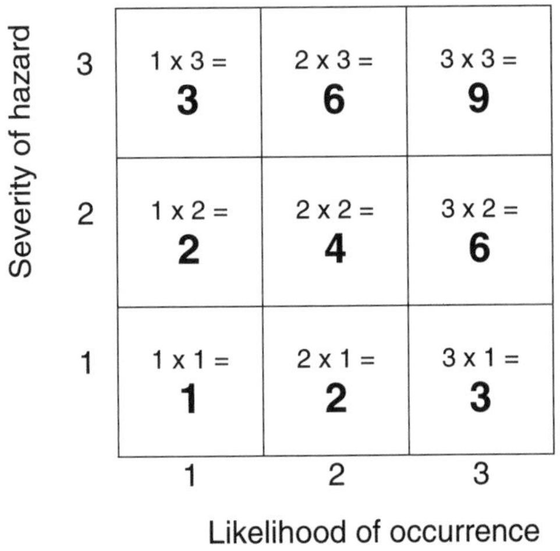

*Severity of hazard*

- *Level 3.* Major—death or major injury or illness causing long-term disability
- *Level 2.* Serious—injuries of illness causing short-term disability
- *Level 1.* Slight—all other injuries or illness

*Likelihood of occurrence*

- *Level 3.* High/probable—where it is certain that harm will occur
- *Level 2.* Medium/possible—where harm will often occur
- *Level 3.* Low/improbable—where harm will seldom occur

The following hierarchy of risk actions are taken from the European Directive (89/391/
EEC) by Griffith and Howarth (2001) but have international applicability:

- "avoiding risks;
- evaluating the risks which cannot be avoided;
- combating the risks at source;
- adapting the work to the individual, especially as regards the design of workplaces, the choice of work equipment and the choice of working and production methods, with a view, in particular, to alleviating monotonous work and work at a pre-determined work rate and to reducing their affect on health;
- adapting to technical progress;

- replacing the dangerous by the non-dangerous or the less dangerous;
- developing a coherent overall prevention policy which covers technology, organization of work, working conditions, social relationships and the influence of factors relating to the working environments;
- giving collective protective measures priority over individual protective measures; and
- giving appropriate instructions to employees."

Recent work at Loughborough University (ConCA, 2002) studying 100 construction accidents has found that many risk assessments are virtually useless, in that they have little or no effect on the actual task operation itself. Too often the risk assessment is done as a "tick-box" exercise rather than a thoughtful assessment of the risk. Frequently the style, language, and length of the documents is such that they are not accessed at the workface by the operatives and supervisors but are retained in the site office "gathering dust." There is a real need for task-based risk assessments. A few organizations have started to address this shortfall. For example, the Channel Tunnel Rail Link (CTRL) project in England included a task risk evaluation as part of the supervisors' briefing and discussion with operative gangs each morning. Other cultures, such as the Japanese, include daily orientation as part of a start of the day routine for all workers. This can provide the opportunity for specific health and safety aspects to be raised and dealt with.

## Designer's Role

"Construction worker safety is impacted by the designer's decisions" (Hinze, 1998). The European Directive, leading to the Construction (Design and Management) Regulations (CDM) in the United Kingdom, have formalized the requirements for designers to consider health and safety in their designs. While not mandatory outside of Europe, this strategy has realized support from researchers and industry leaders worldwide (Gibb, 2000; Hinze and Gambatese, 1996; Tenah, 1996; Oluwoye and MacLennan, 1996). However, despite market leaders emphasizing the importance of designing-in safety and health over many years (e.g., CII, 1996), the take-up of the strategies where not driven by legislation has been very limited.

In the United Kingdom, the CDM regulations require designers to

- inform clients/owners of the CDM regulations;
- apply the hierarchy of risk control to their designs;
- cooperate with other designers;
- cooperate with the "planning supervisor" (who is charged with coordinating H&S effort, particularly during design—the similar EC role is called design phase coordinator); and
- provide information about their design for inclusion in the health and safety file (a document that should form the central core of the health and safety management of a project).

While environmental issues are not covered in the CDM legislation, many projects take the opportunity to deal with them in the same manner as health and safety. In fact, many designers are more comfortable addressing environmental challenges than those associated with health and safety, which are often seen as the responsibility of the construction team alone.

Recommendations from the early concept stage risk assessments (HAZCON 1 or similar) will be made available to the design team and should influence site layouts, detailed design drawings, schematics, and specification. ECI (1995) stresses that these design assessments "must include identification of design errors, ambiguities and/or omissions. Questions of ambiguity and omission are especially important since the definition of design work and its separation from the construction phase is not always clear. In some disciplines, for example structural engineering, parts of the design are not fully detailed by the designer but are subsequently completed or amplified during fabrication and construction." Designers will often need to obtain advice from other domain experts in order to adequately assess the risks, and the contractual arrangements must facilitate this dialogue and knowledge exchange. It is at this stage that SHE benefits from integrated teams can be realized. Effective design risk assessment is still in its infancy, but various guidance documents have been published to assist (e.g., Cooks et al., 1995; CIRIA, 1999; Ove Arup, 1997). One of the challenges for these documents is how to guide a process such as design in a way that is both effective and does not stymie the design creativity.

Most designers will first of all consider the SHE issues of the permanent works, and this is not inappropriate; however, the risks present during construction must also be specifically addressed. Such risks should be systematically identified and removed or reduced during the design phase. A flowchart strategy is described in Figure 6.3

Designers can affect SHE on-site in a number of areas, for example site access. Here they will consider access to site for delivery, offloading, collection, and disposal of materials; access across site to facilitate safe movement of materials and personnel to and from the workplace; and plant/people separation during all construction activities. Another example would be hazardous materials, where designers should ensure that these are used only where necessary and that all materials are classified, with data sheets produced showing all associated risks, including delivery, use, and disposal. The ConCA project (2002) has identified that increased use of preassembly is one of the ways that designers can best improve SHE performance on-site.

## Sustainability during Design

Sustainability is a broad subject dealing with the impact of the built environment on the environment as a whole. It is also often a politically motivated concept, with organizations and even countries playing games with statistics to defend their particular viewpoint. A full discussion of this important subject is clearly outside the scope of this chapter, and this section concentrates only on the issues relating to the design phase of the project. Key considerations for designers are embodied energy of the building elements (covering the energy used to extract, form. and fashion the elements; deliver them to site; install them; and ultimately dispose of them), energy consumption in use, emissions, hazardous materials, and ultimate demolition and disposal of the elements that make up the building. Designing

# FIGURE 6.3. FLOWCHART FOR SYSTEMATIC IDENTIFICATION AND REDUCTION OF RISKS.

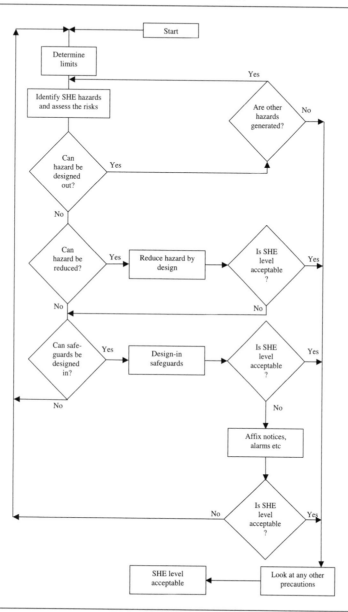

*Source:* Adapted from Pilz GmbH (1993).

for sustainability also covers broader issues such as site location (near public transport to reduce car use, on a previously used "brownfield" site) and stimulating the sustainable community (use of local labor, etc.).

According to Halliday (1998) "materials and products within buildings should not

- endanger the health of building occupants, or other parties, through exposure to pollutants, the use of toxic materials or providing host environments to harmful organisms;
- cause damage to the natural environment or consume disproportionate amounts of resources during manufacture, use or disposal;
- cause unnecessary waste of energy, water or materials due to short life, poor design, inefficiency or less than ideal manufacturing, installation and operating procedures;
- create dependence on high impact transport systems with their associated pollution; and
- further endanger threatened species or environments."

She claims that these are "issues of pollution and toxicity and a strategic approach starting at inception of a project is required to create a truly healthy environment".

Nath et al. (1998) have produced a useful book covering the methods and "tools" of environmental management, with chapters by individual specialist contributors. The first volume commences with the global aspects of environmental management and goes on to cover environmental planning, standards, exposure, and ecological risk assessment, and topics such as environmental risk assessment, life cycle assessment environmental auditing, and environmental accounting. Later sections cover economic and financial instruments for environmental management. The book contains summaries of international, European, and American environmental law, and finishes with chapters on environmental communication and education. Gibb et al. (2000) have produced a glossary of publications on the subject, particularly covering construction implications. The Construction Industry Research and Information Association have also published much on the subject (www.ciria.org.uk). Interested parties are advised to consult these other texts.

## SHE Plan and SHE File

This section is drawn from the European practice where a specific plan for health and safety is central to the effective health and safety management (with larger organizations often including environmental issues as part of SHE) as shown in Figure 6.4. This plan is formally presented as the SHE file, which, as a document, evolves throughout the project process until it is handed over to the end user as a record that tells those who might be responsible for the structure (or facility/building) in the future about the risks that have to be managed during maintenance, repair, or renovation (HSE, 1994).

ECI (1995) explain that a SHE plan is required

- "to fulfill the statutory duty;
- to ensure tenderers take SHE into account and explain their proposals for managing SHE and that clients/owners provide their objectives and background information for the project;

## FIGURE 6.4. CENTRALITY OF THE HEALTH AND SAFETY PLAN IN EUROPEAN PRACTICE.

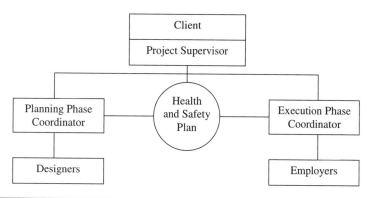

*Source:* Adapted from the EC Directive 92/57/EEC (1992).

- to ensure that all persons involved with the project (client, designers, planning supervisors, principal contractors and subcontractors) provide information to the plan and agree to the SHE management controls;
- to reduce the risk of accidents/incidents both during construction and for the lifetime of the facility;
- to reduce the losses associated with accidents/incidents;
- to protect the health of all project personnel and subsequent employees; and
- to reduce pollution and protect the environment."

The plan covers all construction work that, in Europe at least, is deemed to include maintenance and demolition. The plan is initially developed from information from the client/owner and designers and then developed in detail by the construction team, resulting in one plan rather than two separate documents. The level of detail will depend on the size and nature of the project and the procurement route adopted. Inputs to the plan from the various parties must be carefully coordinated. In Europe, this is a formal role, performed by the planning supervisor/design phase coordinator. ECI (1995) outline the main components of the SHE plan, as shown in Table 6.2.

## Method Statements

Typically, method statements are required by the contract rather than legislation. They are, however, often confused with risk assessments and used interchangeably. Furthermore, as with risk assessments, it is essential that the target readership is acknowledged in the style and delivery of the material—too often the method statements just stay "on the shelf." Clarke (1999) explains the benefits of an effective method statement:

### TABLE 6.2. MAIN COMPONENTS OF A SHE PLAN.

| Section | Subsection |
|---|---|
| Project summary | Objectives |
| | Management organization and responsibilities |
| | Schedule of activities |
| | Existing environment |
| | Contract strategy |
| Design plan | SHE information |
| | Organization and responsibilities |
| | Hazard identification |
| | Designers' risk assessment |
| Procurement plan | Material hazards |
| | Construction risks |
| | Selection of principal contractor and key suppliers |
| Construction plan | Management organization and responsibilities |
| | Selection of contractors, subcontractors, and other suppliers |
| | Site rules and procedures |
| | Welfare arrangements |
| | Training |
| | Hazard identification/risk assessments/method statements |
| | Environmental control |
| | Handover of documents and SHE file |
| | Monitoring, auditing, and review |

*Source:* Adapted from ECI (1995).

- "in getting people to write things down, it encourages them to think about the task in hand;
- it encourages them to commit to what they are writing;
- it helps communicate the planner's thoughts and intentions to operatives;
- it serves as a basis for coordination with other activities and for planning; and,it establishes an audit trail."

## Procurement Strategy

Clarke (1999) cites the following procurement issues as impacting on health and safety management:

- "lowest price mentality of clients;
- competitive tendering;
- dutch auctioning (adversarial leverage to knock down tender prices);
- adversarial contracts;
- subcontracting; and
- design separation".

Citing "experienced commentators," Clarke (1999) claims that the "extensive and increasing use of self-employment" (especially labor-only subcontracting) in construction is an important factor in its poor safety record in construction. Other commentators add that the adversarial nature of many construction contracts also makes cooperation on SHE issues more problematic. Integrated teams are better placed to address the challenges together from a project-wide or even business-wide perspective.

Whatever the procurement strategy, the contract documents must adequately and unambiguously address SHE issues. Risks, rights, and obligations should be clearly spelled out. Efforts to hide important requirements within pages of text goes against the cooperative culture supported by this book and will ultimately lead to SHE problems either during construction or through the facility's life cycle. In most countries there will be specific legislation relating to SHE issues and construction contracts—for instance, in Europe, legislation is explicit about the roles of clients/owners, designers, planning supervisors (coordinators), principal contractors, and other contractors. Any contract strategy must be consistent with the relevant legislation. A number of other countries are considering strategies to draw the owner and designer into this process; however, there is considerable resistance to this move, with some being keen to retain the full responsibility for SHE issues with the contractor, who they argue is the organization best placed to solve the problems. Whereas this may be valid regarding the ability to control risk, the opportunity to remove or reduce the risk is best taken before work starts on site, and the preconstruction team should play a major role in this.

## Assessment of Competence and Resources

"Competence" is an important concept in the recent legislation emanating from the European Union. According to this legislation, for European projects, key staff must have a knowledge and understanding of the work involved in the management and prevention of risk and of relevant SHE standards. They must also have the capacity to apply this knowledge and experience to their role on the project. The client/owner has a duty to ensure that all parties employed on a construction project are competent to perform their duties under the legislation. The client also establishes the extent and adequacy of the resources that have been, or will be, allocated. To assess competence, the key personnel need to be identified at an early stage. This may be hard for some organizations and may require a change in culture, away from the "day-to-day" approach often adopted in construction staff allocation. Where deficiencies exist, they may be addressed by further training. The specific requirements listed here are obviously only legally required on European projects; however, project managers are advised to take this model seriously in their considerations regarding project personnel and resources.

In addition to individuals, each company should be assessed for competence and any deficiencies in their organization and administration arrangements identified. Screening arrangements may include questionnaires, evaluation of previous experience, general reputation within the industry sector, SHE policy review, and specific service provision. Sample competence questionnaires are provided by ECI (1995).

## SHE Training And Education

Training is an essential part of effective project management, both preconstruction and for site-based personnel of all types. A detailed discussion on training is outside of the scope of this chapter; nevertheless, following the assessment of competence, training is often needed to address the identified shortfalls and inadequacies. Designer training rarely moves beyond a cursory coverage of the necessary legislation, and this situation must be changed if improvements are to be made. Construction training will include, but not be limited to, inductions for all personnel, toolbox talk addressing topical issues, and strategic training based on a personal development plan to increase the base level of knowledge and expertise.

There is increasing pressure to include SHE issues in the education of all construction-related professionals. However, in the United Kingdom, progress is slow, confounded by a lack of knowledge of most educators and difficulties with knowing how to include extra information into an already crowded curriculum. A recent survey concentrating on health and safety (Carpenter, 2001) showed that, although there are some exceptions, many higher education establishments have still not begun to address the issues.

# Construction Action Plan

## Planning

Market leaders take a "planned and systematic approach to implementing a SHE policy through an effective SHE management system" (adapted from HSE, 1997), where planning is a continuum throughout the project life cycle. SHE planning starts with a general, high-level plan at concept stage and develops in detail as the availability of detailed information increases. At the start of the construction phase, the initial SHE plan is reviewed and updated, as it is important to build on the foundation already laid and benefit from the knowledge gained by the design team. Once again, integrated teams will achieve this more easily, and the earlier that the construction team becomes involved in the planning process, the better the plan will be. In UK practice, the role of the planning supervisor, who has been coordinating health and safety matters during design, will overlap with the principal contractor, who is responsible for the construction phase. Typically, on most large projects, they will develop and expand the SHE plan jointly as the design is finalized and the construction methods are decided.

In an ideal world the design would have been completed prior to the start of construction. However, in reality, there is always a degree of overlap, and effective management of continued design development during the construction phase is essential for the success of the overall project. This is equally the case in SHE matters.

The SHE plan will be sufficiently complete and detailed to cover the part of the construction work that is to be executed and should be completed as soon as possible. However, planning does not stop with the production of the overall project SHE plan. Individual contractors and subcontractors work is also planned, with special consideration given to the interfaces between the packages (Pavitt and Gibb, 2003).

## Management, Leadership, and Organization

Changes in European legislation in the 1990s have brought the client/owner into the safety and health management process, and ultimately management and leadership starts with the owner. This view is supported by observation, where high-profile clients have achieved much improved SHE performance on their projects. This has also proved to be the case with client/owners such as DuPont, bringing strategies and culture from the hi-tech manufacturing sector to apply pressure on construction. The "zero-accidents" drive that was very prevalent in the 1990s was initiated by informed and influential client/owners (CII, 1993b) and is still prevalent today (CII 2003a, CII 2003b). While client/owners do not do the design or construction work, they clearly produce the brief and requirements and set the overall project culture, and these have a major affect on SHE performance.

Obviously, the "sharp–end" of SHE management is met by site-based managers and supervisors. CIRIA has produced an excellent site safety handbook targeted at site managers (Bielby and Read, 2001). It has also produced many publications on environmental issues for site managers and are now planning an occupational health manual (for more information see www.ciria.org.uk). Beilby and Read (2001) have produced a useful diagram providing a framework for individuals charged with the management of site safety (see Figure 6.5). This shows the effect on an individual's actions of the overlapping requirements of legislation, company policy, specific site rules for safe systems of work (that may be influenced by client/owner requirements), and professional codes of conduct and ethics. The same framework can be applied to health and environmental issues.

Figure 6.6 has been adapted from Griffith and Howarth (2001) to show the outline organization of the project health and safety management for a principal contractor (the main organization responsible for the on-site construction work). This figure shows the roles

## FIGURE 6.5. INFLUENCE FRAMEWORK FOR MANAGING SITE SAFETY.

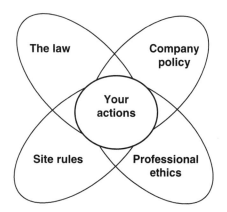

*Source:* Adapted from Beilby and Read (2001).

# FIGURE 6.6. PRINCIPAL CONTRACTOR'S OUTLINE ORGANIZATION FOR PROJECT HEALTH AND SAFETY MANAGEMENT.

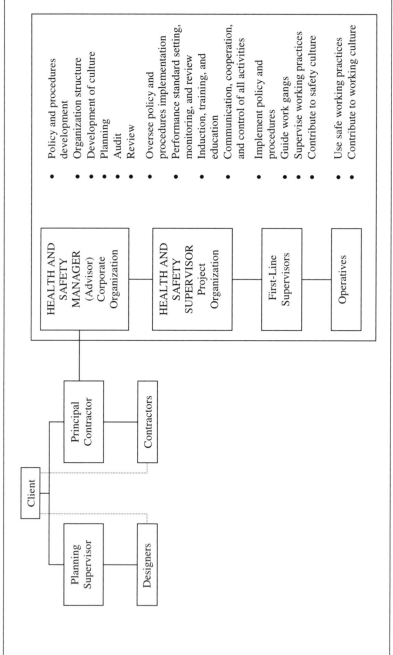

**HEALTH AND SAFETY MANAGER** (Advisor) Corporate Organization

- Policy and procedures development
- Organization structure
- Development of culture
- Planning
- Audit
- Review

**HEALTH AND SAFETY SUPERVISOR** Project Organization

- Oversee policy and procedures implementation
- Performance standard setting, monitoring, and review
- Induction, training, and education
- Communication, cooperation, and control of all activities

**First-Line Supervisors**

- Implement policy and procedures
- Guide work gangs
- Supervise working practices
- Contribute to safety culture

**Operatives**

- Use safe working practices
- Contribute to working culture

Client

Planning Supervisor

Principal Contractor

Designers

Contractors

*Source:* After Griffith and Howarth (2001).

of the main players in the principal contractor's team. In most cases in the United Kingdom, the operatives and possibly the supervisors would actually be employed by the contractors or subcontractors rather than the principal contractor.

## Sustainability during Construction

This section concentrates on sustainability issues relating to the construction phase of the project. One of the most significant areas of environmental management for construction organizations is waste management. In many countries, governments have sought to put pressure on industry to reduce waste by taxing its removal and disposal. When the "landfill" tax was applied in the United Kingdom, the hire of a typical rubbish skip used on construction sites increased from around £25 (US$42) to more than £70 (US$117). Thus, waste management becomes an important factor in the overall financial success of the project.

There are many useful publications on environmental management during construction, and Gibb et al. (2000) have produced a glossary of publications on the subject. Of particular note is a manual by Coventry et al. (1999) covering general site rules; managing materials; water; waste; noise and vibration; dust, emissions, and odors; ground contamination; the natural environment; and archaeology. There is also a companion book and training video for site staff.

## Working Procedures

The method statements developed earlier in the process must be brought down to working procedures such that they can be implemented. Unfortunately, this will be done down to a certain level but rarely taken, in an integrated manner, to the level of the workplace and operative instructions. As a result, the actual impact at the "sharp end" is significantly reduced.

Procedures should be in place to ensure that all contractors and subcontractors comply with the SHE plan and allocate necessary resources. A site layout plan should be developed showing temporary accommodation, storage space, access routes for vehicles and pedestrians, preassembly areas, and emergency access/egress routes. Specific SHE hazards must be identified, following a review of the initial risk assessment and procedures developed for addressing the construction risks including, but not limited to, those shown in Table 6.3.

## Audits

Audits should be part of any management process, and this is equally true for SHE issues. Through audits and reviews, the "organization learns from all relevant experience and applies the lessons" (HSE, 1999). Clarke (1999) states that "the performance of all systems, and of people, changes over time. It usually deteriorates, unless something is done to maintain it." He adds that the purpose of auditing is to "maintain performance and ensure relevance and effectiveness." Watkins (1997) stresses that regular auditing of management systems is vital to sustaining those systems, together with the policies and performance. The

**TABLE 6.3. PRELIMINARY LIST OF TOPICS FOR DEVELOPING PROCEDURES FOR CONSTRUCTION WORKS.**

| Area | Primary Impact | | |
| --- | --- | --- | --- |
| | Safe | Health | Environment |
| Abrasive wheels | X | X | |
| Asbestos | | X | X |
| Cartridge-operated tools | X | X | |
| Cladding and the building envelope | X | X | |
| Confined spaces | X | | |
| Contaminated ground | | X | X |
| Crane operation | X | | X |
| Demolition | X | X | X |
| Diving | X | X | |
| Drainage | X | X | X |
| Electricity | X | | |
| Ergonomics and human factors | X | X | |
| Excavations and groundworks | X | X | X |
| Explosives | X | | X |
| Falsework | X | | |
| Fit out and finishes | X | X | |
| Flammable materials | X | | X |
| Hazardous materials | | X | X |
| Heavy lifts | X | X | |
| Hoists | X | | |
| Lead burning | | X | X |
| Lifting gear | X | | |
| Noise | | X | X |
| Pressure testing | X | X | |
| Radiography | | X | X |
| Roof work/work at height | X | X | |
| Structural frame work | X | X | |
| Transport | X | | X |
| Woodworking machinery | X | X | |
| Work over water | X | X | |
| Work within/near live facilities | X | X | X |

Taylor Woodrow approach mentioned earlier is based on periodic audits of key issues, carried out by visiting auditors.

Another reason for audits is to ensure that the systems devised keep up with the needs and challenges of a changing society. Watkins (1997) explains that "if it were possible to establish the perfect system today, by tomorrow it would begin its long descent into obsolescence. Slowly at first, almost imperceptivity, but steadily. The world moves on. New work practices emerge, legislation is superseded, people change. Unless your systems move along with the rest of the world they will inevitably fall out of step with the demands of the law. Auditing is one of the ways to guard against this".

# Life Cycle Issues

## Operation, Maintenance, and Facilities Management

"Attention to SHE issues during design does not only provide safer construction but will result in more efficient operation, safer maintenance and facility management" (ECI, 1995). This aspect of "construction" varies dramatically depending on the nature of the built facility. Process plants will, by their nature, require more consideration for their operation than, say, speculative office blocks, in that the severity of unplanned events from process plants will be much more serious for health and safety of those in the vicinity as well as for the environment as a whole. *Human Factors in Industrial Safety* (HSE, 1999) stresses the important role that design should play. Reason (1990) describes some of the well-known disasters that have involved human error during operation and/or maintenance—for instance, Bhopal in 1984 or Chernobyl in 1986. In all cases, operational systems should be "fail-safe" and must take into account human error. ECI (1995) provide a list of key considerations for maintenance, particularly for process plants:

- "analysis of the operator-critical tasks and risks of failure;
- evaluation of decisions to be made between automatic and physical controls;
- consideration of emergency actions required and the display of process information;
- arrangement for maintenance access; and
- provision of working environment for lighting, noise and thermal considerations."

ECI (1995) goes on to explain that "the maintenance criteria may be on a routine preventative basis or left to a breakdown/replacement regime. If frequent access to plant controls is required then access can be permanently designed for the facility. If breakdown maintenance is accepted then equipment installed to assist safe and fast turnaround is the designer's consideration".

The SHE issues for other construction projects, such as offices or schools, may appear less crucial when compared to the process sector; however, they are still important. A particular safety issue is maintenance and cleaning access, especially for the building envelope. On the environmental side, emissions from buildings and use of energy are requiring more serious consideration, as are the ultimate demolition and disposal of the elements that make up the building. As already noted, the designer's role in achieving a good SHE performance throughout the life cycle of the project is critical.

One factor that has changed the typical approach toward maintenance issues, at least in the United Kingdom, is the increased use of private/public partnerships (see the chapters by Turner and by Ive). In these projects, the constructing consortium is also responsible for maintenance and operation of the road or hospital or prison for a considerable period after the completion of construction. This does not alter the legal situation, nor should it affect the moral obligation to care for maintenance workers, but it does provide a clearer feedback loop on maintenance issues to designers and constructors.

As explained earlier, the SHE file, prepared by the design and construction team, should be available, identifying SHE implications for maintenance. It is essential that the format and usability of this document is carefully considered to ensure that it can be effectively used throughout the life cycle.

## Demolition and Decommissioning

Demolition and decommissioning are explicitly included as "construction" activities by the European Directives on health and safety issued since the early 1990s. Nevertheless, it has taken some time for designers to address this aspect of design risk assessment. Environmental life cycle strategies, as the sector responds to the sustainability lobby, now commonly have to include demolition and final disposal or, ideally, reuse of the materials from the completed building or facility.

An additional challenge for the construction sector is that most of the built environment has been designed before these considerations were even suggested. This has resulted in a major legacy issue for construction SHE. For instance, the ubiquitous and uncontrolled use of asbestos in all forms of construction now presents one of the biggest challenges for all societies. The health issues for its removal and the long-term environmental risk are leading many building owners to just cover up and leave it in place, perhaps hoping for some miracle solution to be developed. However, all that is happening is that the problem is just being stored up for a future generation. The industry must ensure that an equivalent catastrophe cannot occur in the future.

# Driving Change and Measuring Success

## Driving Change

This chapter has argued that there is a real need to drive change in the SHE performance of construction sector. No one party can deliver this change alone: it requires buy-in of all the stakeholders. If the client/owner is not committed to it, then there will not be enough resources allowed in the brief to adequately manage the risks. The designers have a major influence, and all this previous effort will come to naught unless the construction team, including suppliers and subcontractors, have ownership of the SHE solutions.

## Measuring Success

Measurement is essential to maintain and improve performance. There are two ways to generate information on performance (adapted from HSE 1997):

* Reactive systems that monitor accidents, ill health, and incidents
* Active systems that monitor the achievement of plans and the extent of compliance with standards.

***Reactive Measurement: Quantitative Lagging Indicators.*** The most common form of health and safety performance measures are quantitative, lagging indicators. These are reactive and form the basis of most governmental measurement systems. Laufer (1986) suggested that "safety measuring methods are characterized primarily by the manner in which they relate to the criteria of safety effectiveness, the events measured and the method of data collection." Kunju-Ahmad and Gibb (2003) explain that the "frequency element of the undesirable event usually splits up into four categories:

1. Lost day cases—cases which bring absence from work;
2. Doctor's cases—non-lost workday cases that are attended by a doctor;
3. First aid cases—non-lost workday cases requiring only first aid treatment; and
4. No-injury cases—accidents not resulting in personal injury but including property damage or productivity disruption."

There are a number of additional problems with this approach—for example, the practice of citing only directly employed (and usually office-based) staff in statistics returned, rather than including all the people involved in the project. As most of the people who are injured or suffer ill health are "workers," and many of the owner organizations do not directly employ the workers, this can produce very misleading project statistics. The practice should be to include *all* personnel involved in the project and generally exclude home-office staff from project figures to avoid skewing the statistics. Another dilemma is that where safety culture is poor, there is a tendency to heavily underreport. This leads to the issue of dealing with a perceived increase in incidents once the safety culture starts to improve. These are often caused simply by an increase in the number of incidents being recorded, which may then mask an actual decrease in the incidents themselves.

Environmental performance for specific projects is sometimes also measured, often when a client/owner wants to use the score as a business marketing advantage. In the United Kingdom, the BREEAM technique, developed by the Building Research Establishment (BRE) is typically used. BREEAM assesses the performance of buildings in the following areas:

- *Management.* Overall management policy, commissioning site management, and procedural issues
- *Energy use.* Operational energy and carbon dioxide ($CO_2$) issues
- *Health and well-being.* Indoor and external issues affecting health and well-being
- *Pollution.* Air and water pollution issues
- *Transport.* Transport-related $CO_2$ and location-related factors
- *Land use.* Greenfield and brownfield sites
- *Ecology.* Ecological value conservation and enhancement of the site
- *Materials.* Environmental implication of building materials, including life cycle impacts
- *Water.* Consumption and water efficiency

Developers and designers are encouraged to consider these issues at the earliest opportunity to maximize their chances of achieving a high BREEAM rating. Credits are awarded in each area according to performance. A set of environmental weightings then enables the credits to be added together to produce a single overall score. The building is then rated on a scale of PASS, GOOD, VERY GOOD, or EXCELLENT, and a certificate is awarded that can be used for promotional purposes. More information on this technique can be found at http://products.bre.co.uk/breeam.

***Active Measurement: Behavior, Culture, and Process Management.*** Kunju-Ahmad and Gibb (2003) argue that "proactive measures should be used to evaluate SHE performance rather than backward-looking techniques. These techniques concentrate on evaluating behavior, culture and process management. An industry-wide technique is a potential vision for the future, however, difficulties in applying a single tool to construction remain and, for the foreseeable future, individual organizations are likely to continue to develop their own systems. These individual organizations can derive considerable benefits internally despite being unable to accurately compare their performance with others."

Lingard and Rowlinson (1994) describe behavioral safety management as a "range of techniques which seek to improve safety performance by setting goals, measuring performance and providing feedback." This concentration on behavior is also supported by other research such as Duff et al. (1994). Cameron (1998) describes an audit system as a means to develop goals, implement checks, and provide ongoing feedback.

The United Kingdom's Health and Safety Executive (HSE 1999) describe three aspects of human factors that influence human behavior:

*   *Individual.* Competence, skills, personality, attitudes, risk perception
*   *Organization.* Culture, leadership, resources, work patterns, communications
*   *Job.* Task, workload, environment, display and controls, procedures.

In its work, described earlier in the chapter, the Loughborough ConCA has developed the causality model shown in Figure 6.7. This clearly shows the various levels of influences on a particular accident, and these can be adapted to suit an ill-health or environmental event. The basic point here is that the effective project management approach will address issues much further back up the process chain, rather than leaving all the responsibility for SHE management to those who inherit the problems on-site.

# Summary

This chapter has argued the need to consider SHE from an early stage as part of project management from a moral, legal, financial, and cultural standpoint. It has identified requirements for SHE policy and objectives, SHE concept, initial risk assessment, and the SHE plan. It has described SHE actions required during the design and preconstruction phases, namely risk assessment and risk avoidance, designers actions, sustainability assessment, SHE plan and SHE file development, method statements, procurement strategy, assessment of competence and resources, and SHE training and education. The chapter then introduced the key aspects of the construction action plan, including planning; management, leadership, and organization; sustainability; working procedures; and audits. It has raised issues for the life cycle of the project such as operation, maintenance, and facilities man-

## FIGURE 6.7. FACTORS IN ACCIDENT CAUSALITY.

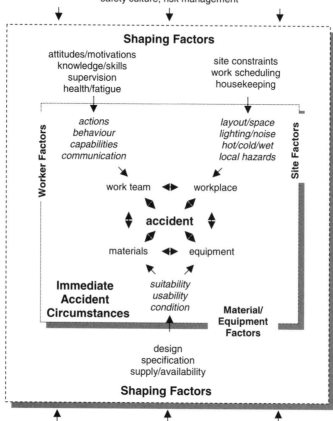

**Originating Influences**

client requirements, economic climate, construction education

permanent works design, project management, construction processes
safety culture, risk management

**Shaping Factors**

attitudes/motivations
knowledge/skills
supervision
health/fatigue

site constraints
work scheduling
housekeeping

**Worker Factors**

*actions
behaviour
capabilities
communication*

*layout/space
lighting/noise
hot/cold/wet
local hazards*

**Site Factors**

work team — workplace

**accident**

materials — equipment

**Immediate
Accident
Circumstances**

*suitability
usability
condition*

**Material/
Equipment
Factors**

design
specification
supply/availability

**Shaping Factors**

permanent works design, project management, construction processes
safety culture, risk management

client requirements, economic climate, construction education

**Originating Influences**

*Source:* ConCA (2002).

agement, as well as demolition and decommissioning. Finally, the chapter looked at driving change and measuring success, using both reactive and proactive techniques.

# References

Beilby, S. C., and J. A. Read. 2001. *Site safety handbook*. 3rd ed. London: Construction Industry Research and Information Association (CIRIA). ISBN: 0-86017-800-5.

Cameron, I. 1998 Pilot study proves value of safety audits. *Construction Manager*, Chartered Institute of Building. London: Thomas Telford.

Carpenter, J. 2001. Identification and management of risk in undergraduate courses. Contract Research Report 392/2001. London: HSE Books.

Casals, M. 2001. Costs and benefits related to quality and safety and health in construction. *Proceedings of the CIB W99 International Conference*, Barcelona, October.

CII. 1993a. *Zero injury economics*. Document No. SP32-2. Austin, TX: Construction Industry Institute.

———. 1993b. Zero injury techniques. Document No. RS32-1. Austin, TX: Construction Industry Institute.

———. 1996. Design for safety. Document No. 101-2. Construction Industry Institute.

———. 2001. Small projects toolkit. Document No. 161-2. Austin, TX: Construction Industry Institute.

———. 2003. The owners' role in construction safety. Document No. RS190-1. Austin, TX: Construction Industry Institute.

———. 2003b. Safety plus: Making zero accidents a reality. Document No. RS160-1. Austin, TX: Construction Industry Institute.

CIRIA. 1998. CDM training pack for designers. Pub. C501. London: Construction Industry Research and Information Association.

———. 2000. Integrating safety, quality, and environmental management. Pub. C509. London: Construction Industry Research and Information Association.

Clarke, T.. 1999. *Managing health and safety in building and construction*. Oxford, UK: Butterworth-Heinemann.

Cooks, J. et al.. 1995. CDM regulations: Case study guidance for designers. Pub. R145. London: Construction Industry Research and Information Association (CIRIA).

ConCA and Loughborough University. Forthcoming. Study of 100 construction accidents to identify causal relationships. Funded by the Health and Safety Executive. Final report awaiting publication. For more information contact a.g.gibb@lboro.ac.uk.

Coventry, Woolveridge, and Kingsley. 1999. Environmental good practice: Working on site. C502 (hardback manual), C503 (pocket book), C525V (training video). London: Construction Industry Research and Information Association (CIRIA),

Croner. 1994. Croner's *management of construction safety*. Surrey, UK: Croner Publications.

DTI. 1998. *Rethinking construction*, Department of Trade and Industry (formerly DETR), Construction Task Force. London: The Stationery Office.

Duff, A. R., I. T. Robertson, R. A. Phillips, and M. D. Cooper. 1994. Improving safety by the modification of behavior. *Construction Management and Economics* 12(6):67–78.

EC. 1989. *Directive concerning the introduction of measures to encourage improvements in the health and safety of workers at work*. Directive 89/391/EEC, European Commission. London: The Stationery Office.

———. 1992. *Directive concerning temporary and mobile construction sites*. Directive 92/57/EEC, European Commission; the basis of the United Kingdom's Construction (Design and Management) Regulations (CDM). London: The Stationery Office.

ECI. 1995. *Total project management of construction safety, health and environment.* 2nd ed., ed. D. Tubb and A. G. F. Gibb. European Construction Institute. London: Thomas Telford.

Gibb, A. G. F., ed. 2000. Designing for safety and health. *Proceedings of the CIB W99/ECI International Conference.* Various papers on designing for safety and health. European Construction Institute. London, June.

Gibb, A. G. F., and A. I. Ayode. 1996. Integration of quality, safety, and environmental systems. In *Proceedings of the First International Conference of CIB Working Commission W99, Portugal, September: Implementation of Safety and Health on Construction Sites,* ed. L. M. Alves Dias, and R. J. Coble. pp. 11–20. Rotterdam: A. A. Balkema.

Gibb, A. G. F.. D. E. Gyi, and T. Thompson., eds. 1999. *The ECI guide to managing health in construction.* London: Thomas Telford. ISBN: 0-7277-2762-1.

Griffith, A., and T. Howarth, T. 2001. *Construction health and safety management.* Pearson Education.

Gibb, A. G. F., J. Slaughter, and G. Cox. 2000. The ECI guide to environmental management in construction. Interactive CD-ROM. Loughborough, UK: European Construction Institute.

Griffith, A.. 1994. *Environmental management in construction.* Basingstoke, UK: Macmillan.

Griffith, A., and T. Howarth. 2001. *Construction health and safety management.* London: Pearson Education. ISBN: 0-582-41442-3.

Halliday, S. 1998. Construction health and safety: Materials impact. In *Proceedings of the International Conference of CIB Working Commission W99: Environment, Quality and Safety in Construction,* ed. Alves, Dias, and Coble. pp. 9–20. Lisbon, Portugal.,

Hide, S., A. G. F. Gibb, R. A. Haslam, D. E. Gyi, S. Hastings, and R. Duff. 2002. ConCA: Preliminary results from a study of accident causality. *Proceedings of the Triennial International Conference of CIB Working Commission W99,* ed. Rowlinson. pp. 61–68, CIB Pub. 274. Hong Kong, May.,

Hinze, J. 1998. Addressing construction worker safety in the design phase. *Proceedings of the International Conference of CIB Working Commission W99: Environment, Quality and Safety in Construction,* ed. Alves, Dias, and Coble. pp. 46–54. Lisbon, Portugal.

———. 1996. Quantification of the indirect costs of injuries. In *Safety and health on construction sites.* CIB Working Commission W99. Pub. 187, ed. Coble, Issa, Elliott. pp. 307–321. ISBN: 1-886431-04-03.

Hinze, J. 1991. *Indirect costs of construction accidents.* Source Document No. 67 Austin, TX: The Construction Industry Institute.

Hinze, J., and L. Appelgate, L. 1991. Costs of construction injuries *Journal of Construction Engineering and Management* 117 (3, September).

Hinze, J., and J. Gambatese. 1996. Design decisions that impact construction worker safety. In *Safety and health on construction sites,* ed. Coble, Issa, and Elliott. pp. 219–231. CIB Working Commission W99. Pub. 187. ISBN: 1-886431-04-03.

HSE. Human factors in industrial safety. HSG 48. *Health and Safety Executive.* London: The Stationery Office.

———. 1994. CDM regulations: How the regulations affect you. *Health and Safety Executive.* London: The Stationery Office.

———. 1997. Successful health and safety management. HSG 65. *Health and Safety Executive.* London: The Stationery Office.

———. 1999. Reducing error and influencing behavior. HSG 48. *Health and Safety Executive.* London: The Stationery Office.

ISO. 1996. ISO 14001: Environmental management systems: Specification with guidance for use. Geneva: International Standards Organization.

Kunju-Ahmad, R., and A. G. F. Gibb. 2003. Towards effective safety performance measurement: Evaluation of existing techniques and proposals for the future. In, ed. S. Rowlinson. London: Spon.

Laufer, A. 1986. Assessment of safety performance measurement at construction sites. *Journal Construction Management and Economics* 112(4):530–542.

Lingard, H., and S. Rowlinson. 1994. Construction site safety in Hong Kong. *Construction Management and Economics* 12(6):501–510.

Nath et al., eds. 1998.*Instruments for environmental management.* Vol.1 of *Environmental management in practice.* London: Routledge.

Nelson, E. J. 1993. *Zero injury economics.* Special Pub. 32-2. Austin, TX: Construction Industry Institute.

Oluwoye, J., and H. MacLennan, H. 1996. Pre-planning safety in project buildability. In *Safety and health on construction sites,* ed. Coble, Issa, and Elliott, pp. 239–248. CIB Working Commission W99. Pub. 187.

Ove Arup and Partners. 1997. *CDM regulations: Work sector guidance for designers.* Pub. R166. London: Construction Industry Research and Information Association (CIRIA).

Pavitt, T. C., and A. G. F. Gibb. 2003. Interface management within construction: in particular the building façade. *Journal of Construction Engineering and Management.* American Society of Civil Engineers. Vol. 129, No. 1: 8–15. ISSN: 0733-9364.

Pilz GmbH and Co. 1997. *Guide to machinery standards.* p. 57.

Reason, J. 1990. *Human error.* Cambridge University Press.

Rwelamila, P., and J. Smallwood. 1996. Total Quality Management (TQM) without safety management? In *Safety and health on construction sites,* ed. Coble, Issa, and Elliott. pp. 83–100. CIB Working Commission W99. Pub. 187.

Smallwood, J. 2001. Total Quality Management (TQM)—the impact?, Costs and benefits related to quality and safety and health in construction. *Proceedings of the CIB W99 International Conference* pp. 289–297. Barcelona, October.

Stubbs, A. 1998. *Environmental law for the construction industry.* London: Thomas Telford.

Tenah, K. 1996. Incorporating safety mechanisms into engineering design. In *Safety and health on construction sites,* ed. Coble, Issa, and Elliott, pp. 249–259. CIB Working Commission W99, Pub. 187.

Watkins, G. 1997. *The health and safety handbook.* London: Street and Maxwell.

## CHAPTER SEVEN

# VERIFICATION

## Hal Mooz

*"Proof of compliance with specifications. Verification may be determined by test, inspection, demonstration, or analysis."*

MOOZ, FORSBERG, COTTERMAN (2002)

When you are managing projects, it is usually necessary to prove that the solution satisfies both the specifications and the users. The process called *verification* develops this proof. Verification encompasses a family of techniques and can be applied irrespective of whether the project is completely hardware, software, a combination of both, or an operations-only solution. While verification methods may differ according to project disciplines, some method of verification is usually required, ranging from full measured compliance of every aspect to the random sampling of production units. It is critical that the verification approach is developed early in the project cycle in conjunction with requirements determination and represents consensus between the solution provider and the customer.

## The Context

Verification is closely associated with other project management terms that address the proof that solutions satisfy one or more requirements. The family of terms includes verification; validation; qualification; certification; integration; independent verification and validation (IV&V); integration, verification, and validation (IV&V); and independent integration, verification, and validation (IIV&V).

As with all of project management and system engineering communication, it is imperative that these terms are well understood and are properly communicated among the team members to avoid unintended consequences. The following definitions are the baseline for the remaining discussions of this chapter (Mooz, Forsberg, and Cotterman, 2002).

> *Validation.* Proof that the user is satisfied.
>
> *Qualification.* Proof that the design will survive in its intended environment with margin. The process includes testing and analyzing hardware and software configuration

items to prove that the design will survive the anticipated accumulation of acceptance test environments, plus its expected handling, storage, and operational environments, plus a specified qualification margin. Qualification testing often includes temperature, vibration, shock, humidity, software stress testing, and other selected environments.

*Certification.* To attest by a signed certificate or other proof to meeting a standard.

*Integration.* The successive combining and testing of system hardware assemblies, software components, and operator tasks to progressively prove the performance and compatibility of all components of the system.

*Independent verification and validation (IV&V).* The process of proving compliance to specifications and user satisfaction by using personnel that are technically objective and managerially separate from the development group. The degree of independence of the IV&V team is driven by product risk. In cases of highest risk, IV&V is performed by a team that is totally independent from the developing organization.

*Integration, verification, and validation (IV&V).* The combining of system entities, the proving the system works as specified, and the confirming that the right system has been built and that the customers/users are satisfied.

*Independent integration, verification, and validation (IIV&V).* The integration, verification, and validation sequence conducted by objective personnel separate from the development organization.

To clarify the interrelated contexts of integration, verification, and validation, system development Vee model illustrations will be used (Forsberg, Mooz, and Cotterman, 2001; Forsberg and Mooz, 2001).

## The Vee Model

Any phased project cycle is composed of three aspects.

* The business aspect represents the pursuit of the business case.
* The budget aspect represents the pursuit and management of the funding.
* The technical aspect represents the technical development strategy.

Projects start with high-level conversations with users/sponsors about the problem to be solved and the tangible proof needed at acceptance to prove that the problem has been solved. The technical process proceeds from those high-level discussions down through solution decomposition with progressively lower-level concepts and designs, and then ascends up to operations and final high-level discussions with the users/sponsor relative to their satisfaction with the solution. The image of this technical aspect of the cycle is best depicted as a Vee where the elaboration of the evolving solution baseline forms on the core of the Vee (see Figure 7.1).

This Vee format accurately illustrates levels of decomposition, from solution requirements and concepts down to the lowest replaceable unit in the left Vee leg and then upward consistent with fabrication and integration of the solution elements into the completed system

## FIGURE 7.1. PROJECT CYCLE VEE+ MODEL.

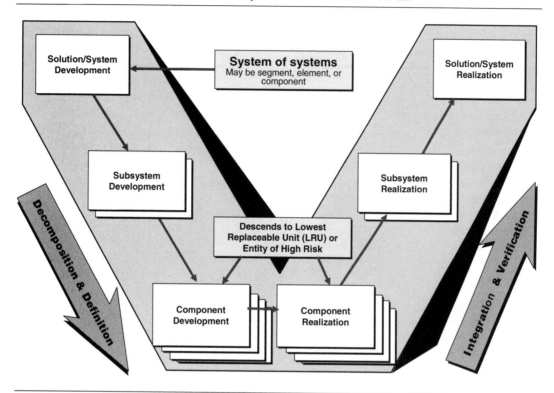

in the right Vee leg. The thickness of the Vee increases downward to reflect the increasing number of elements as a single system or solution is decomposed to its many individual subsystems and their lowest replaceable units (LRU).

At each decomposition level, there is a direct correlation between activities on the left and right sides of the Vee. This is deliberate. For example, the method of integration and verification to be used on the right must be determined on the left for each set of requirements and entities developed at each decomposition level (see Figure 7.2).

This minimizes the chances that requirements are specified in a way that cannot be measured or verified. It also forces the early consideration and preparation of the verification sequence, methods, facilities, and equipment required to meet the verification objectives as well as schedule and cost targets.

Verification facilities may become a task on the critical path requiring stakeholder approval. For example, mechanisms to be deployed in the weightlessness of space may require construction of a large float pool to demonstrate deployment using floatation devices to compensate for gravity. Similarly, a software system might require acquisition of special verification hardware, development, and loading of a verification database, or development of specialized verification drivers. Figure 7.3 illustrates verification planning at one level of

## FIGURE 7.2. PROJECT CYCLE VEE+ MODEL WITH IV&V PLANNING.

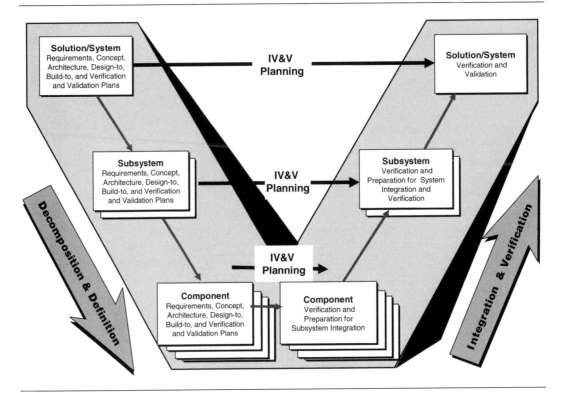

decomposition. This figure also illustrates the investigation of opportunities and their risks to whatever decomposition level is appropriate together with the affirming of the resultant baseline at the user level. However, these aspects of the Vee model are not relevant to this chapter and are not further explained.

There are four key steps in planning for integration, verification, and validation (see Figure 7.4) involving three user types (see Figure 7.5). The ultimate user is the end user of the system. The direct user is up one level in the decomposition from the item being validated. An associate user is any other user potentially impacted by the item being validated and usually exists at the same decomposition level. The three types are further clarified with examples later in this chapter.

- *Step 1.* Determine the integration sequence for combining the entities.
- *Step 2.* Determine how to prove that the solution when built is built right and satisfies both the design-to and build-to specifications.
- *Step 3.* Determine how to prove that the solution when verified is the right solution for both the direct user and the ultimate user.

**FIGURE 7.3. INTEGRATION, VERIFICATION, AND VALIDATION PLANNING.**

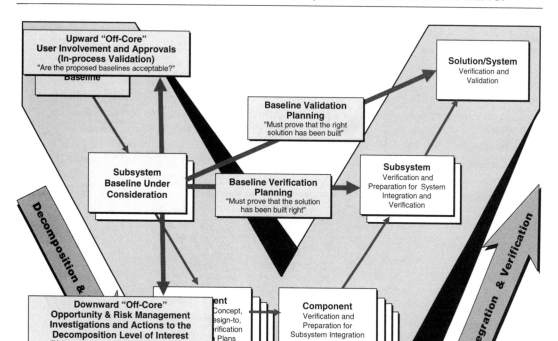

- *Step 4.* Determine if the concept as proposed and the associated proposed integration, verification, and validation approaches are acceptable to the associate, direct, and ultimate users of the solution (see Figure 7.5).

## Risk: The Driver of Integration/Verification Thoroughness

Some projects are human-rated—that is, they must work flawlessly, as human lives are at stake. Some projects are quick-reaction attempts of a concept or an idea, and if they don't work, it is less serious compared to a human-rated project. Human-rated projects require extreme thoroughness, while the quick-and-dirty projects may be able to accept more risk. It is important to know the project risk philosophy as compared to the opportunity being pursued. This reward-to-risk ratio will then drive decisions regarding the rigor and thoroughness of integration and the many facets of verification and validation. There is no standard vocabulary for expressing the risk philosophy, but it is often expressed as "quick and dirty," or "no single point failure modes," or "must work," or "reliability is 0.9997,"

## FIGURE 7.4. FOUR INTEGRATION, VERIFICATION, AND VALIDATION PLANNING STEPS.

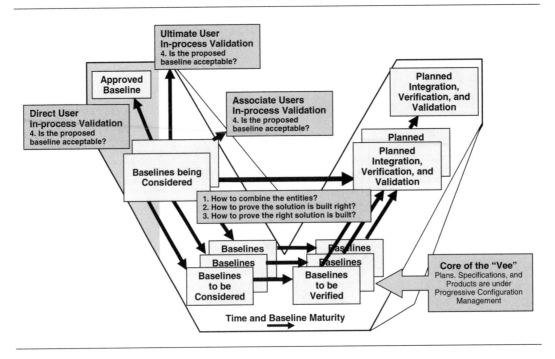

## FIGURE 7.5. EXPANSION OF FIGURE 7.4 DETAILING THREE TYPES OF USERS.

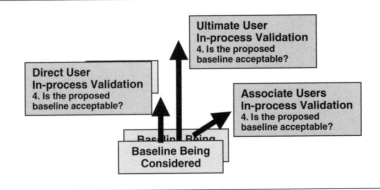

or some other expression or a combination of these. The risk philosophy will determine whether all or only a portion of the following will be implemented.

## Integration

Preparation for integration, verification, and validation begins with planning for integration. The product breakdown structure (PBS) portion of the work breakdown structure (WBS) should reveal the integration approach but often does not. Integration planning must determine the approach so that the interfaces and intrafaces can be provided for, managed, and verified. Figure 7.6 illustrates four possible sequences to integrating four entities into the same higher-level combination.

Each approach reaches the same end result, but for each option, the interfaces are different and must be appropriately managed, followed by verification of both the interfaces and the entity performance before combining into higher-level combinations.

Interface management to facilitate integration and verification should be responsive to the following:

1. The product breakdown structure (PBS) portion of the work breakdown structure (WBS) should provide the roadmap for integration.
2. Integration will exist at every level in the product breakdown structure except at the most senior level.
3. Integration and verification activities should be represented by tasks within the work breakdown structure (see Figure 7.7).

### FIGURE 7.6. FOUR INTEGRATION OPTIONS FOR A SYSTEM.

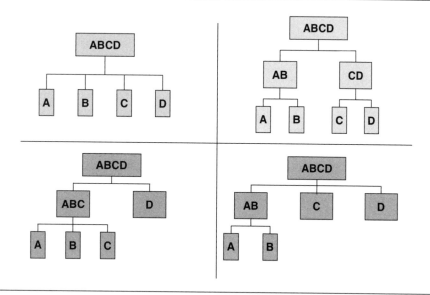

## FIGURE 7.7. RELATIONSHIPS AMONG A SYSTEM, A PRODUCT BREAKDOWN STRUCTURE, AND A WORK BREAKDOWN STRUCTURE.

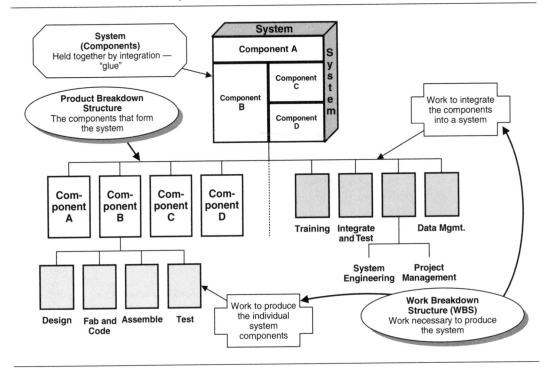

4. The work breakdown structure is not complete without the integration and verification tasks and the tasks to produce the intermediate work products (e.g., fixtures, models, drivers, databases) required to facilitate integration.
5. Interfaces should be designed to be as simple and foolproof as possible.
6. Interfaces should have mechanisms to prevent inadvertent incorrect coupling.
7. Interfaces should be verified by low-risk (benign) techniques before mating.
8. "OK to install" discipline should be invoked before all matings.
9. Peer review should provide consent-to authorization to proceed.
10. Haste without extra care should be avoided.

Integration and verification can be performed in a methodical incremental sequence by adding entities one at a time and proving the combined performance; or, all entities can be combined and then verified as a group in the "Big Bang" approach; or any combination between these two extremes can be used. In the sequential approach, anomalies are usually quickly resolved to the last entities; combined and corrective action can be swift. In the Big Bang approach, anomalies are more difficult to resolve, as there may be multiple causes working together to produce an undesired result. While the Big Bang approach, if it works, can result in substantial cost and time savings, it rarely works on newly developed systems that have not been adequately debugged. Other incremental variations, especially in software integration, though not limited to software, include the top-down, bottom-up, thread, and mixed approaches (see Figure 7.8). Note that prior to initiating any of these integration approaches each component or software module should have been verified against its specification.

Hindsight and lessons learned can be beneficial to the avoidance of future problems and to the development of improved methods. The following are valuable lessons learned related to the integration and verification of solution entities:

- Make sure names and identifiers are consistent and correct across entities being integrated.
- Ensure the correct versions of the entities are being integrated.
- Ensure no changes to external interfaces during integration.
- Be aware that logical integration problems are subtle. (They don't emit smoke.)
- Ensure that software and hardware baselines are compatible.
- Use peer reviews, software walk-throughs, and inspections to confirm compatibility.
- Verify software modules incrementally and resolve discovered anomalies.
- Use mechanical mock-ups to verify space, access, clearances and to practice the installation process.
- Use thermal models to confirm thermal predictions.
- Use an electrical/electronic simulator to verify functionality on both sides of the interface before mating.
- Enforce power off during connector mating.
- Ensure frame ground is common with power ground.
- Use connector keying and clocking to prevent incorrect mating.
- Use a single supplier for both halves of mating connectors.

## FIGURE 7.8. INCREMENTAL INTEGRATION APPROACHES.

| Technique | Features |
|---|---|
| Top - Down | • Control logic testing first<br>• Modules integrated one at a time<br>• Emphasis on interface verification |
| Bottom - Up | • Early verification to prove feasibility and practicality<br>• Modules integrated in clusters<br>• Emphasis on module functionality and performance |
| Thread | • Top down or bottom up integration of a software function or capability |
| Mixed | • Working from both ends toward the middle<br>• Choice of modules designated top-down vs. bottom-up is critical |

- Use "OK to install" discipline to make sure everything is perfect prior to each and every mating.
- Examine all connectors for debris, pushed and bent pins, and correct clocking.
- If it doesn't mate easily, STOP. Don't force it.
- Use "OK to power" discipline before applying power.
- Compare results to predictions; then identify and resolve discovered anomalies.

# Validation and Validation Techniques

Validation is proof that the users are satisfied regardless of whether the specifications have been satisfied or not. Occasionally a product meets all specified requirements but is rejected by the users and does not validate. Famous examples are the Ford Edsel, IBM PC Junior, and more recently, Iridium and Globalstar. In each case the products were exactly as specified but the ultimate users rejected them, causing very significant business failures. Conversely, Post-It-Notes failed verification to the glue specification, but the sticky notes then catapulted into our lives because we all loved the failed result. The permanently temporary or temporarily permanent nature of the glue was just what we were looking for, but it hadn't been specified.

Traditionally, validation occurs at the project's end when the user finally gets to use the solution to determine the level of satisfaction. While this technique can work, it can also cause immense waste when a project is rejected at delivery. Too many projects have been relegated to scrap or a storage warehouse because of user rejection. Proper validation management can avoid this undesirable outcome. When considering the process of validation, recognize that except for the top product level having just the ultimate or end user, there are direct users, associate users, and ultimate users at each decomposition level and for each entity at that level, all of whom must be satisfied with the solution at that level. Starting at the highest system level, the ultimate user is also the direct user. At the outset, the ultimate users should reveal their plans for their own validation so that developers can plan for what the solution will be subjected to at delivery. A user validation plan is valuable in documenting and communicating the anticipated process.

Then within the decomposition process, as the solution concept and architecture is developed, the ultimate users should be consulted as to their satisfaction with the progression of proposed concepts. The approved concepts then become baselined for further decomposition and rejected concepts are replaced by better candidates. This process is called *in-process validation* and should continue in accordance with decomposition of the solution until the user decides the decisions being made are transparent to his or her interface and use of the system. This on-going process of user approval of the solution elaboration and maturation can reduce the probability of user dissatisfaction at the end to near zero. Consequently, this is a very valuable process to achieve and maintain user satisfaction throughout the development process and to have no surprise endings.

Within the decomposition process, validation management becomes more complex. At any level of decomposition, there are now multiple users. The ultimate user is the same. However, there is now a direct user that is different from the ultimate user, and there are associate users that must also be satisfied with any solution proposed at that level of decomposition. Consider, for instance, an electrical energy storage device that is required by the power system within the overall solution. The direct user is the power system manager, and associate users are the other disciplines that must interface with the storage device's potential solutions. If a chargeable battery is proposed, then the support structure system is a user, as is the thermodynamic system, among others. In software, a similar situation exists. Software objects have defined characteristics and perform certain specified functions on request, much like the battery in the prior example. When called, the software object provides its specified service just as the battery provides power when called. Associate users are any other element of the system that might need the specified service provided by the object.

All direct and ultimate users need to approve baseline elaboration concepts submitted for approval. This in-process validation should ensure the integration of mutually compatible elements of the system. In eXtreme and Agile programming processes, intense user collaboration is required throughout the development of the project to provide ongoing validation of project progress.

Ultimate user validation is usually conducted by the user in the actual user's environment, pressing the solution capability to the limit of user expectations. User validation may incorporate all of the verification techniques that follow. It is prudent for the solution developer to duplicate these conditions prior to delivery.

## Verification and Verification Techniques

As stated at the outset, verification is proof of compliance with specifications. Verification may be determined by test, inspection, demonstration, or analysis. The following four techniques should be applied as appropriate to the verification objectives.

*Verification by test.* Direct measurement of specification performance relative to functional, electrical, mechanical, and environmental requirements. (Measured compliance with specified metrics).

*Verification by inspection.* Verification of compliance to specifications that are easily observed, such as construction features, workmanship, dimensions, configuration, and physical characteristics such as color, shape, software language, style, and documentation. (Compliance with drawings, configuration documents)

*Verification by demonstration.* Verification by witnessing an actual operation in the expected or simulated environment, without need for measurement data or post-demonstration analysis. (Observed compliance without metrics)

*Verification by analysis.* An assessment of performance using logical, mathematical, or graphical techniques, or for extrapolation of model tests to full scale. (Predicted compliance based on history)

### Verification Objectives

The definition of verification calls for proof of specification performance. However, since specifications can require nominal performance, design margin, quality, reliability, life, and many other performance factors, the verification plan must be formulated to prove compliance within each of these requirement categories. Figure 7.9 illustrates the context of design margin.

To be conservative, engineers include design margins to ensure that their solution performs its function. Verification may then be designed to prove both nominal performance and a specified design margin with or without deliberately forcing the solution into failure.

The more common verification objectives are outlined in the following paragraphs:

## Design Verification

Design verification proves that the solution's design performs as specified, or conversely, that there are identified design deficiencies requiring design corrective action. Design verification is usually carried out in nominal conditions unless the specification has design margins already built into the specified functional performance. Design verification usually includes the application of selected environmental conditions. Design verification should confirm positive events and the absence of negative events. That is, things that are supposed to happen happen, and things that are not supposed to happen do not. Software modules are often too complex to verify all possible combinations of events, leaving a residual risk within those that have not been deliberately verified.

eXtreme Programming and other Agile methods advocate thorough unit testing and builds (software integration) daily or even more frequently to verify design integrity in-process. Projects that are not a good match for an Agile methodology may still benefit from rigorous unit tests, frequent integrations, and automated regression testing during periods of evolving requirements and/or frequent changes.

## Design Margin Verification: Qualification

Design margin verification, commonly called qualification, proves that the design is robust with designed-in margin, or, conversely, that the design is marginal and has the potential of failing when manufacturing variations and use variations are experienced. For instance, it is reasonable that a cell phone user will at some time drop the phone onto a concrete surface from about four or five feet. However, should the same cell phone be designed to

## FIGURE 7.9. DESIGN MARGIN.

survive a drop by a high lift operator from, say, 20 feet? Qualification requirements should specify the margin desired.

Qualification should be performed on an exact replica of the solution to be delivered. The best choice is a unit within a group of production units. However, since this is usually too late in the project cycle to discover design deficiencies which would have to be retrofitted into the completed units, qualification is often performed on a first unit that is built under engineering surveillance to ensure that it is built exactly to print and as the designers intended. Qualification testing usually includes the application of environment levels and duration to expose the design to the limits that may be accumulated in total life cycle use. Qualification tests may be performed on specially built test articles that simulate only a portion of an entity. For instance, a structural test qualification unit does not have to include operational electronic units or software; inert mass simulators may be adequate. Similarly, electronic qualification tests do not need the actual supporting structure, since structural simulators with similar response characteristics may be used for testing.

The exposure durations and input levels should be designed to envelop the maximum that is expected to be experienced in worst-case operation. These should include acceptance testing (which is quality verification) environments, shipping environments, handling environments, deployment environments, and any expected repair and retesting environments that may occur during the life of an entity. Environments may include temperature, vacuum, humidity, water immersion, salt spray, random vibration, sine vibration, acoustic, shock, structural loads, radiation, and so on. For software, transaction peaks, electrical glitches, and database overloads are candidates.

The qualification margins beyond normal expected use are often set by the system level requirements or by the host system. Twenty-degree Fahrenheit margins on upper- and lower-temperature extremes are typical, and either three or six dB margins on vibration, acoustic, and shock environments are often applied. In some cases, safety codes establish the design and qualification margins, such as with pressure vessels and boiler codes. Software design margin is demonstrated by overtaxing the system with transaction rate, number of simultaneous operators, power interruptions, and the like. To qualify the new Harley-Davidson V Rod motorcycle for "Parade Duty," it was idled in a desert hot box at 100 degrees Fahrenheit for eight hours. In addition, the design was qualified for acid rain, fog, electronic radiation, sun, heat, structural strength, noise, and many other environments. Actual beyond specification field experience with an exact duplicate of a design is also admissible evidence to qualification if the experience is backed by certified metrics.

Once qualification has been established, it is beneficial to certify the design as being qualified to a prescribed set of conditions by *issuing a qualification certification* for the exact design configuration that was proven. This qualification certification can be of value to those that desire to apply this design configuration to other applications and must know the environments and conditions under which the design was proven successful.

## Reliability Verification

Reliability verification proves that the design will yield a solution that over time will continue to meet specification requirements. Conversely, it may reveal that failure or frequency of repair is beyond that acceptable and anticipated. Reliability verification seeks to prove *mean time between failure* (MTBF) predictions. Reliability testing may include selected environments to replicate expected operations as much as possible. Reliability verification tends to be an evolutionary process of uncovering designs that cannot meet life or operational requirements over time and replacing them with designs that can. Harley-Davidson partnered with Porsche to ultimately achieve an engine that would survive 500 hours nonstop at 140 mph by conducting a series of evolutionary improvements.

Life testing is a form of reliability and qualification testing. Life testing seeks to determine the ultimate wear-out or failure conditions for a design so that the ultimate design margin is known and quantified. This is particularly important for designs that erode, ablate, disintegrate, change dimensions, and react chemically or electronically, over time and usage. In these instances the design is operated to failure while recording performance data. Life testing may require acceleration of the life process when real-time replication would take too long or would be too expensive. In these instances acceleration can be achieved by adjusting the testing environments to simulate what might be expected over the actual life time. For instance, if an operational temperature cycle is to occur once per day, forcing the transition to occur once per hour can accelerate the stress experience.

For software, fault tolerance is the reliability factor to be considered. If specified, the software must be tested against the types of faults specified and the software must demonstrate its tolerance by not failing. The inability of software to deal with unexpected inputs is sometimes referred to as "brittleness."

## Quality Verification

In his book Quality is Free, Phillip Crosby defines quality as "conformance to requirements" and the "cost of quality" as the expense of fixing unwanted defects. In simple terms, is the product consistently satisfactory, or is there unwanted scrapping of defective parts? When multiple copies of a design are produced, it is often difficult to maintain consistent conformance to the design, as material suppliers and manufacturing practices stray from prescribed formulas or processes. To detect consistent and satisfactory quality—a product free of defects—verification methods are applied. First, process standards are imposed and ensured to be effective; second, automatic or human inspection should verify that process results are as expected; third, testing should prove that the ultimate performance is satisfactory. Variations of the process of quality verification include batch control, sampling theory and sample inspections, first article verification, and nth article verification. Quality testing often incorporates stressful environments to uncover latent defects. For instance, random vibration, sine sweep vibration, temperature, and thermal vacuum testing can all help force latent electronic and mechanical defects to the point of detection. Since it is difficult to apply all of these environments simultaneously, it is beneficial to expose the product to mechanical environments prior to thermal and vacuum environments where extended power-on testing can reveal intermittent malfunctions.

## Software Quality Verification

The quality of a software product is highly influenced by the quality of the individual and organizational processes used to develop and maintain it. This premise implies a focus on the development process as well as on the product. Thus, the quality of software is verified by verifying that the development process includes a defined process based on known best practices and a commitment to use it; adequate training and time for those performing the process to do their work well; implementation of all the process activities, as specified; continuous measurement of the performance of the process and feedback to ensure continuous improvement; and meaningful management involvement. This is based on the quality management principles stated by W. Edwards Deming that "Quality equals process—and everything is process."

## -ilities Verification

There are a host of -ilities that require verification. Figure 7.10 provides a list of common -ilities.

Verification of -ilities requires careful thought and planning. Several can be accomplished by a combined inspection, demonstration, and/or test sequence. A verification map can prove to be useful in making certain that all required verifications are planned for and accomplished.

## Certification

Certification means "to attest by a signed certificate or other proof to meeting a standard." Certification can be verification of another's performance based on an expert's assurance. In the United States, the U.S. Food and Drug Administration grades and approves our meat to be sold, and Consumer Reports provides a "Best Buy" stamp of approval to high value products.

Certification often applies to the following:

- *The individual.* Has achieved a recognized level of proficiency
- *The product.* Has been verified as meeting/exceeding a specification
- *The process.* Has been verified as routinely providing predictable results

### FIGURE 7.10. OTHER -ILITIES REQUIRING VERIFICATION.

| | | |
|---|---|---|
| Accessibility | Efficiency | Reusability |
| Adaptability | Hostility | Recyclability |
| Affordability | Integrity | Securability |
| Compatibility | Interoperability | Survivability |
| Compressability | Liability | Scalability |
| Dependability | Mobility | Testability |
| Degradeability | Manageability | Usability |
| Distributability | Producibility | Understandability |
| Durability | Portability | Variability |

In all cases certification is usually by independent assessment or audit to a predefined standard. When material is "certified," it should arrive at the user's facility complete with a pedigree package documenting the life history of the contents and a signed certification with associated test or other verification results substantiating that the contents of the container are as represented.

Certification is becoming more and more popular as professional organizations promote organizations and individuals to improve their performance capability and to be recognized for it. Individual or organizational certification such as ISO (International Standards Organization; see Figure 7.11), Carnegie Mellon's Software Engineering Institute's CMM® (see Figure 7.12) or CMMI,® and the Project Management Institute's PMP® Project Management Professional (see Figure 7.13) designation are achieved by demonstrated compli-

**FIGURE 7.11. ISO 9000 QUALITY STANDARD.**

| | |
|---|---|
| Quality Policy Statement | ✓ |
| Quality Organization | ✓ |
| Management Quality Reviews | ✓ |
| Quality System Procedures and Planning | ✓ |
| Contract Review Procedures | ✓ |
| Design Control Procedures | ✓ |
| Document and Data Control System | ✓ |
| Purchasing Control System | ✓ |
| Control of Customer-Supplied Products | ✓ |
| Product Identification and Traceability System | ✓ |
| Process Controls | ✓ |
| Inspection and Testing Procedures | ✓ |
| Control of Inspection, Measuring, and Test Equipment | ✓ |
| Inspection and Test Status System | ✓ |
| Control of Nonconforming Products | ✓ |
| Corrective and Preventive Action Procedures | ✓ |
| Handling, Storage, Packaging, Presentation, and Delivery Procedures | ✓ |
| Quality Record Control System | ✓ |
| Internal Quality Audit Procedures | ✓ |
| Training in Quality | ✓ |
| Procedures for Servicing | ✓ |
| Statistical Techniques | ✓ |

90437

## FIGURE 7.12. SEI CMM; CAPABILITY MATURITY MODEL.

### Implications of Advancing Through CMM Levels.

| | Level 1 | Level 2 | Level 3 | Level 4 | Level 5 |
|---|---|---|---|---|---|
| **Processes** | Few stable processes exist or are used. | Documented and stable estimating, planning, and commitment processes are at the project level. | Integrated management and engineering processes are used across the organization. | Processes are quantitatively understood and stabilized. | Processes are continuously and systematically improved. |
| | "Just do it" | Problems are recognized and corrected as they occur. | Problems are anticipated and prevented, or their impacts are minimalized. | Sources of individual problems are understood and eliminated. | Common sources of problems are understood and eliminated. |
| **People** | Success depends on individual heroics. | Success depends on individuals; management system supports. | Project groups work together, perhaps as an integrated product team. | Strong sense of teamwork exists within each project. | Strong sense of teamwork exists across the organization |
| | "Firefighting" is a way of life. | Commitments are understood and managed. | Training is planned and provided according to roles. | | Everyone is involved in process improvement. |
| | Relationships between disciplines are uncoordinated, perhaps even adversarial. | People are trained. | | | |
| **Technology** | Introduction of new technology is risky. | Technology supports established, stable activities. | New technologies are evaluated on a qualitative basis. | New technologies are evaluated on a quantitative basis. | New technologies are proactively pursued and deployed. |
| **Measurement** | Data collection and analysis is ad hoc. | Planning and management data used by individual projects. | Data are collected and used in all defined processes. | Data definition and collection are standardized across the organization. | Data are used to evaluate and select process improvements. |
| | | | Data are systematically shared across projects. | Data are used to understand the process quantitatively and stabilize it. | |

95-357 drw 12B

## FIGURE 7.13. PMI/PMP KNOWLEDGE AREAS.

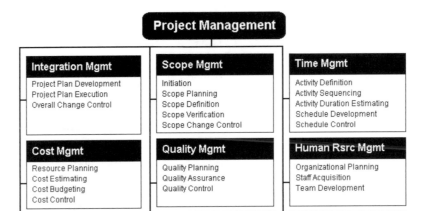

ance to a recognized and controlled set of standards. It has become increasingly common for buyers of services to include these certifications in their buying decision criteria.

In some cases, certifications take the form of licenses to do business, such as a certified public accountant and for lawyers who are required to pass a bar exam to practice law.

Many other individual certifications support the project management discipline. Most are in the quality discipline, such as:

- Certified Quality Manager
- Certified Quality Engineer (CQE)
- Certified Quality Auditor (CQA)
- Certified Reliability Engineer (CRE)
- Certified Quality Technician (CQT)
- Certified Mechanical Inspector (CMI)
- Certified Software Quality Engineer (CSQE)

In addition, in 2004 the International Council on Systems Engineering initiated certification of systems engineers as Certified Systems Engineering Professionals.

The objective of organizational and personal certification is to ensure that the required level of individual and organizational competency exists throughout a project's internal and supplier organizations so as to achieve the project's objectives the first time and every time. Product and material certification is evidence that results are consistently being achieved at delivery.

The ultimate project certification is the system certification provided by the chief systems engineer that the solution provided to the customer will perform as expected. This testimonial is based on the summation of the verification history and the resolution of all anomalies. Figure 7.14 is an example certification by a chief systems engineer.

### Verification and Anomaly Management

In the management of verification, it is important to keep latent biases removed from the process as much as possible. To achieve maximum objectivity, the verifiers should be independent of both the developers and the verification planners. Figure 7.15 illustrates a candidate organization structure.

## Verification Management

The management of verification should be responsive to lessons learned from past experience. A few are offered for consideration:

1. A Requirements Traceability and Verification Matrix (RTVM) should map the top-down decomposition of requirements to their delivering entity and should also identify the integration level and method for the verification. For instance, while it is desirable to verify all requirements in an all-up systems test, there are many requirements that cannot be verified at that level. There may be stowed items at the system level that cannot and will not be deployed until the system is fielded. In these instances, verification of these entities must be provided at a lower level of integration. The RTVM should

### FIGURE 7.14. CSE SYSTEM CERTIFICATION EXAMPLE.

Date: _____
I _____ certify that the _____ system delivered on _____ will perform as specified. This certification is based on the satisfactory completion of all verification and qualification activities. All anomalies have been resolved to satisfactory conclusion except two that are not repeatable. The two remaining are:
1. _____
2. _____
All associated possible causes have been replaced and regression testing confirms specified performance. If either of these anomalies occurs during the operational mission there will not be any effect on the overall mission performance.

Signed _____
Chief Systems Engineer (CSE)

## FIGURE 7.15. ORGANIZATION FOR VERIFICATION.

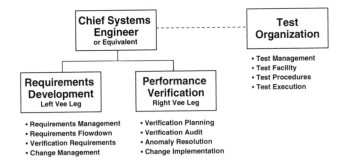

- **Requirements for verification and proof of verification should be organizationally separate, if possible**
  May be in same organization or independent

ensure that all required verification is planned for, including the equipment and faculties required to support verification at each level of integration.

2. The measurement units called out in verification procedures should match the units of the test equipment to be used. For example. considerable damage was done when thermal chambers were set to degrees Centigrade when the verification procedure called for degrees Fahrenheit. A perfectly good spacecraft was destroyed when the range safety officer, using the wrong flight path dimensions, destroyed it during ascent thinking it was off course. Unfortunately, there are many other examples that caused perfect systems to be damaged in error.

3. Red-line limits are "do not exceed" conditions, just as the red line on a car's tachometer is designed to protect the car's engine. Test procedures should contain two types of red-line limits. The first should be set at the predicted values so that if they are approached or exceeded the test can be halted and an investigation initiated to determine why the predictions and actuals don't correlate. The second set of red-line limits should be set at the safe limit of capability to prevent system failure or injury. If these limits are approached the test should be terminated and an investigation should determine the proper course of action. One of the world's largest wind tunnels was destroyed when the test procedures that were required to contain red-line limits did not. During system verification, the testers unknowingly violated engineering predictions by 25 times, taking the system to structural failure and total collapse.

4. A Test Readiness Review (TRR) should precede all testing to ensure readiness of personnel and equipment. This review should include all test participants and should dry-run the baselined verification procedure, including all required updates. Equipment used to measure verification performance should be confirmed to be "in calibration," projected through the full test duration including the data analysis period.

5. Formal testing should be witnessed by a "buyer" representative to officially certify and accept the results of the verification. Informal testing should precede formal testing to

## FIGURE 7.16. REQUIREMENTS TRACEABILITY AND VERIFICATION MATRIX—BICYCLE EXAMPLE.

| Level | Rev | ID | Name | Make or Buy | Requirement | | Predecessor | Verification | | Auditor | Date |
|---|---|---|---|---|---|---|---|---|---|---|---|
| 0 | 0 | 0.0 | Bicycle System | M | 0.0.1 | "Light Wt" - <105% of Competitor | "User Need" Doc ¶ 1 | 0.0.1 | Assess Competition | | |
| 0 | 0 | 0.0 | Bicycle System | M | 0.0.2 | "Fast" - Faster than any other bik | "User Need" Doc ¶ 2 | 0.0.2 | Win Tour de France | | |
| 1 | 0 | 1.1 | Bicycle | M | 1.1.1 | 8.0 KG max weight | 0.0.1, Marketing | 1.1.1 | Test (Weigh bike) | | |
| 1 | 0 | 1.1 | Bicycle | M | 1.1.2 | 85 cm high at seat | Racing rules ¶ 3.1 | 1.1.2 | Test (Measure bike) | | |
| 1 | 0 | 1.1 | Bicycle | M | 1.1.3 | 66 cm wheel dia | Racing rules ¶ 4.2 | -- | *Verif at ass'y level* | | |
| 1 | 0 | 1.1 | Bicycle | M | 1.1.4 | Carry one 90 KG rider | Racing rules ¶ 2.2 | 1.1.4 | Demonstration | | |
| 1 | 0 | 1.1 | Bicycle | M | 1.1.5 | Use advanced materials | Corporate strategy ¶ 6a | -- | *Verif at ass'y level* | | |
| 1 | 0 | 1.1 | Bicycle | M | 1.1.6 | Survive FIVE seasons | Corporate strategy ¶ 6b | 1.1.6 | Accelerated life test | | |
| 1 | 0 | 1.1 | Bicycle | M | 1.1.7 | Go VERY fast (>130 kpm) | 0.0.2 | 1.1.7 | Test against benchmark | | |
| 1 | 0 | 1.1 | Bicycle | M | 1.1.8 | Paint frame Red, shade 123 | Marketing | 1.1.8 | Inspection | | |
| 1 | 0 | 1.2 | Packaging | B | 1.2.1 | Packaged for Shipment | 0.0.4, Marketing | | | | |
| 1 | 1 | 1.2 | Packaging | B | 1.2.1 | Photo of "Hi Tech" Wheel on Box | 0.0.4, Marketing | | | | |
| 1 | 0 | 1.2 | Packaging | B | 1.2.2 | Survive 2 m drop | Industry std | | | | |
| 1 | 1 | 1.3 | Documentation | M | 1.3.1 | Assembly Instructions | 0.0.4 | | | | |
| 1 | 1 | 1.3 | Documentation | M | 1.3.2 | Owner's Manual | 0.0.4 | | | | |
| 2 | 0 | 2.1 | Frame Assembly | B | 2.1.1 | Welded Titanium Tubing | 1.1.5, 1.1.6 | | | | |
| 2 | 0 | 2.1 | Frame Assembly | B | 2.1.2 | Maximum weight 2.5 KG | 1.1.1, allocation | | | | |
| 2 | 0 | 2.1 | Frame Assembly | B | 2.1.3 | Demo 100 K cycle fatigue life | 1.1.6 | | | | |
| 2 | 0 | 2.1 | Frame Assembly | B | 2.1.4 | Support 2 x 90 KG | 1.1.4, 1.1.6 | | | | |
| | | | • | | | • | | | | | |
| | | | • | | | • | | | | | |
| | | | • | | | • | | | | | |

discover and resolve all anomalies. Formal testing should be a predetermined success based on successful informal testing.

6. To ensure validity of the test results the responsible tester's or quality control's initials should accompany each data entry.

7. All anomalies must be explained including the associated corrective action. Uncorrected anomalies must be explained with the predicted impact to system performance.

8. Unrepeatable failures must be sufficiently characterized to determine if the customer/users can be comfortable with the risk should the anomaly occur following operations.

## Anomaly Management

Anomalies are deviations from the expected. They may be failure symptoms or may just be un-thought-of nominal performance. In either case, they must be fully explained and understood. Anomalies that seriously alter system performance or that could cause unsafe conditions should be corrected. Any corrections or changes should be followed by regression testing to confirm that the deficiency has been corrected and that no new anomalies have been introduced.

The management of anomalies should be responsive to the past experience lessons learned. A few are offered for consideration:

1. Extreme care must be exercised to not destroy anomaly evidence during the investigation process. An effective approach is to convene the responsible individuals immediately on detecting an anomaly. The group should reach consensus on the approach to investigate the anomaly without compromising the evidence in the process. The approach should err on the side of care and precaution rather than jumping in with uncontrolled troubleshooting.

2. When there are a number of anomalies to pursue, they should be categorized and prioritized as Show Stopper, Mission Compromised, and Cosmetic. Show stoppers should be addressed first, followed by the less critical issues.

3. Once the anomaly has been characterized, a second review should determine how to best determine the root cause and the near- and long-term corrective actions. Near-term corrective action is designed to fix the system under verification. Long-term corrective action is designed to prevent the anomaly from ever occurring again in any future system.

4. For a one-time serious anomaly that cannot be repeated no matter how many attempts are made, consider the following:
   a. Change all the hardware and software that could have caused the anomaly.
   b. Repeat the testing with the new hardware and software to achieve confidence that the anomaly does not repeat.
   c. Add environmental stress to the testing conditions, such as temperature, vacuum, vibration, and so on.
   d. Characterize the anomaly and determine the mission effect should it recur during any phase of the operation. Meet with the customer to determine the risk tolerance

**TABLE 7.1. IV&V ARTIFACTS.**

| Artifact | Purpose |
| --- | --- |
| System Engineering Management Plan | The technical strategy, including the overall approach to integration, verification, qualification, and validation. |
| Interface and Intraface Specifications | The requirements for entities to properly combine into higher assemblies. |
| Validation Plan | The approach to in-process and final validation. |
| Validation Procedures | The step-by-step actions required to accomplish validation. |
| Verification Plan | The approach to verification including environments imposed. |
| Verification Procedures | The step-by-step actions required to accomplish the various types of verification. |
| Verification Data | The raw data produced by verification activity. |
| Discrepancy Report | The characterization of an anomaly. |
| Failure Analysis Report | The results of failure analysis and the recommended corrective action. |
| Qualification Certificate | The summation of evidence and certification that a configuration item has survived a defined set of environmental and operational conditions. |
| Verification Report | A summation of the verification history and the verification results. |
| Requirements Traceability and Verification Matrix (RTVM) | A map of verification results against their requirements to ensure completeness and adequacy of verification. |

of the using community and whether deployment with the risk, as quantified, is preferred over abandoning the project.

## IV&V Artifacts

The integration, verification, and validation process is managed by an integrated set of artifacts. Table 7.1 summarizes the most popular artifacts and their purpose.

## Summary

Following is a summation of the important points of this chapter:

Integration should be planned early and should be reflected in the product breakdown structure, the work breakdown structure, and the tactical project network.

Interfaces should be designed to be simple to facilitate integration and to simplify verification of those interfaces.

An "okay to install" discipline should be imposed for all integrations and matings.

At each level of decomposition, determine how the solution users will determine their satisfaction (validation).

Practice in-process validation with both the direct and ultimate users, being aware that, at some point in the decomposition, decisions may be transparent to the user, since they won't be impacted. For instance, the ultimate user probably won't be concerned if slotted or Phillips head screws are used as fasteners. The direct user who must apply the fasteners will care.

When new customers and users emerge, re-baseline their validation expectations.

One of the systems engineer's jobs is to reduce the expectations of the customer and users back to the approved baseline. Customers and users often expect more without adding funds or schedule.

Verification must prove performance of design, design margin, quality, reliability, and many other "-ilities."

Verification incorporates various combinations of testing, demonstration, inspection, and analysis.

Anomalies must be characterized and resolved without destroying the evidence.

For adequate qualification, all life cycle environments must be understood and planned for including the possibility of multiple environmental retests of a unit that has failed several times and has been repaired and retested each time. Certification demonstrates a level of capability and competency attested to by an authority.

Individual and organizational certification is available and expanding.

# References

Ambler, S. W., and R. Jefferies. 2002. *Agile modeling: Effective practices for extreme programming and the unified process.* New York: Wiley.

Beck, K. 1999. *Extreme programming explained: Embrace change.* Reading, MA: Addison-Wesley.

Buede, D. M. 2000. *The engineering design of systems.* New York: Wiley.

Carnegie Mellon Software Engineering Institute Capability Maturity Model (CMM) and Capability Maturity Model Integrated (CMMI). www.sei.cmu.edu/cmm/cmms/cmms.integration.html.

Crosby, P. 1992. *Quality Is Free.* Denver: Mentor Books.

Cockburn, A. 2001. *Agile software development.* Reading, MA: Addison-Wesley.

Forsberg, K., H. Mooz, and Cotterman, H. 2001. *Visualizing project management.* New York: Wiley.

Forsberg, K. and H. Mooz. 2001a. Visual explanation of development methods and strategies including the Waterfall, Spiral, Vee, Vee+, and Vee++ models. INCOSE 2001 International Symposium.

Grady, J. O. 1998. *System validation and verification.* Boca Raton, FL: CRC Press.

International Organization for Standardization (ISO). www.iso.ch/iso/en/ISOOnline.openerpage.

Martin, R. C. 2002. *Agile software development: Principles, patterns, and practices.* Upper Saddle River, NJ: Prentice Hall.

Mooz, H., K. Forsberg, and H. Cotterman. 2002. *Communicating project management.* New York: Wiley.

Project Management Institute (PMI). www.pmi.org/info/default.asp.

International Council On Systems Engineering (INCOSE). www.incose.org/.

Walton, M., and W. E. Deming. 1988. *Deming management method.* New York: Perigee

Stevens, R., P. Brook, K. Jackson, and S. Arnold. 1998. *Systems engineering.* Upper Saddle River, NJ: Prentice Hall.

CHAPTER EIGHT

# MANAGING TECHNOLOGY: INNOVATION, LEARNING, AND MATURITY

Rodney Turner, Anne Keegan

In this chapter we take a slightly wider view of the management of technology than is usually taken, and in two ways. First, rather than viewing "technology" just as the engineering skills an organization uses to do its projects, we are going to focus on all the skills it brings to bear to do its projects. We suggest those skills exist on three levels:

1. The ability of an organization to manage projects in general. This is the organization's general competence or maturity at managing projects, stored in its collective wisdom, in standards, and elsewhere. Turner (1999) has described this as the "projectivity" of an organization.
2. The ability of an organization to manage a given project. This skill reflects the organization's ability to recognize the relevant success criteria and appropriate success factors for a given project, and to take its standards and develop a strategy for this project to deliver it successfully. It also reflects the organization's ability to learn what works and what does not work in given situations, and to manage the risks so that it does not keep on repeating the same mistakes.
3. The ability of an organization to use its technology ("engineering" skills) to build its assets as efficiently and as effectively as possible, and thereby obtain the best value (whatever that may mean for them).

Second, rather than just considering how an organization uses those skills to repeat previous performance, we focus on how an organization gets better at doing its projects. We describe how an organization learns from its previous experience to get better at what it does, and how it innovates to introduce new ideas to get better still. Surprisingly, project-based organizations often do not provide an environment supportive of innovation. This chapter looks at why that is and what can be done to overcome it. It also considers linear-rational and organic approaches to innovation management.

# A Four-Step Process of Innovation and Learning

Turner, Keegan, and Crawford (2003) developed a four-step process of innovation and learning, based on earlier studies from management learning (Miner and Robinson, 1994), adapted for project-based organizations:

1. *Variation.* The organization experiments with new ways of delivering its projects.
2. *Selection.* It chooses those innovations that work and that it wants to adopt.
3. *Retention.* It stores the selected innovations in its collective wisdom.
4. *Distribution.* The organization then needs to distribute those new ideas to new projects and ensure they are used to improve project performance. This step is not needed in functional organizations; it is specific to project-based ways of working.

In a functional organization, only the first three steps are necessary, and all take place within the line organization. The function experiments with new ways of working, selects those it wishes to retain, and stores them in the functional organization where they are immediately accessible to people working in the function. In a project-based organization, variation occurs on one project, which comes to an end. Selection occurs in the post-completion review process, and ideas then need to be retained in a central project management function. But those ideas are not used there, so they need to be distributed to new projects about to start, and different project managers must be encouraged to use them. There are, however, two problems in the project-based context, the issues of deferral and attenuation:

## Deferral

We have suggested (Keegan and Turner, 2001) the process can build in a delay at each step. The emphasis is on *post*-completion reviews. Assuming a two-year delay at each step, it can be eight years between a new idea being generated to its being used on a future project. We adopted the idea of viscosity of information. Some information oozes through the organization like treacle, with eight years from idea generation to its becoming widely adopted. To overcome this, people suggest the use of intranet-based technologies. Then information zips through the organization like gas in a vacuum. Yesterday's hearsay becomes today's perceived wisdom. Later in the chapter we discuss how to achieve a balance and discuss the use of internal project reviews to assess the technology used on a given project.

## Attenuation

Cooke-Davies (2001) has measured the loss of information at each step. Of the worthwhile new ideas generated, about 70 percent are selected; about 70 percent of those get retained; about 70 percent of those get distributed to the organization; and about 70 percent of those get reused on future projects. The project-based organization gets to reuse about a quarter of the worthwhile new ideas it generates. (We are surprised the numbers are as high as they are.)

# Creating an Environment Supportive of Innovation

Innovation is vital for technological development within organizations, developing new engineering and project management process skills to enhance project performance. It occurs at the first of the preceding four steps. There are many things an organization can do to create an environment supportive of variation and innovation (Keegan and Turner, 2003), including:

- Creating channels for formal and informal communication
- Blurring organizational boundaries, using integrating boundary spanners
- Creating flexible roles and multidisciplinary teams
- Allowing some stress and ambiguity
- Facilitating projects rather than rigidly controlling them

## Information and Communication

Communication is essential, to ensure the right people have the right information. When establishing a project, the manager needs to ensure that everybody who may have knowledge or information or opinions about the design and specification of the end product is properly consulted, and that those people who need to know are informed. But just as important, he or she does not want to be overwhelmed with conflicting opinions from people who have no real input to make. Further, if everybody on the project is informed about every decision, those people who need to know may ignore the bit of information targeted at them.

Informal channels of communication are also important for innovation. New things happen and new ideas are generated because people who do not normally talk to each other make new contacts, and through those contacts generate the new ideas that underpin technological development (Keegan and Turner, 2003). The Nobel laureates Sir Alexander Fleming and Niels Bohr encouraged creative, informal contacts and communication (Larsson, 2001). Later in this section, when discussing organic approaches to innovation, we cover how to encourage new contact. Having people work in cross-discipline teams is also essential to that.

Management can often be viewed as a series of dilemmas, and information and communication on projects can be no different:

- Communication can be too formal, stifling cross-functional working, innovation, and people's ability to perform (see Figure 8.1).
- Or communication can be too informal, with too many conflicting opinions, too many siren voices, and too much hype about the possibilities for the project.
- The project team can talk to too many people, again with too many conflicting opinions and informing too many people about progress so that nobody bothers to listen.
- Or they can talk to too few, behaving in a cloak-and-dagger fashion, so that nobody knows what they are doing and nobody cares.

Information and communication on projects needs to be carefully managed to achieve the desired innovations.

## Cross-Functional Working with Boundary Spanners

Innovation requires cross-functional working. It is about creating change, and change involves people working together in novel ways. We have just seen that people working across functions can lead to new ideas through new contacts. But it also needs people to be empowered to make progress together; a project must be a rugby scrum, not a relay race. The

### FIGURE 8.1. EXCESSIVELY FORMAL COMMUNICATION.

Rodney Turner wrote a case study on an IS/IT project in the UK's Civilian Aviation Authority. One of the people interviewed had just joined from the private sector. He said he was working on a project involving people from other departments, but found it impossible to make progress. If he needed to write a memo to somebody in another department to make a decision, he had to write a draft of the memo to give to his boss to critique. He would then revise the memo, and it would go to his boss's boss to critique, and so on until it got to the lowest common boss, when it would be sent to the relevant person. The reply would come back the same way, and the whole process would take weeks. He said you cannot make progress on a cross-functional project working in this way.

You can understand why managers are doing this. If a plane crashes, the media will be on a witch hunt, and managers want their stamp on decisions they will be held accountable for. But in the temporary organization that is a project, you have to empower people to make decisions to make progress (see the chapter by Huemann, Turner, and Keegan).

The next person spoken to was a Royal Air Force Officer on secondment to the CAA. Asked if it was true, he said he was afraid it was. What he did was send the first draft of the memo to the person he was trying to communicate with and they got on with it, while the official memo did the rounds. You can see the military is used to empowering people. In the heat of battle you cannot refer decisions up the line; there isn't time. Projects are like the heat of battle; you need to empower people (within firm guidelines).

communication methods described in Figure 8.1 result in projects being artificially extended, as they go from one department to another. People must work in flexible, cross-discipline teams, with decentralized authority. People can be given roles to act as boundary spanners—people whose role is to bridge gaps between people and get them working together as a team. Figure 8.2 (adapted from Fong, 2003) illustrates the need, bridging boundaries between disciplines, hierarchies, and areas of expertise to achieve new thinking and innovation.

## Stress and Ambiguity

Surprisingly, or perhaps not, some stress and ambiguity can lead to better innovation. It is well known that people work better under reasonable levels of stress than they do under no stress at all. But this is another dilemma. Too much stress can be deleterious. Likewise, ambiguity can encourage innovation. It is recognized by psychologists that it is not risk that people fear, but ambiguity and the chance of loss (Bernstein, 1998). People will respond to risk if they see it as a chance of gain. But if it is unclear whether the risk will result in loss of gain, they fear that. Thus, where ambiguity exists, people will try to eliminate it, leading

### FIGURE 8.2. BOUNDARY SPANNERS.

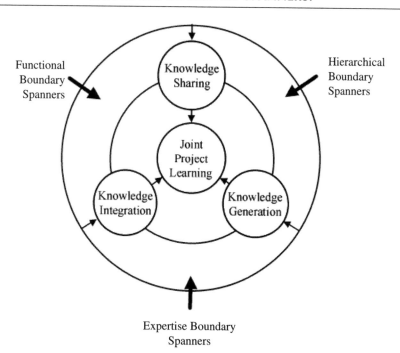

to innovative solutions to their problems. Reasonable levels of stress and ambiguity should be encouraged on projects to find the most innovative solutions.

## Facilitation vs. Rigid Control

Projects involve stress and ambiguity, they involve cross-discipline teams with boundary spanners, and they involve communication among team members. They should therefore encourage innovation, right? Unfortunately not. The project environment often kills innovation through too-rigid control. On projects we can do all the other things but still kill innovation. We consider below the need to achieve a balance between the rigid linear-rational approach usually adopted on projects and the organic approach sometimes adopted. But let us first consider some of the problems of too-rigid control.

## Competency Traps

Projects are more exposed to competency traps than routine operations. A competency trap is where there is an established way of working, which may not be optimal (Levitt and March, 1995). But people fear trying an alternative for risk of failure. In a routine environment it is easier to try something new. If it does not work on the first attempt, it may be improved next time, and if it doesn't work at all, the old way can be reinstated. It is easier to experiment in a routine environment. Projects (in their pure form) are only done once. If they do not work the first time, they do not work at all. There may be a way of doing a project that is twice as good as the preferred way, but with, say, a 20 percent chance of failure. So if it were done several times, on average it would be 60 percent better. In a routine environment, people can experiment, find the flaws, and get it right second time around. But on a project, if it is only done once, people prefer the certain, though less efficient, way. They are trapped in the inferior way of working.

## Rigid Evaluation Criteria

Project management has developed evaluation criteria for assessing the value of projects and their contribution to corporate wealth, including techniques such as net present value (NPV) and internal rate of return (IRR) (Turner, 1995; Lock, 2000). However, these do not properly evaluate IS/IT and innovation projects (Akalu, 2003). For such projects a technique known as option pricing gives a better view, but unfortunately is more difficult to apply. Akalu (2003) showed that most organizations are reduced to applying qualitative assessment criteria to innovation projects and IS/IT projects. This can create problems in firms wanting to compare the IRR of all of their projects and applying strict hurdle rates that are not appropriate to all the projects they do. Artto and Dietrich, Archer and Ghasemzadeh, and Thiry describe benefits management and the evaluation of projects by linking their outcomes to business objectives in their chapters in this book.

## Rigid Resource Utilization

Standard project management techniques also suggest tight assignment of resources, allocating the precise number required to do the job. This is not always appropriate for some projects. Projects are risky, and some flexibility is required to deal with uncertainty. But it is especially inappropriate for innovation projects. People with time to think develop much more innovative, creative solutions. Innovation projects require creativity, coupled with high uncertainty. That is stress enough, without adding additional stress by making the project team work to tight resource limits.

## Rigid Control

Traditional project management also suggests tight control (see the chapter on changes by Cooper and Reichelt). On innovation projects this may be appropriate at the development stage, but not at the research stage. Innovation projects should still be managed, but at the research stage more organic approaches are appropriate, emphasizing facilitation and co-ordination of the people working on the project. At the development stage there can be more rigid deadlines—for instance, the new product needs to be delivered to market by a certain date, or the Web space needs to be online by a certain date. Thus, the appropriate form of control, organic or linear rational, depends very much on the type of project and the stage it is at.

## Rigid Contract Management

Traditional contract management procedures can also be a block to innovation (see Figure 8.3). Innovative contracting techniques, such as partnering and alliancing (Scott, 2001) and appropriate sharing of risk on conventional contracts (Turner, 2003) are necessary to achieve

### FIGURE 8.3. RIGID CONTRACT MANAGEMENT DISCOURAGING INNOVATION.

A client wanted a vessel to be shot-blasted and painted. They drew up a specification of how the job should be done, and asked potential contractors to bid competitively to do the job on a fixed-price basis. A contractor won the bid process, but as they were about to sign the contract, the contractor said they could do the job for half the price, but they would only tell the client how under two conditions:

- The client would not reopen the bid process.
- The contractor would earn the same absolute profit, not the same percentage profit.

You can see the contractor's concern. The client was so stuck in rigid contract management procedures that they could not award the job without compulsory competitive tendering, and they only wanted their contractors to make a certain profit margin; the client wouldn't share their increased profit with the contractor.

In this case the client agreed. They got the job done for 55% of what it would have cost them, and the contractor made their same absolute profit and a higher percentage profit.

innovation on projects. (See the chapters by Venkatarman, Langford and Murray, Lowe, and others in this book.)

# Achieving Innovation in Project-Based Firms

In this section we consider how innovation can be achieved in a project context.

## Encouraging Variation

*"If you always do it the way you have always done it, you will always get what you have always got."*

SIR MICHAEL LATHAM.

To improve the performance of their projects, organizations need to innovate and try things new. So how do people do that? The organic approaches we describe in the following help encourage the creation of new ideas. Other techniques that our research into innovation and learning have identified include the following.

## Obtaining Senior Management Support

First we mention the importance of senior management support. This was identified as a key part of improving the management of information services (IS) projects in the Research and Development Department in SmithKline Beecham by Gibson and Pfautz (1999). Without senior management support, junior people will either fear making changes or not take the initiative.

1. A manager in IBM suggested junior people may avoid making honest reports in project reviews for fear of upsetting middle managers. Organizations must learn not to shoot the messenger, and the support of senior management helps junior people to make honest reports. The nature of the organization also has an impact here. If the organization has a blame culture, nobody will give honest reports, either for fear of attracting blame to themselves or through fear of damaging their immediate colleagues, particularly their immediate superior. A learning organization, on the other hand, will welcome honest reviews and treat them as opportunities to improve, rather than a basis for witch-hunts.

2. In many organizations in the construction industry, including government procurement departments, it is the "jobsworths" at junior levels who block the adoption of new contracting practices that would lead to improved effectiveness as suggested by Latham (1994). They fear that if they try something new and it goes wrong, they will get the blame. "The risk of that going wrong is more than my job's worth." In reality, this fear is often imagined and is an excuse for making their lives comfortable by doing what they have always done (and getting what they have always got). More than senior management support is needed here. Junior managers need to recognize it's more than their job's worth not to adopt new practices.

## Involving Construction and Operations People at the Design Stage

We saw previously the importance to innovation of multidisciplinary teams and cross-functional working. On a project, one way of achieving that is to involve construction and maintenance people in the design process. Without being stimulated to think new thoughts, design people may apply their traditional ways of thinking. Figure 8.4 contains examples of obtaining improved value by bringing other experiences into the design review process. Thiry further describes in his chapter on value management how value can be improved by involving a broad range or project participants in value management workshops during the design stage.

## Researching New Techniques at the Feasibility Stage

The project manager from the main contractor on one of the early alliance contracts in the Netherlands told us that he extensively researched alliance contracts as early as the bid stage. He assigned members of the potential project team to research different elements of alliance contracts. That helped the firm to learn a new approach to both contract management and in the area of project team building.

## Managing Innovation Projects

There are two opposing approaches to managing innovation projects:

- *Linear-rational approach.* Emphasizes rigid evaluation criteria, rigid resource utilization, rigid control, and the following of a strict process
- *Organic approach.* Emphasizes more fluid and flexible approaches.

The linear-rational approach is about keeping to the straight and narrow, moving as briskly as possible to the final objectives. It suggests tight, rigid control. On the other hand, the

### FIGURE 8.4. INVOLVING MAINTENANCE IN BUILDING DESIGN.

The British Airports' Authority, BAA, was planning an extension to one of its airports, and held workshops with its maintenance people during the design stage. In a building project, operation and maintenance costs are five times construction costs. Considerable improvements in value can be obtained by reducing operation and maintenance costs.

A window cleaner pointed out that if the windows were made of frosted glass they wouldn't need cleaning so often. And another member of maintenance staff pointed out that if gray grout was used in the bathrooms, the floors and walls would be easier to clean.

Both suggestion were adopted, to give considerable saving in maintenance costs at no change in capital cost organic approach is more pagan, following the seasons through cycles of development to the end objectives. It may be less efficient, but in the right circumstances is far more effective. Which is appropriate for a given innovation project depends on the nature of the project and the stage of development the product is at. If the project is in the research phase, organic approaches may be more appropriate, as it is here the new ideas must be generated. If it is in the final product development stage, more rigid control may be appropriate, as time-to-market is now significant (Turner, 1999).

## The Linear-Rational Approach

The linear-rational approach is usually based on a version of the life cycle, and on toll-gates or stage-gates. Projects move through a series of go/no-go decisions, being evaluated against strict criteria at each stage-gate before being allowed to proceed. Wheelwright and Clark (1992) suggested a funnel as a metaphor, indicating that the surviving projects become fewer and fewer at each stage-gate. (See the chapter by Cooper, Aouad, Lee, and Wu.) The PRINCE2 process, shown in Figure 8.5 (OGC, 2002), is in essence a Stage-Gate model, where the number of stage-gates can be varied to meet the needs of the project. At the completion of each stage, the project is evaluated against business and project control criteria.

Table 8.1 is the Stage-Gate model used by a financial services supplier to assess data products they supply over their own network and over the Web. They suggest that at each stage-gate, half the surviving projects fall. Thus, 64 product proposals need to start for one to be released to market. Even at the last stage-gate, one out of two products falls, two have

## FIGURE 8.5. THE PRINCE2 PROCESS.

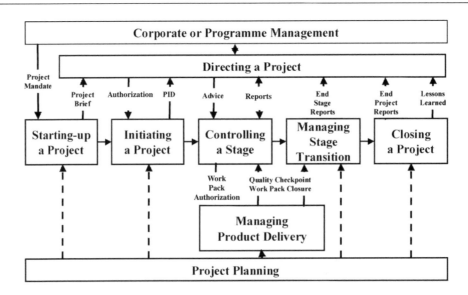

## TABLE 8.1. THE STAGE-GATE MODEL FOR A FINANCIAL SERVICES SUPPLIER.

| Stage | Time | Action |
|---|---|---|
| Identify market opportunity | One week | Sales prepares a brief description of the customer's requirement. |
| Initial product description | One month | Marketing appoints a product manager who develops a product description and initial project plan. |
| Project portfolio committee | Three months | An the resource requirements are quantified and the product portfolio committee prioritizes products for development-based resource availability. |
| Requirements definition | | Functional and system requirements are defined. |
| Assess risks, and reconcile project success criteria | | The project is planned in detail, a formal risk analysis is conducted, and success criteria and factors are evaluated. |
| Approve release | | The product is ready for release to market. Likely maintenance costs and profitability are assessed. |

made it all the way through the product development process, and only one is released to market.

At an innovation conference in Ireland a few years ago a speaker described the 63 projects that do not make it as "failures." They are not failures. To be an innovative organization, it is necessary to try out many things in order to have the one successful one. The challenge is to shut projects down as quickly as possible. Table 8.1 illustrates that seven-eighths of the projects should be shut down within three months, the time of the third stage-gate, and that is the role of the project portfolio committee. The same speaker said that in his company, out of every four products released to market, two made a loss and only one made a profit. The financial services agency tried to stop that with their last stage-gate, so every product released at least breaks even.

Table 8.1 and Figure 8.5 illustrate that project reviews are held at the completion of each stage. Thus, they may be held at the completion of initiation, feasibility, and design. One purpose of project reviews is to ensure the right technology is being used in the form of the project management process and appropriate engineering skills, as well as checking that the project meets required business and operational criteria. This is discussed again at the end of the chapter. (See the chapters by Thiry on value management and by Huemann on quality reviews.)

The advantages of the linear-rational approach are as follows:

- It provides clear no-go decisions, encouraging the development of a business plan and allowing the project to be matched to company strategy.
- It allows ideas to be tried and tested; closure is not seen as a failure.

- It provides clear, strict control, through the stage-gates and through milestone planning.
- It helps manage risk; ideas are tested before progressing, and the stage-gates are clear review points.
- It creates a system that all employees are familiar with and provides a systematic methodology for evaluation across disciplines or functional areas.

The disadvantages of the linear-rational approach are as follows:

- It favors efficiency and control over creativity and effectiveness
- It can artificially extend projects if you insist that one stage-gate is passed before work on the next begins.

## The Organic Approach

The organic approach overcomes these weaknesses, but at the expense of efficiency and control. It is still possible to provide vision and direction for the project, and coordinate the input of resources. But the organic approaches favor more flexible management. Through our research we have identified the following organic approaches to innovation on projects.

***Deliberate Redundancy.*** Advertising agencies, when starting work on a new account, create three or four teams to work independently to develop ideas. They then sample those ideas, to come up with an overall proposal for the client. We wonder how many project-based organizations would consider having four teams work independently in the feasibility stage to think of different ideas. (BAES does.) Many high-tech companies often have two people do a job where strictly one will do. The advantage is two people working together come up with more creative solutions to the client's need, and when the job is over, two people have the new skill, improving learning in the organization. We call this the creation of Nellies. In traditional companies, the new recruit learns the work of the company by serving an apprenticeship, sitting next to an experienced person, "Nellie." There are no Nellies in high-tech companies because the technology is changing so fast, so firms create Nellies in the way described. Some people ask how can they afford this inefficiency. But if they obtain more creative solutions, with better learning for the organization, then the cost is repaid. Unisys, Intel, and Hewlett-Packard have all told us they adopt this approach. For innovation on their projects, organizations should consider having several people or teams work independently during the early stages and then choose the best solution. Those people can deliver more creative solutions and communicate those solutions to the rest of the organization. People might say they cannot afford the cost. Well, they need to compare the value of the creative solution to the additional resource input, and sometimes (not always) this approach provides better value.

***Sampling.*** Having generated several different solutions to the problem, the best solution needs to be selected. That is what advertising agencies do. But the best solution may not be one of the proposals, but a mixture of them all. So the sampling needs to be more a blending process, where the best solution emerges as a mixture of the proposals. In the financial services company mentioned previously, the one product that emerged at the end was not one of the 64 that started but a mixture of them. This approach led to a tension between the product development personnel and senior management. Senior management wanted to shut half the projects down at each stage-gate, but the product development personnel wanted to keep them alive, as they could not know which ones would contribute to the final product. Somebody described it as being like blending whisky. Sampling from the different casks, they could not know they had the right blend, and what it would comprise, until they had it. The problem here is that the decision to keep projects going on the off chance they may be needed or that unanticipated synergies will emerge can be a cost-prohibitive move. The main reason for review gates within most organizations is cost control. But sometimes (not often) this approach can lead to better value solutions. Advertising agencies have a strictly limited time for the parallel working, and the financial services agency relied heavily on the intuition of the product development experts, and usually they were reliable.

*Chance Encounters.* We said previously that innovation comes through people making contacts that have not existed previously. New ideas come from old ideas reforming in new ways. An Irish advertising agency decided to try to reduce office rent and commuting costs for their employees by having them work from home. All their creative people were given laptops and an ISDN link at home. Creativity plummeted! Sir Alexander Fleming, who discovered penicillin, deliberately left petri dishes lying around to see if something unexpected happened, and it did (Larsson, 2001). We spoke above about the need for cross-discipline teams and boundary spanners. For innovation on projects, do not let the design people work in isolation.

*Creative Communications.* Chance encounters lead to creative communications and vice versa. The Danish hearing aid company Oticon encourages people to chat at the water fountain. At Henley Management College, morning and afternoon tea are something of a ritual, but creative communication occurs at them.

*Creative Tensions.* Difference between people, rather than being avoided, should be encouraged, as they too can lead to new ideas by reforming old ones in new ways. In advertising agencies, the tension between the suits and non-suits (businesspeople and creative people) is encouraged. On projects, rather than avoiding differences between engineers and marketing people, it should be encouraged to find the best solution (Graham, 2003).

The linear-rational and organic approaches are quite different, but not entirely incompatible. Clearly, deliberate redundancy is incompatible with efficiency, but not with effectiveness. Further, it is possible to adopt organic and linear-rational approaches at different stages of the life cycle. At the early stages organic approaches are best. Here the new ideas need to be generated, so cross-functional working with deliberate redundancy will be used through the first and second stage-gates in Table 8.1. Then ideas can be sampled at the second stage-gate, and then more strict management processes applied from then on, when costs begin to increase and time-to-market becomes significant (Turner, 1999).

***Viewing Uncertainty as an Opportunity.*** Some project managers, in a desire to achieve certainty and strict control, try to squeeze risk and uncertainty out of their projects as quickly as possible. They follow Path A in Figure 8.6. However, in the process they lock themselves into high-cost solutions at an early stage. It has been suggested by Latham (1994) and Egan (1998) that in construction following Path B can lead to a 30 percent reduction in cost, by allowing more innovative, cheaper solutions to be found. It has even been suggested that following Path C can lead to a further 30 percent reduction in cost.

The problem with Path B, and more so with Path C, is that they are not compatible with predictability and certainty. They are not compatible with conventional project management thinking, which likes rigid control, following the straight and narrow, linear-rational approach. They are not compatible with normal relations between clients and contractors, based on compulsory competitive tendering for fixed-priced contracts and confrontational relationships.

However, there is now a growing body of case study evidence that Paths B and C can achieve what is promised of them, allowing options to be explored for a longer period and innovative, cheaper solutions to be found. The application of strategies B and C is based on at least two requirements, though:

1. The use of strict configuration management (Turner, 1999, 2002; see also the chapter by Kidd and Burgess) to manage the reduction in uncertainty and track the various options as they are explored, merged, or discarded
2. The use of modern contracting techniques that encourage collaborative working between clients and contractors, who view the project as an opportunity to work together toward

## FIGURE 8.6. STRATEGIES FOR RISK REDUCTION.

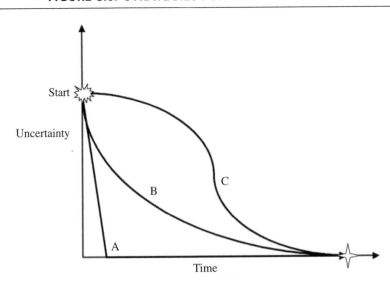

a common objective, with appropriate sharing of risk (Turner, 2003; Scott, 2001; see also the chapter by Langford and Murray).

# Retaining and Using Technological Developments

Innovations occur at the first of the four steps we identified earlier in the in the section on variation. Having made new technological developments, the organization must decide which are worthwhile for further use (selection), record them so that people can draw on the knowledge (retention), and pass the knowledge on to people working on other projects so that they can use them (distribution). The organization must also ensure that the right technological solutions are being used on a given project. We describe here practices adopted for the selection, retention, and distribution of technological developments at the three levels described previously, and for checking the technological solutions on a given project. We have observed the use of four practices:

1. Systems and procedures
2. Project reviews
3. Benchmarking
4. Project management communities

We describe these four practices and then show how they support learning and maturity.

## The Four Practices

*Systems and Procedures.* This is where the organization formally stores its technological knowledge, in "written" systems, procedures, and standards. They can take many forms:

- Procedures manuals
- Engineering standards
- Computer-based project management information systems
- Virtual project office in the intranet

An organization's competence can be described as its collective knowledge or wisdom. The systems and procedures are the concrete evidence of its collective wisdom. Systems, procedures, and standards are a key way organizations capture knowledge and experience. They are the collective representation of the firm's experiences.

The procedures and standards should be treated as flexible guidelines, tailored to the needs of each project. Every project is different, and so requires a unique procedure. Standard procedures represent captured experience and best practice, but they must be tailored project by project. Hopefully the tailoring is marginal, but it must done. It is part of a project manager's tacit knowledge that enables the manager to know how the procedures need to be tailored to individual projects. People who have the lack of maturity that makes them want to follow procedures to the letter are not yet ready to practice as project managers. The United Kingdom's Office of Government Commerce in its maturity model overtly states that part of maturity level 3 (of 5) is the ability to tailor procedures. A main contractor from the engineering construction industry reported that new project personnel are told to follow procedures strictly on their first project (when they are in a support role— sitting next to Nellie). On subsequent projects, they can reduce the amount they refer to the standards, as they internalize the firm's good practice. They are encouraged to adapt the procedures to individual projects as their experience grows.

Ericsson requires that its PROPS process should be used on all projects, although it is not mandatory. PROPS is designed to be tailored to the needs of individual projects. It represents good practice in Ericsson, but that good practice is flexible enough to be adapted to the size and type of project. PROPS is also continually updated to reflect new experiences, and the changing technology and nature of projects. The same is true for the PRINCE2 process produced by the United Kingdom's Office of Government Commerce (OCG, 2002). PRINCE2 certification is becoming mandatory to bid for many projects in both the public and private sector in the United Kingdom. In this way the government is contributing to the increasing competence of public sector projects, and to the increasing project management competence of the society. Organizations that have not captured their experience in project procedures are able to use industry-standard procedures, such as PRINCE2, ISO 10,006 (ISO, 1997), the PMI Guide to the Body of Knowledge (PMI, 2000), or other bodies of knowledge.

The emphasis on procedures, both as a learning medium and as a measure of maturity, does tend to emphasize process over outcome and intent (Levitt and March, 1995). However, both process and outcome should be emphasized on projects, and an emphasis on one is not mutually exclusive with an emphasis on the other (ISO, 1997). Project managers need to learn to emphasize both. The emphasis on procedure can lead to redundancy of experience and competency traps. However, the need to develop project-specific procedures for each project helps to ensure new processes are developed and tried. This encourages variation, although many project-based organizations do tend to be very conservative (Keegan and Turner, 2003).

***Reviews.*** Reviews can fulfil two purposes:

1. They can be conducted internally throughout the project, to check that the project's requirements are properly defined (see the chapter by Davis, Hickey, and Zweig) and the right technologies (project management process and engineering skill) have been selected for the project
2. They can be conducted at the end of the project, so that the organization can learn how well it did and capture its success and learn from its failures.

Huemann describes both types of reviews in her chapter on quality reviews, though she reserves the word "review" for the first case and uses the word "audit" for the second.

Internal reviews may be conducted at the completion of project initiation, feasibility, design, and other project stage transitions. They were described previously in the discussion on project stage-gates.

Post completion reviews, or audits, play a vital part in capturing experience. PRINCE2 and ISO 10,006 suggest a review be conducted at the end of every project, and company procedures updated to reflect that learning. Pinto (1999) reported that one contributing factor for the failure of many IS projects was a failure of the organization to review its performance on previous projects and learn from experience. People working on failing projects are met with a strong sense of déjà vu: "We have been here before and can see we are locked on a path to failure." Many project-based organizations continually bench-mark their procedures and processes, gathering data about project performance, storing that as historical data to help plan future projects, and thereby improving overall project performance.

However, many firms report less than satisfactory use of project audits. They find the practice difficult to enforce, and where it is enforced, it is a meaningless box-ticking exercise. An ICT contractor reported that post-completion reviews were an essential part of their quality assurance procedures, but there was no check on the quality of the outputs. Further, where reviews are conducted, it can be difficult to transmit the learning to the organization, because of the problems of deferral and attenuation identified earlier.

## Benchmarking

The organization compares its performance on projects to projects elsewhere (Gareis and Huemann, 2003). It may compare its performance with the following:

- Earlier projects it undertook, to track performance improvement
- Projects undertaken by other parts of the same organization
- Projects undertaken by other organizations if it can access the data

Benchmarking is also essential to increasing project management performance, but is not something that is well done by many organizations. The European Construction Institute and the American Construction Industry Institute are benchmarking projects in the engineering industry in the two continents. There are also many benchmarking communities in Europe, the Far East, and Australasia.

## Project Management Communities

The fourth learning practice adopted by many organizations is the maintenance of a project management community. The importance of the project management community is mentioned by many authors, for example, Gibson and Pfautz (1999) and Pinto (1999). The last step of the innovation and learning cycle is distribution, and project management communities help achieve that. Specific practices used by organizations through their project management communities to distribute innovations and technological knowledge are as follows:

- Regular (quarterly) seminars and conferences
- Mentoring of project management professionals
- Career committees, and support for individual competence and career development

- Overseas postings
- Centers of excellence
- The use of the intranet

*Seminars and conferences:* Many organizations, especially from high-technology industries, have regular, quarterly meetings of project management professionals, where they can network and share experiences. These can range form informal to extremely formal. The Dutch bank ABN-Amro has a quarterly meeting of its project managers from its information services (IS) department. This lasts a couple of hours, during which they have one or two lectures, followed by a *borrel.* In the Dutch army, the meeting lasts all day. Other organizations have a more formal conference one to four times a year.

As Huemann, Turner, and Keegan show in their chapter on human resource management (HRM) that most industries maintain industry-wide communities. These may be in the form of professional associations for individuals, such as the Project Management Institute or International Project Management Association, or professional institutes for companies, such as the European Construction Industry and Construction Industry Institute. Such associations hold seminars and conferences and provide other networking opportunities in the industry, rather than in individual companies.

*Mentoring:* Pinto (1999) reports that another contributing factor to the failure of IS projects in many organizations is a failure to mentor new project managers. They do not serve an apprenticeship; they do not spend time "sitting next to Nellie."

*Career committees:* These are an essential HRM practice to manage the learning and development of individual project managers, which in itself is critical to increasing performance in the organization.

*Overseas postings:* Moving people around the organization, through the spiral staircase career, is a way of spreading technological competence and learning throughout the organization. People take their learning with them and pick up new learning as well.

*Centers of Excellence:* Many project-based firms maintain centers of excellence for retaining learning and disseminating it throughout the company. They may maintain the company procedures and offer consultancy advice and training within the firm. The Office of Government Commerce is the UK government's Centre of Excellence in project management, maintaining the PRINCE2 process. It is also establishing satellite offices in all government departments.

*The Intranet:* Many firms are also using the Intranet to support organizational learning. However, experience is patchy, and the main risk is totally inviscid information. Yesterday's hearsay becomes entered in the system without being tested and proven, and becomes today's perceived wisdom. Some companies suggest the use of gatekeepers to monitor the entering of information, but then cannot afford the cost.

# The Four Practices and Four Steps of Innovation and Learning

Table 8.2 shows how the four practices contribute to the development and distribution of technological competence throughout the organization through variation, selection, retention, and distribution. The four practices can also be related to organizational maturity and learning in organizations.

## The Four Practices and Project Management Maturity

The purpose of innovating—of developing new technologies for project delivery, new project management processes, and engineering skills—is to improve project performance. The organization aims to get better at doing its projects, to increase its competence at project delivery. The jury is still out on what the project management competence of organizations should be called and how it should be measured. In the late 1980s, Turner labeled it *projectivity* (Turner, 1999). More recently it has been called *capability* and *maturity* (Fotis, 2002).

Cooke-Davies offers a challenging review of the application of the maturity concept to project management in his chapter later in this book but if we take the five levels of the Organizational Project Management Maturity Model, OPM3, we can see how the four practices contribute to increasing maturity, so defined.

*Level 1—Initial.* There is no guidance or consistency in the organization's approach to project management.

*Level 2—Repeatable.* The organization begins to pick off individual project management processes (scope, quality, cost, time, risk) and defines how those should be managed. It begins to write *procedures* for individual processes, the most often used, and begins to give minimum guidance to its project managers on how to use those through the embryonic *project management community*.

### TABLE 8.2. THE ROLE OF THE FOUR PRACTICES IN INNOVATION AND LEARNING.

| Learning Process | Contributing Themes |
|---|---|
| Variation | Project management communities |
| | Benchmarking |
| Selection | Benchmarking |
| | Reviews |
| Retention | Reviews |
| | Systems and procedures |
| Distribution | Systems and procedures |
| | Project management communities |

## TABLE 8.3. ORGANIZATIONAL PROJECT MANAGEMENT MATURITY MODEL, OPM3.

| No. | Level | Theme | Attainment |
|---|---|---|---|
| 1: | Initial | Procedures | Ad hoc processes |
| | | Review | |
| | | Benchmarking | |
| | | Community | No guidance, no consistency |
| 2: | Repeatable | Procedures | Individual processes for the most often used |
| | | Review | |
| | | Benchmarking | |
| | | Community | Minimum guidance |
| 3: | Defined | Procedures | Institutionalized processes across the board |
| | | Review | |
| | | Benchmarking | |
| | | Community | Group support |
| 4: | Managed | Procedures | Processes measured |
| | | Review | Experiences collected |
| | | Benchmarking | Metrics collected |
| | | Community | |
| 5: | Optimized | Procedures | Continuous improvement |
| | | Review | Defects analyzed and patched |
| | | Benchmarking | Data collected |
| | | Community | Continuous improvement |

*Level 3—Defined.* The organization begins to formalize the individual processes into a coherent, integrated *project management procedures*. It offers group support, through the *project management community*, mentoring apprentice project managers in the use of the company procedures

*Level 4—Managed.* *Review* and *benchmarking* become formalized, and the systems and *procedures* are measured as a basis for *benchmarking* and performance improvement.

*Level 5—Optimized.* The organization moves into continuous improvement. Data is collected, and defects are analyzed and patched to achieve that continuous improvement. *Procedures, reviews, benchmarking, and the project management community* are practiced to achieve *variation, selection, retention, and distribution* of new technological knowledge so the organization moves into a permanent state of innovation and performance improvement:

1. Improvement in the performance of its project management systems and procedures
2. Improvement in the performance individual projects
3. Improvement in the performance of its technological and engineering skills and in the efficiency and effectiveness of its products

## The Four Practices and Organizational Learning

To increase its project management competence, in order to better use its technology, the organization needs to learn how to better use its project management processes and engineering skills. Learning is considered formally by Bredillet and by Morris in their chapters in this book. However, we wish to show here that the four practices described previously for selecting, retaining, and distributing technological knowledge do contribute to organizational learning using a model developed by Nonaka and Takeuchi (1995), two authors whose work is described more fully in the chapters on learning by Bredillet and by Morris. Figure 8.7 illustrates the organizational learning spiral postulated by Nonaka and Takeuchi (1995). It shows an organization and the people in it learning by cycling between explicit and tacit knowledge:

> *Explicit knowledge*—Codified knowledge as reflected in the technological systems, project management procedures, and engineering standards used by the organization.
>
> *Tacit knowledge*—Inherent knowledge reflected in the combined wisdom of the project management community.

Nonaka and Takeuchi suggest that organizations move clockwise through this cycle to improve explicit and tacit knowledge and so enhance organizational learning. We can see as

## FIGURE 8.7. NONAKA AND TAKEUCHI'S LEARNING CYCLE.

| | | To | |
| --- | --- | --- | --- |
| | | Tacit knowledge | Explicit knowledge |
| From | Tacit knowledge | *Socialization* Sharing-creating tacit knowledge through experience | *Externalization* Articulating tacit knowledge through reflection |
| | Explicit knowledge | *Internalization* Learning and acquiring new tacit knowledge in practice | *Combination* Systematizing explicit knowledge and information |

an organization moves through this cycle, it follows the four-step process of variation, selection, retention, and distribution, using the four practices of standards, reviews, benchmarking, and community. This is best described starting at the selection (review) step, socialization:

*Socialization.* The *project management community* consolidates its tacit knowledge through reflection and *review*. It *selects* the tacit knowledge considered valuable for further use.

*Externalization.* Through further reflection, itarticulates that tacit knowledge and converts it into explicit knowledge. It decides what should be *retained* in its systems and *procedures*. It compares how it is doing by *benchmarking* its performance internally and externally.

*Combination.* Itsystematizes that explicit knowledge into systems and *procedures*, *retaining* it for further use. It can now be *distributed* to the organization through the *project management community*.

*Internalization.* The *project management community* can now use the explicit knowledge, and through use convert it into tacit knowledge. It can also try new ideas through a process of *variation* and thereby acquire new tacit knowledge.

Returning to socialization thus we see that the four practices suggested for selecting, retaining, and distributing technological knowledge contribute to increasing project management competence of organizations through a process of learning.

## Summary

Organizations achieve superior project performance through the effective use of the technological knowledge available to them. However, they can either try to repeat past performance or they can try to improve their performance through the development and use of new technological knowledge. This chapter looked at how organizations can do that.

First the scope of technological knowledge was widened. Technological knowledge includes engineering skills, but also includes an organization's ability to manage projects—that is, its overall competence at project management, as well as its ability to manage specific projects through its ability to identify appropriate success criteria and key performance indicators, and to identify and manage risk effectively. A four-step process was introduced for the development and use of new technological knowledge: variation, selection, retention, and distribution. It was first shown what organizations can do to encourage innovation through variation and manage innovation and development projects effectively. Four practices were then introduced for the selection, retention and distribution of technological skills in project-based organizations. These practices are as follows:

- The use of systems, procedures, and standards
- Internal and cost completion reviews on projects
- Benchmarking project performance internally and externally
- The maintenance of project management communities

It was shown how these four practices contribute to increasing project management competence and maturity by supporting organizational learning.

# References

Bernstein, P. L. 1998. *Against the gods: The remarkable story of risk.* New York: Wiley. ISBN: 0-471-29563-9.

Cooke-Davies, T. 2001. Project close-out management: More than just "good-bye" and move on. In *A project management odyssey: Proceedings of PMI Europe 2001,* ed. D. Hilson and T. M. Williams. London: Project Management Institute, UK Chapter.

Crawford, L. 2003. Assessing and developing the project management competence of individuals. In *People in project management,* ed J. R. Turner. Aldershot, UK: Gower. ISBN: 0-566-08530-5.

Egan, J. 1998. *Rethinking construction.* Construction Task Force report, London.

Fong, P. S. W. 2003. Knowledge creation in multidisciplinary project teams: An empirical study of the processes and their dynamic interrelationships. *International Journal of Project Management* 21(6). To appear Fotis, R. 2002. Maturity. *PM Network* 16(9):39–43.

Gareis, R., and M. Huemann. 2003. Project management competences in the project-oriented company. In *People in project management,* ed. J. R. Turner. Aldershot, UK: Gower.

Gibson, L. R., and S. Pfautz. 1999. Re-engineering IT project management in an R&D organization—a case study. In *Managing business by projects: Proceedings of the NORDNET Symposium,* ed K. A. Arrto, K. Kähkönen, and K. Koskinnen. Helsinki: Helsinki University of Technology.

Graham, R. G. 2000. Managing conflict, persuasion, and negotiation. In *People in project management,* ed. J. R. Turner. Aldershot, UK: Gower.

ISO. 1997. *ISO 10,006: Quality management: Guidelines to quality in project management.* Geneva: International Standards Organization.

Keegan, A. E., and J. R. Turner. 2001. Quantity versus quality in project based learning practices. *Management Learning* (special issue on project-based learning) 32(1):77–98.

———. 2003. The management of innovation in project based firms. *Long Range Planning* Larsson, U., ed. 2001. *Cultures of creativity: the Centennial Exhibition of the Nobel Prize.* Canton, MA: Science History Publications.

Latham, M. 1994. *Constructing the team: Final report of the government/industry review of procurement and contractual arrangements in the UK construction industry.* London: The Stationery Office.

Levitt, B. and J. G. March. 1995. Chester I Barnard and the intelligence of learning. In *Organization theory: From Chester Barnard to the present and beyond.* Exp. ed., Oliver E. Williamson. New York: Oxford University Press.

Lock, D. 2000. Project appraisal. In *The Gower handbook of project management.* 3rd ed, J. R. Turner and S. J. Simister. Aldershot, UK: Gower.

Miner, A., and D. Robinson. Organizational and population level learning as engines for career transition. *Journal of Organizational Behaviour* 15:345–364.

Nonaka, I., and H. Takeuchi. 1995. *The knowledge-creating company.* New York: Oxford University Press.

OGC. 2002. *Managing successful projects with PRINCE2.* 3rd ed. London: The Stationery Office.

Pinto, J. K. 1999. Managing information systems projects: regaining control of a runaway train. In *Managing business by projects: Proceedings of the NORDNET Symposium,* ed. K. A. Arrto, K. Kähkönen, and K. Koskinnen. Helsinki: Helsinki University of Technology.

Project Management Institute. 2000. *A guide to the Project Management Body of Knowledge.* Newtown Square, PA: Project Management Institute.

Scott, R., ed. 2001. *Partnering in Europe: Incentive based alliancing for projects.* London: Thomas Telford.

Turner, J. R., ed. 1995. *The commercial project manager.* London: McGraw-Hill.

Turner, J. R. 1999. *The handbook of project-based management.* 2nd ed. London: McGraw-Hill.

———. 2002. Configuration management. In *Project management pathways,* ed. M. Stevens. High Wycombe, UK: Association for Project Management.

———. 2003. Farsighted project contract management. *In Contracting for project management,* ed. J. R. Turner. Aldershot, UK: Gower.

Turner, J. R., A. E. Keegan, and L. Crawford. 2003. Delivering improved project management maturity through experiential learning. In *People in project management.* ed. J. R. Turner. Aldershot, UK: Gower.

Wheelwright, S. C., and K. B. Clarke. 1992. *Revolutionizing new product development.* New York: Free Press.

# CHAPTER NINE

# INTEGRATED LOGISTIC SUPPORT AND ALL THAT: A REVIEW OF THROUGH-LIFE PROJECT MANAGEMENT

David Kirkpatrick, Steve McInally, Daniela Pridie-Sale

Traditionally, project management has been associated with the activities of an organization creating new products and has therefore focused on the early phases of a project—from concept through design and development to production, up to the point of sale to a customer, who is most generally an end user. Relatively little attention has been given to later phases in the project's life, perhaps because the operation and support of a product in service require different skills from those used earlier in its design and production, and perhaps because the sale is seen to mark a significant transfer of responsibility from the supplier to the customer.

During the early phases of a project, a variety of pressures on the project manager tend to encourage a short-term approach—seeking to solve immediate problems without due regard for the consequences that solution will impose on later phases. For example, the development of the Tornado attack aircraft was truncated because of short-term budgetary constraints; consequently, the aircraft in service initially provided an unduly low level of availability for active duty and required many expensive design changes to rectify problems that should have been resolved in development (UK Parliamentary Select Committee on Defence, 2000).

In the private sector, a short-term approach is promoted by the need to maintain the organization's profitability and its share price, in order to satisfy shareholders' expectations. In the public sector, politicians and their officials face a chronic shortage of resources immediately available to meet limitless demands for public services, so they may be tempted by a policy that matches supply and demand in the short term but that might create problems some years ahead. In both sectors project decisions should ideally be guided by a process of investment appraisal that takes account of all the resulting costs and benefits through the life of a project, but in practice managers often pay more attention to immediate problems and neglect through-life issues.

An emphasis on the early phases of a project may be justified in those cases where the transfer of a product from supplier to customer is a purely financial transaction, involving virtually no exchange of information, or where the product's (short) life after its transfer to the end user absorbs only insignificant resources. However, an emphasis on the early phases is quite inappropriate in those cases where the costs of the later phases (operations, support, and disposal) constitute the larger fraction of the project's through-life costs. This latter category includes many defense equipment projects, so the UK Ministry of Defence (for example) has repeatedly exhorted its project managers to adopt a through-life approach (Ministry of Defence, 1998).

Furthermore, a good project manager should be aware that unsatisfactory performance of a product in service could damage the organization's reputation and its future sales; inadequate performance could even subject it to crippling litigation, and the manager to prosecution, if the product adversely affects the health and welfare of customers. In many countries an increasing body of legislation insists that a product being used for its designed purpose should not damage the environment and that it can later be safely recycled. For instance, in response to social and legislative pressures carpet fiber manufacturers DuPont Antron have developed a carpet reclamation initiative as part of a life cycle management methodology (DuPont Antron, 2002).

Thus, today's project managers must address all the phases in a project's life in an integrated manner, to ensure that all phases meet their targets of performance, timescale, and cost.

## Life Cycle Phases of a Project

The term project can be applied in at least two ways depending on one's point of view. For simple products such as pencils and personal computers, the term project would usually refer to the creation process in the early part of the life cycle that brings about the new product, system, or equipment, as in a *design project*. This point of view is reflected in the definition that the *Oxford English Dictionary* provides for a project (*Oxford English Dictionary*, 2000):

A co-operative enterprise, often with a social or scientific purpose.

For more complex and costly products and systems such as spacecraft, hospitals, and aircraft carriers, the term project is generally synonymous with the whole life cycle of those products and systems. Given that this chapter is primarily concerned with more complex and costly products, the second interpretation that a *project* is concerned with all life phases applies in this chapter.

Different industries use different nomenclatures to describe various life phases. Figure 9.1 describes the life phases from three different perspectives.

## FIGURE 9.1. COMPARISON OF LIFE CYCLES FOR DIFFERENT PRODUCTS AND SYSTEMS.

NASA

| Mission Feasibility | Mission definition | System definition | Preliminary Design | Final design | Fabrication and Integration | Prepare for deployment | Deployment and Operations verification | Mission Operations | Disposal |
|---|---|---|---|---|---|---|---|---|---|

CADMID

| Concept | Assessment | Demonstration | Manufacture | In -service | Disposal |
|---|---|---|---|---|---|

BS7000

| Trigger, Product Planning, and Feasibility | Development and Production | Installation, Commissioning, Operation, and Use | Disposal and Recycle |
|---|---|---|---|

- The National Aeronautics and Space Administration (NASA) life cycle model (Shishko, 1995) reflects the complex nature of space flight projects. NASA retains responsibility and ownership throughout a spacecraft's life, so all life phases, from Mission Feasibility through to Disposal, are included.
- The CADMID cycle adopted by the United Kingdom's Ministry of Defence (MoD) is applied to military equipment in all shapes and sizes. Like NASA, but unlike manufacturers of simpler products, UK MoD retains responsibility throughout the equipment's life; hence, the whole life cycle is viewed as a project.
- The British Standard BS7000- 1:1999 Guide to Managing Innovation (BSI, 1999) life cycle is simpler and is intended primarily to provide guidance for the development of mass-produced products.

For the purposes of illustrating ILS, this chapter will follow the MoD's CADMID designation.

It should be noted that in the MoD the start of the Assessment and Demonstration phases of the CADMID cycle must be formally authorized by the allocation of appropriate funding, provided that the preceding phases have produced satisfactory results. Other organizations adopt less formal procedures and may allow some overlap in the timescale of the project's phases.

## Responsibility for Project Phases: Civil Sectors

In the civil sector, the supplier and the customer take or share responsibility for the various phases of a project. The supplier would typically be responsible for project phases up to the point of sale—for instance, through Product Planning & Feasibility and Development & Production phases of BS 7000, with the customer taking ownership after the Point of Sale (BS 7000 Operation and Disposal phases). However, the allocation of responsibility can be different in different industries, as demonstrated in the following examples.

In many consumer goods industries, the early Concept, Assessment, Demonstration, and Manufacture phases are the exclusive responsibilities of the supplier, guided by market surveys and by related insights on latent customer demand. The supplier then transfers ownership to the user and, at the point of sale (often by a retailer or agent), provides only simple, if any, instructions. The later In-Service and Disposal phases are the exclusive responsibility (and sometimes irresponsibility) of the customer.

The suppliers of expensive, durable products recognize that their reputations, and their hope for future business, depend on the continued acceptable performance of those products; they therefore provide an extensive set of detailed instructions and a guarantee to repair or replace the product if it fails through ordinary use within a specific period. In the automobile industry, for example, the suppliers often seek to retain responsibility for repair and maintenance activities in the in-service phase to ensure, as far as possible, that their cars remain safe and reliable and their owners remain satisfied. Similarly, organizations supplying capital equipment to commercial customers often undertake a contractual responsibility to provide repair and maintenance through the equipment's service life.

In the civil engineering and building sectors, a customer organization may take an active role alongside one or more suppliers in the early Concept and Assessment phases, assign total responsibility for the Demonstration and Manufacturing/Construction to a chosen supplier under an agreed contract, and later take full responsibility for the project's operation and support. In some cases, however, the supplier may retain responsibility for rectifying problems that arise within an agreed period.

Under the UK government's policy for the provision of public services and infrastructure under the Private Finance Initiative (PFI; see the chapter by Ive), it is now usual for a prime contractor to undertake the construction of a school or a hospital, and later to undertake its operation and support to deliver over an agreed period an agreed volume of educational or medical services. Similarly, under PFI, the contractor responsible for building a motorway may assume responsibility for its repair and maintenance to a satisfactory standard for an agreed period, as well as for its design and construction. In these and similar cases, the customer is involved in the Concept phase, which captures the requirement and considers alternative options; thereafter the customer adopts a detached supervisory role, providing the funds agreed and monitoring the supplier's performance via an appointed regulator.

## Responsibility for Project Phases: Defense Sector

In the special case of the defense sector, the Armed Forces have a unique and exclusive knowledge of the realities of military operations and of any developing shortfalls in the capabilities of their current equipment. Accordingly, they take a leading role in the Concept phase of a new defense equipment project, although potential suppliers may, in this phase, offer advice on alternative options that might provide the required increment in capability. As the project passes through the Assessment, Demonstration, and Manufacturing phases, the Armed Forces and the relevant MoD branches and agencies play a less active role, and the chosen prime contractor takes a progressively greater share of responsibility. As the new equipment enters service, the Armed Forces take full responsibility for the operation of front-line equipment but may assign to contractors the operation of equipment located in "benign" rear areas. Responsibility for in-service support may be allocated according to circumstances; in a nuclear submarine on extended patrol all repair and maintenance during that period must be done by the crew, but a squadron of aircraft operating near the contractors' facilities can easily draw on their expertise and stocks of spares. It follows that in some cases suppliers are involved in the day-to-day support of their equipment, but in other cases they are involved only in any major refit or refurbishment.

In former times it was customary for the Armed Forces to maintain and repair their own equipment whenever practicable. This activity gave Service personnel a greater knowledge of the equipment's strengths and weaknesses and a greater ability to repair battle damage or to improvise modifications in a crisis. Today, by contrast, support arrangements vary widely between different nations and between different classes of equipment. In the United Kingdom it is perceived to be more cost-effective to rely on contractors to provide wherever practicable the in-service support for peacetime training and for expeditionary operations, except in the front line. While on some projects the contract for support may be negotiated separately from the contract for procurement, it is increasingly common for the equipment supplier to be given a portmanteau contract covering design, production, and support within an agreed payment schedule.

In the Disposal phase, the Armed Forces have full responsibility of ensuring either that sale of the equipment to a third party does not breach national arms control policies or that the equipment is safely destroyed according to current environmental legislation.

## The Need for Through-Life Management

These examples indicate that there are many sectors of industry where project managers must adopt a through-life approach, taking full account of their product's durability, reliability, maintainability, and repairability in the In-Service phase of its life cycle and of the need for safe and efficient Disposal.

# Integrated Logistic Support (ILS)

## The Need for ILS

Modern military equipment requires many inputs to keep it operational. It may need regular supplies of fuel and ordnance. It undoubtedly needs regular attention from skilled artificers to undertake scheduled maintenance and unscheduled repair, and these activities will require technical documentation describing the equipment and its potential faults in useful detail. Specialist tools and test equipment, spare components or assemblies to replace those found to be damaged or faulty, and a logistic chain designed to provide supplies when they are needed (or at worst soon afterwards) are also required. In many cases the cost of operating and supporting equipment through its service life equals or exceeds the cost of its procurement.

Furthermore, a modern expeditionary force (and its associated equipment) needs a large and consistent inflow of supplies to maintain its effectiveness in a remote theater of operations. For example, in the Gulf War of 1991, the UK armoured division required some 2000 tons of supplies per day before the period of active operations, and triple that volume during the land campaign (White, 1995). The provision of this large quantity of supplies demands considerable planning and resources, and hence must be organized efficiently.

## What Is ILS?

Usually applied to defense systems, ILS is a disciplined, structured, and iterative approach to ensure that all the inputs required by each item of defense equipment are provided where and when they are required and that the cost of providing them is minimized. During the life cycle of a defense equipment project, the principal aims of the ILS process are to

- analyze the through-life requirement for logistical support;
- formulate plans to provide sufficient support resource;
- influence the equipment design; and
- deliver the support resources when required.

The U.S. Department of Defense provides a useful and succinct definition for ILS (U.S Department of Defense, 1983):

> ILS is a structured management approach aimed at influencing the design of the asset and ensuring that all the elements of design are fully integrated to meet the client's requirements and asset's operational and performance, including availability, reliability, durability, maintainability, and safety at minimum whole life cost.

One of the most important features of ILS practice is the notion that it should be closely integrated with procurement and development cycles.

> The basic management principle of the ILS process is that logistic support resources must be developed, acquired, tested, and deployed as an integral part of the materiel acquisition process."
>
> From the *US DoD Integrated Support Manager's Guide* (U.S. Army, 1998)

Throughout that process the overall objective of the ILS is to maximize the cost-effectiveness of the equipment by striking a balance between its logistic requirements for resupply, its reliability and maintainability, the scale and the organization of the support resources, and the equipment's life cycle cost (LCC). It must also satisfy the dual objectives of being economical in the peacetime environment, which is familiar and well understood, and also being effective in the strange and demanding environment of conflict.

ILS was originally developed and applied in the USA in the early 1980s with the introduction of US DoD directive 5000.39 (U.S. Department of Defense, 1983) and became a compulsory part of all UK MoD projects since the early 1990s, as embodied in Def. Stan 00-60 (UK Ministry of Defence, 2002). The MoD ILS process is defined as:

> Integrated Logistic Support (ILS) provides the disciplines for ensuring that supportability and cost factors are identified and considered during the design stage of an equipment so that they may influence the design, with the aim of optimizing the Whole Life Cost (WLC)."
>
> From the UK MoD ILS Guide (UK Ministry of Defence, 2001).

## ILS in the CADMID cycle

Ideally the ILS process should be used initially as part of the Concept phase and should then be progressively updated during the later phases of the CADMID cycle, incorporating additional project data as it becomes available.

Figure 9.2 illustrates (in a general manner) the relationship between the project life cycle (CADMID) and the support system life cycle (within ILS). At the beginning of CADMID,

**FIGURE 9.2. PHASES IN THE ILS LIFE CYCLE SIMULTANEOUSLY EVOLVE WITH THE CADMID LIFE CYCLE.**

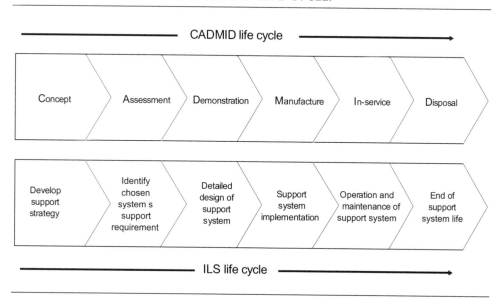

there may be some general information available for the logistic requirements of systems "of this type," but there is little or no validated data for this specific system. As the CADMID cycle progresses, the ratio of vague/specific information decreases. Designers and procurers start to generate meaningful system design and performance data, and that data is communicated to those responsible for ILS analyses, specific guidance as to logistical requirements and constraints is fed back to system designers, and so on.

The interactive relationship between ILS and acquisition activities during the CADMID cycle, as illustrated in Figure 9.3, is fundamentally a learning process.

The majority of the expenditure in a project's life cycle is, by implication, committed by early design decisions, probably before the end of the assessment phase. Because the cost of in-service support of defense equipment is very high, sometimes higher than the cost of its procurement, this in-service stage has to be managed in a disciplined way and using the appropriate tools. It is therefore important that these early decisions are appropriately influenced by the results of ILS studies early in the project. In principle the early application of ILS can have a major influence on the project's through-life management plan, but in practice the ILS analysis is often constrained by predetermined (upper and/or lower) limits on the number of units to be deployed, or on the number of service personnel to be involved, or on the number of planned operating bases. Even constrained ILS studies can, however, favorably influence the initial design of equipment, provided that the study results are both timely and robust, and thus can significantly reduce the project life cycle cost.

**FIGURE 9.3. AT EACH STAGE OF CADMID, THE VOLUME AND QUALITY OF ACQUISITION/DESIGN DATA INCREASES, ILS REQUIREMENTS BECOME CLEARER.**

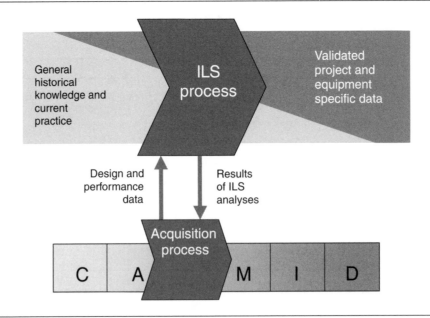

## Components of ILS

ILS is a *framework* of tools and techniques; a *method* for prescribing the use of those tools and techniques; and in execution, a *process* whereby tools and techniques are systematically applied to a particular equipment life cycle. ILS starts from a proposed equipment design and a proposed support arrangement for the planned equipment and then uses a process of modeling and prediction to generate forecasts of equipment availability and life cycle costs. This provides a foundation for the comparative assessment of alternative design features and alternative support arrangements to identify the most cost-effective combination from a through-life perspective.

The ILS framework incorporates three principal activities (see also Figure 9.4):

- Logistic support analysis (LSA)
- Creation of technical documentation (TD)
- Formulating integrated supply support procedures (ISSP)

**Logistic Support Analysis.** The purpose of logistic support analysis is to identify the repair and maintenance tasks likely to be involved in the support of a new project and to plan how those tasks can most efficiently be accomplished. The results of this analysis can identify costs drivers in the proposed design and can stimulate trade-offs in which the design is refined to reduce its support costs without unacceptable penalties on performance, timescale, or procurement costs.

The LSA includes several discrete but integrated activities:

*Failure modes effects and criticality analysis (FMECA)* for each component in the proposed design determines how it might fail and the consequences of each failure for the equipment's safety and military capability. The FMECA results can guide decisions on component quality standards, duplication, and preventative maintenance.

*Reliability-centered maintenance (RCM)* considers alternative policies on inspection, preventative maintenance, and repair to establish the most cost-effective approach. Alternative policies include repairing or replacing items when they fail (with an appropriate level of servicing designed to delay failure); repairing or replacing items when electronic, visual, or other types of inspection reveal damage or deterioration approaching critical levels; and repairing or replacing items on a planned schedule linked to their durability (obtained from calculation or experiment) in order to avoid untimely failures. The optimal policy in each case depends on ease of inspection, the cost of preventative maintenance or repair or replacement, and the consequences of failure.

*Maintenance Task Analysis* considers the timescale and the resources of personnel and equipment required for each of the potential tasks. The personnel may require particular knowledge and skills and the equipment may include specialist tools and test facilities.

*Level of Repair Analysis (LORA)* is the process of determining the most efficient maintenance level for repairing items of equipment. Military organizations often have four levels of repair, with the *first line* in an operational unit, *second line* in a higher-level formation, *third line* in a base workshop, and *fourth line* at the contractor's factory. The

## FIGURE 9.4. ILS INCORPORATING ITS THREE PRINCIPAL ACTIVITIES.

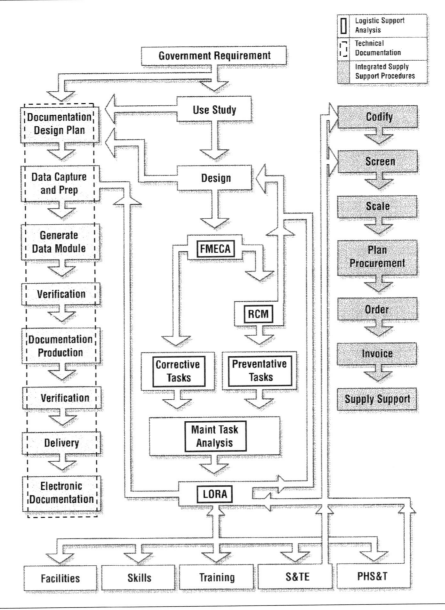

*Source:* Def Stan. 00-60.

number of levels and the arrangements within them varies between service environments and for different types of equipment. The LORA must balance the delay and resources required to transfer faulty equipment between the different levels of maintenance against the cost and risk involved in having skilled personnel, specialized test equipment, and spares holdings available in or close to operational units. The results of the LORA determine which types of spares should be held at each level of repair.

LSA coordinates these activities through five sets of tasks:

1. *Program planning and control* establishes the scale and scope of the analytical tasks and the procedures for ongoing management and review.
2. *Mission and support systems definition* considers how the equipment is to be operated and supported, and thus identifies design changes that would yield significant reductions in its support costs.
3. *Preparation and evaluation of alternatives* assesses detailed design trade-offs to determine the options yielding best value for money.
4. *Determination of logistic support requirements* quantifies the resources needed to support the equipment through its In-Service phase.
5. *Supportability assessment* reviews the effectiveness of the LSA and the lessons to be learned from it.

The data resulting from the LSA activities is assembled in a structured *logistic support analysis record (LSAR)*, which can easily be used by the various government and commercial organizations involved in the project.

**Technical Documentation (TD).** Technical documentation contains all the information necessary to operate, service, repair, and support an equipment project through its service life and to dispose of it afterward. This information includes data on

- system description and operation
- illustrated parts data
- system servicing, maintenance, and repair
- diagnostic support equipment, and so on

and may be held as text or drawings on paper, fiche, text or drawings in electronic format, video, and data to support computer-aided design (CAD). For modern projects, electronic technical documentation (ETD), in the format established by Def. Stan. 00-60 part 10, is generally the most cost-effective option.

**Integrated Supply Support Procedures (ISSP).** ISSP cover the procurement of new spares, the repair and overhaul of defective items, and the administration of these processes. Given the multitude of spares of many types required by one equipment project, and of the overlapping sets of spares required by concurrent projects, it is vitally important that various service units and supporting contractors benefit from early, rapid, and unambiguous exchange of data, using electronic documentation.

The ISSP include the following:

- Codification that assigns to each item used by the Armed Services a unique identifier, using the NATO codification system
- Initial provisioning to provide adequate spares to support an initial period of operations (nominally two years in the UK) within which experience of reliability and maintainability yield definitive data
- Reprovisioning analysis that determines how many spares of each type should be held at each level of maintenance, when an order for replacement should be placed, and the economic size of the order
- Repair and overhaul plans that define how defective items, which cannot economically be replaced, can be restored to serviceability
- Procurement procedures that define how orders and invoices (ideally electronic) will be administered

## Tailoring ILS

Although off-the-shelf procedures are widely used, every project is a unique enterprise, and therefore the ILS process should be tailored according to the realities of each and every project and program. Tailoring establishes which of the tasks and subtasks must be performed, when, and to what depth. A skillfully tailored ILS process can produce more saving than the use of off-the-shelf procedures, which are sometimes preferred by less experienced project managers because of their lower cost and their convenience.

All projects and programs have to accomplish certain core activities according to standards and regulations. When contracting for the ILS, the U.S. Army recommends that

the ILS requirements will be tailored according to the acquisition strategy and included in the solicitation documents. The contractor will be required to define his approach to meeting the stated ILS requirements in the proposal developed in response to the solicitation.

U.S. Department of the Army, 1999

## Management of ILS

Within the ILS activities reviewed previously, it is necessary to take account of the following:

- Provision and upkeep of support and test equipment (S&TE)
- Test and evaluation (T&E) facilities
- Personnel and human factors
- Computing and IT resources
- Training and training equipment
- In-service monitoring
- Packaging, handling, storage, and transportation
- Safe and economical disposal

As described in the previous *Responsibility for Project Phases* sections, some of these tasks will be the sole responsibility of the ILS contractor; others will have to be carried out in conjunction with the customer or/and with other contractors.

The role of the ILS management process is to facilitate the development and integration of these elements. It is vital that ILS is integrated into the overall system development process in order to ensure the best balance between a system design, its operation, and its related support. The development of the ILS elements must be done in coordination with the system engineering process and with each other. When you are trying to achieve a system that fulfils all the desired criteria—performance, affordable, operable, supportable, sustainable, transportable, and environmentally sound—within the resources available, it is often necessary to have trade-offs between all these elements.

The ILS management process requires a demanding and rigorous approach to the development of a through-life management plan, requiring close attention to forecasts of the cost and duration of the successive phases of the project life cycle, and appropriate trade-offs of overall performance, cost, and timescale. The through-life management plan and detailed cost forecasts provide a good basis for a disciplined monitoring of the actual progress of the project. Furthermore, the logistic support analysis process supports a more precise forecast and assessment of the design costs and of the effect on costs of the changes that occur during the life of a project.

# "ILS" in the Civil Sectors

Processes similar to ILS are used in the civil sectors of industry alongside a variety of techniques and methodologies that focus on identifying, analyzing, and optimizing with reference to issues that may emerge during the life of a product or system; for instance, systems engineering (SE), concurrent engineering (CE) and integrated product and process development (IPPD) are all through-life approaches. All these approaches, as well as certain proprietary life cycle management methodologies, have been successfully applied in the medical, automotive, nuclear, construction, and manufacturing industries for many years. All focus on analyzing and planning for a whole life cycle, and even for subsequent life cycles of replacement products. The motivation for developing and applying whole-life cost analysis and design techniques is born of a number of economic, environmental, and legislative factors.

## Systems Engineering and ILS

The discipline of systems engineering was developed in response to the problems of managing complexity and reducing risk of failure in the design of large-scale, technology-driven systems such as information system, civil engineering, and aerospace development projects (see the chapters by Davis et al. on requirements management, Harpum on design management, and Mooz on verification). Systems engineering in its broadest interpretation includes a variety of concepts, models, techniques, and methods, including many or all of the concepts found in concurrent engineering, project management, integrated product and process development, as well as ILS (see the chapters by Thamhain, Cooper et al., and others). The central body for systems engineering, the International Council On Systems Engineering (INCOSE), defines a through-life approach as key to systems engineering as (INCOSE, 1999):

> Systems Engineering is an interdisciplinary approach and means to enable the realization of successful systems. It focuses on defining customer needs and required functionality early in the development cycle, documenting requirements, then proceeding with design synthesis and system validation while considering the complete problem: Operations; Performance; Test; Manufacturing; Cost & Schedule; Training & Support; Disposal. Systems Engineering integrates all the disciplines and specialty groups into a team effort forming a structured development process that proceeds from concept to production to operation. Systems Engineering considers both the business and the technical needs of all customers with the goal of providing a quality product that meets the user needs.

Although SE and ILS are two different concepts, in practice they are in some ways interdependent. SE is concerned with designing systems specifically with the emerging through-life considerations in mind, for instance, design for supportability. In his paper discussing the relationship between SE and ILS in the design of military aircraft, Strandberg describes logistic support analysis as the activity that bridges both ILS and SE (Bergen, 2000).

The concepts of SE and ILS are further integrated within international standard ISO 15288 Life Cycle Management—System Life Cycle Processes (International Organization for Standardization, 2002). The standard is intended to offer guidance for acquiring and supplying hardware, software systems, and services, but it also claims to offer a framework for the assessment and improvement of the project life cycle.

Although SE is concerned primarily with exploring and solving complex technical problems, it has a complementary relationship with project management. As Hambleton (Hambleton, 2000) puts it:

> You can't engineer a complex system without managing it properly and you can't manage a complex system without understanding its engineering. Systems Engineering and Project Management are two sides of the same coin . . ."

SE and ILS are different disciplines within project management. Both provide methodologies for the management of complexity to achieve specific organizational goals (such as optimal cost-effectiveness). SE is an overarching discipline integrating several other project management activities (such as requirement capture and equipment design) as well as ILS. Some of those other activities apply rigorous engineering methodologies, but ILS retains a more pragmatic approach with the methods and techniques applied being adapted to the project circumstances.

## PPP, PFI, DBFO, and ILS

Public Private Partnership (PPP), Private Finance Initiative (PFI), and Design-Build-Finance-Operate (DBFO) initiatives have been key factors in the growth of interest in ILS and ILS-like methods The PPP initiative was introduced in the 1980s, the PFI launched in 1992, and more recently the DBFO initiative were all intended to bring the skills and resources of the public and private sectors together to improve the success of large-scale projects. (See the chapters by Ive and by Turner.) The shift toward a single contractor being responsible for the whole life of project has emphasized the need for contractors to adopt tools, methods, and techniques like ILS to reduce risk and cost in large projects.

For instance, the UK government has established a number of risk-sharing initiatives such as Contractor Logistic Support (CLS), financed by Private Finance Initiative (PFI) or Public Private Partnership (PPP) arrangements that will allow risk sharing for potentially expensive support services, which have traditionally been provided by the government. In order for nongovernmental organizations to provide these services, they need to understand fully how and why a system fails, what are the impacts of each failure, and what maintenance and resources would be required to carry out repairs. Under CLS initiatives, the UK government will no longer pay industry to perform the ILS activity of logistic support analyses and then use their own resources to carry out the work; rather, the industrial contractor will bid for the whole task at the Invitation to Tender (ITT) stage of a project.

According to the UK Confederation of British Industries (CBI), PPP, PFI, and DBFO initiatives have been a great success (Confederation of British Industries, 2002):

> Public Private Partnerships are a crucial element of delivering the government's commitments on improving public services. There is a vast range of PPP models and activities. Private Finance Initiative projects, for example, deliver public sector "capital and service package solutions", e.g. PFI prison service contracts where the private sector designs, builds and operates the prison for, say, 25 years. Over 400 PFI contracts had been signed to date. Investment in public services through the PFI is expected to increase from £1.5 billion to at least £3.5 billion by the end of the current spending round in 2003/4. The range of savings identified is considerable, ranging from less than 5% to over 20%.

In a report examining the value for money for the PFI-financed redevelopment of the West Middlesex Hospital, the National Audit Office (NAO) noted that

> the Trust considered that the unquantifiable benefits of doing this as a PFI deal outweighed the disbenefits (NAO,2001).

However, in a recent report from the UK Audit Commission, its chairman James Strachan indicated (BBC News, 2003)

> schools built by the Private Finance Initiative are "significantly worse" in terms of space, heating and lighting than new publicly-funded schools. . . . The early PFI schools have not been built cheaper, better, or quicker and learning from this early experience is critical.

Given that the application of PFI and PPP for many projects is still in the early stages, the benefits and implications of this type of financing are not yet fully understood. In June 2002 the Audit Commission for Scotland reported on the use of PFI contracts to finance the renewal of 12 schools projects in Scotland (Audit Scotland, 2000). It commented that

> we are at an early stage in the 25–30 year life-span of PFIs, so it is too early to judge their contribution to education. . . . it was not possible to draw overall conclusions on value for money as it is difficult to quantify the benefits associated with PFI. The Report notes that it is important to the whole integrity of the PFI process that councils as clients hold the providers to their contractual commitments.

The challenge for the project manager is that PPP financing significantly increases the scale and complexity of the management task. With PPP, PFI, and particularly with DBFO projects, the scope of the project cycle may well extend beyond the traditional handover point to many years into the future. This implies a need for "ILS" activity to support through-life management. Though it is unlikely that the original project manager would continue to be responsible throughout the complete life of the system being designed, the structure and processes of management must always address the whole life of the project and its associated costs, particularly at the early phases.

## ILS in civil construction projects

In a study of the application of ILS techniques applied in the construction industry (El-Haram et al., 2001), researchers at the University of Dundee's Construction Management Research Unit noted a number of issues:

- PFI was a key motivator in adopting and applying ILS techniques.
- ILS needs to be broadly and thoroughly applied early in the development cycle in order to maximize benefits.
- In the absence of formal guidance as to the order and circumstances that the various ILS techniques and procedures should apply, participant organizations interpreted and adapted ILS to their own specific needs.
- Approximately one-third of the data used in ILS analyses was based on engineering intuition rather than recorded data.
- As ILS was relatively new to the organizations and their particular industry sector, co-ordination between stakeholders (designers, facility managers, manufacturers, and so on) was poor.

The study is part of the Construction Management Research Unit's ongoing research efforts, particularly in developing a framework for capturing and analyzing whole-life data for constructed facilities, and in developing guidance for which ILS techniques will be appropriate used in differing construction projects.

In recent years, the U.S. Department of Energy (DOE) has applied life cycle analysis methodologies to the nuclear industry in response to financial, social, and environmental pressures. In a recent report on the costs of managing nuclear waste, the DOE estimates that the total costs of radioactive waste management will be in excess of $49 billions (US

Department of Energy, 2001). In response to this, the DOE has published its own life cycle cost savings analysis methodology to assist the deployment of new technologies in the nuclear industry (U.S. Department of Energy, 1998), as part of DOE Order 430.1 (U.S. Department of Energy, 1998). Similarly, the Australian Federal Highway Administration (FHWA) has developed a life cycle cost analysis approach to support the choice of materials and design of major highway projects (Hicks and Epps, 2003).

## Continuous Acquisition and Life Cycle Support (CALS) and ILS

The CALS acronym has come to take on various meanings since the term was first coined, for instance:

- Computer-Aided Logistic Support
- Computer-Aided Acquisition and Logistic Support
- Continuous Acquisition and Life Cycle Support
- Commerce at Light Speed

In general, though, all refer to the same fundamental objective: to acquire, store, manage, and distribute design data electronically. CALS is effectively the means by which ILS is implemented on acquisition and design projects.

CALS began life in the 1980s as a U.S. Department of Defense (DoD) initiative. The basic idea was that technical data should be exchanged between government and its contractors in electronic format rather than on paper; as the DoD puts it "a core strategy to share integrated digital product data for setting standards to achieve efficiencies in business and operational mission areas" (Taft, 1985).

In the United Kingdom the initiative was adopted by the Ministry of Defence, which, in 1990, developed its own strategy to implement CALS called CIRPLS (Computer Integration of Requirements, Procurement and Logistic Support), and in 1995 the use of CALS technologies became a common and obligatory strategy for organizations and governments in NATO member countries.

A key concept for CALS is "create data once, use many times." This idea was made feasible by the growth of computerized information networks with the subsequent increased connectivity between enterprises. The problem was that potentially useful technical data was being held in many locations on different systems in different organizations.

The aim of CALS was to allow any authorized individual, from any stake-holding organization, to access the body of data which grows and matures as a project develops. This would have the benefits of

- increasing the rate at which information was exchanged;
- reducing information management overhead costs; and
- allowing information to be reused through all stages of a product's life cycle.

The concept of sharing applied both to individuals and to collaborating organizations. Within a single organization, design engineers, manufacturing staff, and product support staff all need to share design and logistics data right from an early stage in the project, so a strategy that improved information sharing could lead to important gains, particularly in the reduction of product development and manufacturing costs, and in reduced lead times. Additionally, information shared between different organizations in partnering-style relationships reduces the burden of information systems development, populating, and maintenance.

According to the UK MoD's National Codification Bureau (NCB), the body responsible for ILS, CALS, and similar initiatives (Clarke, 2003), CALS and CALS-like strategies are being applied by many companies around the world in a variety of industries, from consumer goods to aircraft, petrochemical plants to building and maintaining a road network.

Since its debut, CALS has continued to evolve in response to political, industrial, and technological changes. According to the U.S. Department of Defense, the term CALS is starting to disappear (U.S. Department of Defense, 2003), not because of any inherent flaw in CALS, but rather by its success. The original concept of information sharing during the system acquisition and design process is evolving into strategies such as the Integrated Digital Environment (IDE), Interactive Electronic Technical Manuals (IETM), and a Common Operating Environment (COE).

The *Integrated Digital Environment* initiative is CALS-like in that it focuses on information sharing, particularly at the enterprise level, and at early project phases. The initiative aims to overcome the barriers to efficient communication caused by program-unique information environments. The aim is to create seamless collaborative digital business environments shared by stakeholders, allowing the right information to be acquired at the right time and leading to fewer formal reviews and the improved quality of analyses. The benefits are improved general visibility throughout the supply chain, online access to technical information, reduced need for a information management infrastructure investment, and reduced cycle time.

The concept of *Interactive Electronic Technical Manuals* has been evolving since the 1970s. The idea is straightforward enough. Shared electronic media replaces technical documentation such as books and manuals with the inherent problems of storage, distribution, and version management.

Technicians and managers are able to consult centrally stored electronic reference information, use that information, and provide immediate feedback if any amendments or updates are required. IETM also provides the opportunity for those who apply the information, and who are also experts on the documented procedures and methods to author new and additional procedures and methods.

The hope is that maintenance tasks can be accomplished quicker with fewer errors, with no opportunity to "lose" pages.

The concept of a *Common Operating Environment*, developed in the early 1990s, is that the various stakeholders involved in procurement and design processes benefit from economies of scale in the development of databases and communication system. The idea is that program cost and risk can be reduced by reusing proven solutions and by sharing common functionality. The benefits of COE are improvements in development times, technical obsolescence, training requirements, and life cycle costs.

The relationship between ILS and CALS is becoming ever more integrated, for instance, as seen in the NATO initiative, as described in detail in the NATO CALS Handbook

(NATO, 2000). The NATO CALS initiative funded by 11 of the 19 NATO member nations was formed in order to improve NATO's ability to exploit information and communications technology, in the acquisition and life cycle support of complex weapons systems. A key follow-on activity to this initiative will be to develop a new international standard based on ISO 10303 for industrial automation systems and integration—Product Data Representation and Exchange (International Organization for Standardization, 1994), also known as the Standard for the Exchange of Product Model Data (STEP), to cover in-service and disposal phases of a system's life cycle. STEP ensures that the information produced in digital form can be read by others, is not hardware- or software-dependent, and has a life cycle dependent on its value.

The continuing trend toward larger and more complex projects involving many organizations in many countries, and the concurrent complementary introduction of CALS and associated strategies and standards, has significant implications for project management in organizations large and small. Project management of large, technically advanced systems is becoming more complex, and project managers will have to access, use, and contribute to information in external as well as internal databases, and to manage the interfaces involved. The manager of projects involving many enterprises must operate a complex information interchange, in which problems may be compounded by language and cultural differences between the participants if the project spans several nations. Project managers in such situations must therefore learn to exploit initiatives like NATO CALS and to operate within standards like ISO 10303.

## Medical systems Life Cycle Management

Medical equipment manufacturers face a number of commercial, technical, legal, and ethical challenges that force them to analyze and plan for a variety issues to emerge during the life of a product. Radiotherapy oncology systems and magnetic resonance imaging systems in particular have long in-service lives; are highly complex; require specialized technical staff to install, commission, operate, maintain, and dispose of; and are expensive to purchase and own. A single radiotherapy or magnetic resonance imaging suite costs a hospital around $10 million to install. Medical systems of this kind typically have design lives of 10 to 15 years and are often in service for 20 or more years. Although medical system manufacturers still compete largely on purchase price, there are increasing pressures from customers and purchasing authorities to identify and minimize costs of ownership.

In response to these pressures, companies such as Philips Electronics have developed and now apply a range of techniques in early project phases to optimize the design for many factors. The phases of Philips' proprietary life cycle model, the Product Creation Process (Sparidens, 2000), is illustrated in Figure 9.5.

The model applies to a wide range of product types from medical systems to manufacturing systems and consumer goods. Depending on the type of product, market and legislative pressures, a number of life cycle analyses and optimization strategies analogous to ILS will be applied, for instance, design for cost, usability, patient and operator safety, serviceability, environmental friendliness, and disposability.

## FIGURE 9.5. KEY PHASES IN THE LIFE OF A PHILIPS PRODUCT.

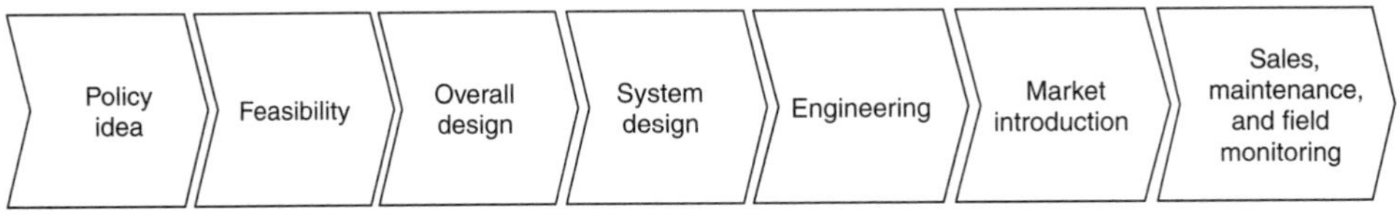

| Policy idea | Feasibility | Overall design | System design | Engineering | Market introduction | Sales, maintenance, and field monitoring |

Though medical equipment manufacturers apply ILS-like methods and approaches, they are not always 100 percent successful. Equipment that incorporates leading-edge technologies makes prediction of life cycle costs and environmental impacts difficult. When Philips Medical Systems developed a lightweight solid-state digital replacement for its heavier glass tube image intensifier system, it was unable to predict all knock-on energy consumption effects in the supporting electronic control systems. Issues such as these seem obvious in hindsight, but at the time of development there was insufficient data on energy consumption and thermal radiation data with which to predict emergent properties. Philips was able to resolve the difficulties once operational data became available, but only with added costs, which seek to be recovered through sales and post sales revenue. Commercial medical equipment suppliers are obliged to seek other opportunities to recover research and development investment costs, through sales of service contracts, spare parts, user training, complementary products, and accessories.

Despite being fundamentally commercial products whose ownership transfers sometime shortly after being delivered to the customer, the manufacturer's responsibility for radiotherapy systems, as with other safety critical systems like aircraft and automobiles, does not cease at the point of sale.

## Difficulties in Implementing ILS

While it is evident that many projects in the military and civil sectors can benefit from the application of ILS and related disciplines, there are some intractable difficulties in implementing ILS. These include a dearth of data on current systems, difficulties of forecasting accurately the characteristics of future systems, the sheer scale and complexity of the arrangements necessary for the logistic support of large projects, and the tendency of decision makers, in defiance of any existing ILS plans, to resolve urgent problems by solutions that are not cost-effective in the long-term.

Organizations have often failed to collect systematically data on the operation and support of equipment now in service. They may, for example, record the delivery of a batch of spare parts but take no account of when (and in what circumstances) these spares are used to repair existing equipment. They may record the delivery of fuel or utilities, but not identify the vehicles that consumed then. Not all organizations have yet been motivated to collect data that would be useful to ILS analyses. Company and service financial systems have been designed to monitor the various purchased inputs, rather than to facilitate input-output analysis linking such inputs to the organization's activities.

The problem of high-quality data varies between industries and product types. In the industries with particular concerns about safety, such as civil air transport and nuclear power generation, there is generally comprehensive data on all aspects of operation and support. In other industries having a large number of similar projects, such as civil engineering, it should be feasible to collect data of reasonable quality and volume. Data on the operation and maintenance of schools, hospitals, and prisons may be relatively easily acquired. Gathering good-quality data for, say, highly-innovative medical equipment, would be much more challenging.

Even when good data on the operation and support of current equipment has been collected, to provide a basis for forecasting the characteristics of future equipment and justifying the ILS policies chosen for it, the process of forecasting the operation and support of future equipment is extremely difficult. It is notorious that many forecasts of equipment reliability during its concept and the initial design stages have proved to be grossly inaccurate (Augustine, 1983), though in recent years a better understanding of the physics of failures has led to improvements in forecasting methodologies. However, when the new equipment incorporates unfamiliar technology or will be used in an unfamiliar environment, the initial estimates of equipment reliability and maintainability cannot be regarded as accurate until they have been confirmed by rigorous and realistic field trials.

In addition to doubts about the characteristics of future equipment, there are additional difficulties involved in forecasting the efficacy of some of the alternative arrangements for logistic support considered in the ILS process, particularly on those arrangements involving unfamiliar contractors and innovative contractual arrangements.

Even if the performance of the equipment itself (and of the organizations involved in its operation and support) could be forecast with confidence, there often remains considerable uncertainty about the employment of the equipment in service and the duration of its service life. Such uncertainty is greatest for military projects and for other capital equipment with long life cycles. The equipment's planned service life may be lengthened or shortened, according to the vagaries of military or corporate policy. It may or may not be subjected to mid-life upgrades or improvements. A military vehicle may be used for training in a benign peacetime environment or may be exposed to the rigors of warfare of various intensities and in different climates. A civil construction project may (during its lifetime) have to withstand more damaging levels of traffic or climatic conditions, or radical changes of use.

Because the future is inherently uncertain, any forecast of a project's life cycle cost is unlikely to be accurate, and hence should be accompanied by upper and lower confidence limits covering a substantiated range of uncertainty. Some project managers, accustomed to precise engineering calculations or auditable balance sheets (depending on their past experience), become demoralized by the distance between realistic confidence limits and cannot for that reason regard ILS as a really important influence on their management plan. Some of these managers may therefore be reluctant to allocate sufficient resources to ILS, when there are many urgent problems to engage the attention of their staff. In fact, many of the future uncertainties apply to all of the alternative design configurations and to all of the alternative logistic support arrangements; so it is it possible to select with confidence the most cost-effective designs or support arrangements based on their relative life cycle costs, even the where the absolute values of life cycle cost are very obscure.

Another inherent difficulty with ILS is the scale and complexity of some of the projects on which it must be used, the number of different organizations that must contribute, and the nature of the interfaces between these organizations. If these interfaces are blocked by mistrust or distorted by perverse contracting, the ILS process is unlikely to be completed satisfactorily. Furthermore, the proliferating multitude of interacting analyses, studies, and plans for the ILS of major projects encourages the growth of management procedures, bureaucracy, jargon, and acronyms, which together obscure the underlying principles of ILS and tend to insulate decision makers from operational realities.

Even when the ILS process has been satisfactorily completed and the most cost-effective strategy has been determined to manage the project through its entire life cycle, it remains difficult to ensure that the stakeholders are always guided by the best long-term policy. The politicians, government officials, and service officers directing military projects may be involved with the project for only a few years before at their respective career paths take them to other responsibilities. Business executives managing commercial projects may have personal goals (such as an annual bonus, stock options, or ambitions for promotion) whose attainment in the short term may not exactly correspond with the optimal policy for the project. In both cases, the stakeholders may take decisions that are attractive in the short term, but that in the long run can prove enormously expensive. The existence of an ILS plan can inhibit such decisions by highlighting and quantifying the scale of their adverse consequences, but it requires an appropriately forceful ILS manager to insist that the ILS management plan is widely understood and acknowledged as a significant factor in decision making.

Although a rational notion, there is a risk that ILS will lead to being unduly conservative in design. One criticism of PFI projects (a key driver for ILS) is that its application may lead to mediocrity of end product. Early consideration of later life cycle issues such as maintainability may stifle creativity and innovation, so that the end products may be maintainable but excessively dull as a result of the compromises made to make them so.

## Summary

It is evident that integrated logistic support (or any similar process under another label) is an essential part of the development of a new product in the defense or civil sectors of industry. The ILS process specifies the facilities and supplying arrangements that are required to maintain and repair the products in service and to achieve the target level of availability. ILS is particularly necessary for large and complex projects that are expected to remain in service for many years, such as major capital items of defense equipment, investment goods, or infrastructure. ILS specifies the resources necessary for equipment support and hence defines their contribution to the equipment's life cycle cost, which is an essential input to its through-life project management plan (including budgeting).

There are many difficulties in implementing the ILS process, and these increase with the scale and complexity of the project considered. ILS involves many stakeholders who may have imperfect understanding of each other's problems and who may offer various levels of cooperation of the ILS process. The information available to support the ILS process is inevitably incomplete, particularly near the start of the product's life cycle, and the process itself is therefore prone to error and inaccuracy.

The ILS process should accordingly be tailored to match the information available and will help to identify critical areas of uncertainty. There are often inadequate resources (human and/or financial) and insufficient time to implement the ILS process as rigorously as would ideally be appropriate, since the project manager must always balance limited resources between ILS and various other activities required in creating a new product.

In poorly managed projects there is the risk that the ILS process is accomplished early in the life cycle only in order to obtain the funding necessary to launch the project but may subsequently be ignored during the Demonstration, Manufacturing, and In-service phases.

Despite these difficulties, ILS is a necessary activity since it provides vital inputs to through-life project management, except in those very rare cases where the supplier bears no accountability whatsoever for outcomes after the point of sale.

# References

Audit Scotland. 2000. Taking the initiative: Using PFI contracts to renew council schools. Report to the Auditor General for Scotland in June 2002. Accounts Commission Scotland. www.audit-scotland.gov.uk/.

Augustine, N. R. 1983., *Augustine's laws*. p. 176. New York: AIAA.

BBC News. 2003, PFI schools criticised by report. BBC news report. Thursday, January 16, 2003, http://news.bbc.co.uk/1/hi/wales/2662999.stm.

Bergen, T. 2000. Supportability: A key to system effectiveness. Conference paper for Norwegian Systems Engineering Council (NORSEC) Annual Symposium, January 2000. www.incose.org/norsec/Dokumenter_og_nedlastbare_filer/NORSEC_moter/20000111/teknisk_referat20000111_2.pdf.

BSI. 1999. BS7000-1:1999. *Guide to managing innovation*. London: British Standards Institution.

Clarke, J. 2003. An Introduction to Codification, Statement by the Director of the National Codification Bureau, Glasgow, March 5. www.ncb.mod.uk/.

Confederation of British Industries. 2002. Making PFI / PPP work. Issue statement from the CBI Information Centre. September 23, 2002.

DuPont, A. 2002. Ensuring sustainability. Sustainability Brochure H93234, published by de Pont de Nemours and Company, United States, http://antron.dupont.com/pdf_files/literature/sustainability.pdf.

El-Haram, M., Marenjak. Horner. 2001. The use of ILS techniques in the construction industry. *Proceedings of the 11th MIRCE International Symposium*. Exeter, December.

Hambleton, K. 2000. Systems engineering: An educational challenge. *Ingenia* (November). Royal Academy of Engineering, London.

Hicks, R. and J. A. Epps. 2003. Life cycle costs analysis of asphalt rubber paving materials. Industry report. The Rubber Pavements Association, Tempe, AZ, May 1. www.rubberpavements.org/library/lcca_australia.

INCOSE. 1999. What is systems engineering? International Council on Systems Engineering. May 1. www.incose.org/whatis.html.

ISO. 2002. ISO/IEC 15288:2002(E). Systems engineering: System life cycle processes. Geneva: International Organization for Standardization/International Electrotechnical Commission.

ISO. 1994. ISO 10303-1:1994. Industrial automation systems and integration: Product data representation and exchange. Part 1: Overview and fundamental principles. Geneva: International Organization for Standardization.

NAO. 2001. The PFI contract for the redevelopment of West Middlesex University Hospital. National Audit Office Press Notice. National Audit Office. London: Stationary Office. ISBN:

NATO. 2000. *NATO CALS Handbook.* Version 2, June 2000. Available at www.dcnicn.com/ncmb/ (accessed May 1, 2003).

*Oxford English Dictionary.* 2000. Oxford, UK: Oxford University Press.

Shishko, R. 1995. *NASA systems engineering handbook.* SP-6105. Washington, D.C.: National Aeronautics and Space Administration.

Sparidens, H. 2000. Purchasing and supplier involvement is the product creation process. Philips Medical Systems corporate communication. Technische University Eindhoven, October 16. Available at www.tm.tue.nl/ipsd/educate/pms-2000.pdf (accessed May 1, 2003).

Taft, W. H. 1985. Computer Aided Logistics Support (CALS). Memorandum for Secretaries of the Military Departments, Defense Logistics Agency, U.S. Department of Defense. Report no. MIL-HDBK-59A. Washington, D. C.

UK Ministry of Defence. 2001. *MoD Guide to integrated logistics support.* Andover, UK MoD Corporate Technical Services. www.ams.mod.uk/ams/content/docs/ils/ils_web/ilsgdef.htm (accessed May 1, 2003).

———. 2002. *Integrated logistic support.* Defence Standard 00-60 Part 0, Issue 5, May. Glasgow: Directorate for Standardisation.

UK Parliamentary Select Committee on Defence. 1998. *The strategic defence review.* HC 138, Eighth Report. Vols. I–III.

———. 2000. *European security and defence* HC 264, Eighth Report. ISBN: 0-10-229400-3.

U.S. Army. 1998. *Integrated logistic support (ILS) manager's guide.* PAM 700-127. Washington, D.C.: United States Army Publishing Agency.

U.S. Department of the Army. 1999. Logistics, Integrated Logistic Support, Army regulation 700-127, November 10, p. 6. Washington DC Department of the Army.

U.S. Department of Defense. 1983. *Acquisition and management of integrated logistical support for systems and equipment.* Directive 5000, 39. November 17. Washington, D.C.: U.S. DoD Directives and Records Division.

———. 1983. Military Standard 1388-1A, Logistics support analysis.

———. 2003. *Integrated Digital Environment Initiative.* Integrated Digital Environment

———. 2003. Performance-centered learning module: IDE relation to CALS. January 13. www.acq.osd.mil/ide/learning_modules/ide/what_is_an_ide/ide_relation_to_cals.htm (accessed May 1, 2003).

U.S. Department of Energy. 1998. *Life cycle asset management.* U.S. Department of Energy order DOE 430.1 A.

———. 2001. *Analysis of the total system life cycle cost of the Civilian Radioactive Waste Management Program.* Report DOE/RW-0533 May 2001. Washington D.C.: U.S. Department of Energy.

———. 1998. *Standard life-cycle costs-savings analysis methodology for deployment of innovative technologies* Washington, D.C.: U.S. Federal Energy Technology Center.

White, M. 1995. *Gulf logistics from Blackadder's war.* London: Brasseys.

CHAPTER TEN

# PROJECT SUPPLY CHAIN MANAGEMENT: OPTIMIZING VALUE: THE WAY WE MANAGE THE TOTAL SUPPLY CHAIN

Ray Venkataraman

During the 1990s, many organizations, both public and private, embraced the discipline of supply chain management (SCM). These organizations adopted several SCM-related concepts, techniques, and strategies such as efficient consumer response, continuous replenishment, cycle time reduction, vendor-managed inventory systems, and so on to help them a gain a significant competitive advantage in the marketplace. Companies that have effectively managed their total supply chain, as opposed to their individual firm, have experienced substantial reductions in inventory- and logistics-related costs, shorter cycle times, and improvements in customer service. For example, Procter & Gamble estimates that its supply chain initiatives resulted in $65 million savings for its retail customers. "According to Procter & Gamble, the essence of its approach lies in manufacturers and suppliers working closer together jointly creating business plans to eliminate the source of wasteful practices across the entire supply chain" (Cottrill, 1997).

While the adoption and implementation of total SCM-related strategies is quite prevalent in retail and the manufacturing industries and their benefits are well understood, project-based organizations have lagged behind in their acceptance and use of such strategies. For instance, the engineering and construction industry worldwide has been plagued by poor quality, low profit margins, and project cost and schedule overruns (Yeo and Ning, 2002). It is estimated that in the construction industry about 40 percent of the amount work constitutes non-value-adding activities such as time spent on waiting for approval or for materials to arrive on project site (Mohamed, 1996). The current project management practices of the construction industry in the areas of resource and materials scheduling would seem to be inefficient and lead to considerable waste. There is an urgent opportunity to adopt the practices of total supply chain management to reduce inefficiencies, improve profit margins, and optimize value. Sir John Egan, who headed a construction task force backed by the British government in 1997, strongly recommended in his report *Rethinking Construction* that the construction industry's performance would dramatically improve if it adopted the partnering approach in its supply chain (Watson, 2001).

Given that there are proven benefits in adopting total supply chain management-related strategies, the challenge then for project managers is to integrate these strategies into their management of projects.

## What Is Supply Chain Management?

Supply chain management is a set of approaches utilized to efficiently and fully integrate the network of all organizations and their related activities in producing/completing and delivering a product, a service, or a project so that systemwide costs are minimized while maintaining or exceeding customer-service-level requirements. This definition implies that a supply chain is composed of a sequence of organizations, beginning with the basic suppliers of raw materials, and extends all the way up to the final customer. Supply chains are often referred to as *value chains*, as value is added to the product, service, or project as they progress through the various stages of the chain. Figure 10.1 illustrates typical supply chains for manufacturing and project organizations. Each organization in the supply chain has two components: an inbound and an outbound component (Stevenson, 2002). The inbound component for an organization may be composed of suppliers of basic raw materials and components, along with transportation links and warehouses, and it ends with the internal operations of the company. The outbound component begins where the organization delivers its output to its immediate customer. This portion of the supply chain may include wholesalers, retailers, distribution centers, and transportation companies, and it ends with the final consumer in the chain. The length of each component of the supply chain depends on the nature of the organization. For a traditional make-to-stock manufacturing company, the outbound or the demand component of the chain is longer than the inbound or supply component. On the other hand, for a project organization, the inbound component is typically longer than the outbound component. These concepts are illustrated in Figure 10.1.

## The Need to Manage Supply Chains

Business organizations in the past have focused only on the performance and success of their individual firms. Such firm-focused approaches, however, will not help companies achieve a competitive edge in the current global business environment. Survival, let alone success, hinges on the ability of companies to manage their total supply chain. There are several reasons that make it necessary for companies to adopt supply chain management approaches.

First, businesses are encountering competition that is no longer regional or national; it is global. Competitive pressure from foreign competitors in both domestic and international markets is intense. Customers increasingly are seeking the best value for their money, and

FIGURE 10.1A. TYPICAL SUPPLY CHAIN FOR A MAKE-TO-STOCK MANUFACTURING COMPANY.

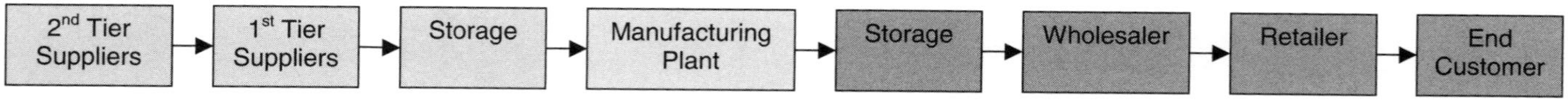

## FIGURE 10.1B. TYPICAL SUPPLY CHAIN FOR A PROJECT ORGANIZATION.

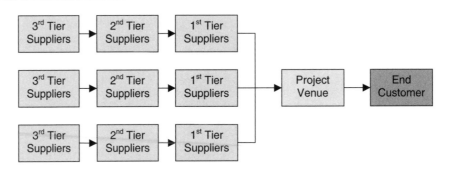

the advances in information technology and transportation have provided them the ability to buy from any company anywhere in the world that will provide that value. To win over these customers, business organizations need to reduce costs and add value, not just for their individual firm, but throughout their supply chain.

Second, inventory is a non-value-adding asset and is a significant cost element for businesses. The increasing variability in demand as we move up in the supply chain, known as the "bull-whip effect", can force some individual members of a supply chain to carry very high levels of inventory that can substantially increase the final cost of the product. Effective supply chain management approaches can enable a business to achieve a visible and seamless flow of inventory, thereby reducing inventory-related costs throughout the supply chain.

Third, the chain of organizations involved in producing and delivering a product or completing and delivering a project is becoming increasingly complex and is fraught with many inherent uncertainties. For example, inaccurate forecasts, late deliveries, equipment breakdowns, substandard raw material quality, scope creep, resource constraints, and so on can contribute to significant schedule and cost overruns for a project organization. The more complex the supply chain, the greater would be the degree of uncertainty and hence the more adverse the impact on the supply chain.

Supply chain management approaches such as partnering, information, and risk sharing can greatly reduce the impact of these uncertainties on the supply chain. Finally, management approaches such as lean production and TQM enabled many organizations to realize major gains by eliminating waste in terms of time and cost out of their systems. New opportunities for businesses to improve operations even further now rest largely in the supply chain areas of purchasing, distribution, and logistics (Stevenson, 2002). In the present-day global environment, because the competition is no longer between individual firms but between supply chains, companies need to better manage their supply chain to remain viable.

While several project-based organizations have adopted SCM-related strategies, evidence indicates their efforts to mitigate project schedule and cost overruns have fallen woefully short of expectations. The reason may be that project supply chain management is considerably more difficult as project supply chains are inherently more complex. For example, many projects typically involve a multitude of suppliers and experience considerable variability in supply delivery lead times and resource constraints, as well as frequent changes to the project scope. Such project supply chain complexities underscore the importance and need for project-based organizations to manage their total supply chain in a more formal and organized manner.

## SCM Benefits

Companies that effectively manage their supply chain accrue a number of benefits. A recent study by Peter J. Metz of the MIT Center for eBusiness found that companies that manage their total supply chain from suppliers' supplier to customers' customer have achieved enormous payoffs, such as 50 percent reduction in inventories and 40 percent increase in on-time deliveries (Betts, 2001). Effective SCM has enabled Campbell Soup to double its inventory turnover rate, Hewlett-Packard reduced its printer supply costs by 75 percent, profits doubled and sales increased by 60 percent for Sport Obermeyer in two years, and National Bicycle achieved an increase in its market share from 5 percent to 29 percent (Stevenson, 2002; Fischer, 1997). Companies such as Wal-Mart that have better managed their supply chain have benefited from greater customer loyalty, higher profits, shorter lead times, lower costs, higher productivity, and higher market share. These benefits are not restricted to traditional manufacturing or retail businesses. Organizations that manage projects can enjoy similar benefits by effectively managing their supply chains.

## Critical Areas of SCM

Effective supply chain management requires companies to focus on the following critical areas: Customers, suppliers, design and operations, logistics, and inventory.

### Customers

Customers are the driving force behind supply chain management. Effective supply chain management, first and foremost, requires a thorough understanding of what the customers want. In a project environment, determining customer requirements and integrating the voice of the customer by working with the customer throughout the project will in all likelihood lead to a satisfied customer and project success. An important mechanism to achieve such a customer focus in projects is the integration of all project activities and participants into the larger framework of supply chain management. However, given that customer expectations and needs are ever-changing, determining these is tantamount to hitting a moving target. In recent years, customer value as opposed to the traditional measures of quality and customer satisfaction has become more important. "Customer value is the measure of a company's contribution to its customer, based on the entire range of products, services, and intangibles that constitute the company's offerings" (Simchi-Levi et al., 2003). Clearly, effective supply chain management is a fundamental prerequisite to satisfying customer needs and providing value. The challenge for project organizations, then, is to provide this customer value by managing the inevitable scope changes without incurring significant project schedule and cost overruns.

## Suppliers

Suppliers constitute the back-end portion of the supply chain and play a key role in adding value to the chain. Their ability to provide quality raw materials and components when they are needed at reasonable cost can lead to shorter cycle times, reduction in inventory-related costs, and improvement in end-customer service levels. Traditionally, the relationship between suppliers and buyers in the supply chain has been adversarial, as each was interested in their own profits and made decisions with no regard to their impact on other partners in the chain. Supplier partnering is vital for effective supply chain management, and without the involvement, cooperation, and integration of upstream suppliers, value optimization in the total supply chain cannot be a reality. For project-based organizations, this issue is even more critical, as the supply or back-end portion of the chain is typically long, and without the total involvement of each and every supplier, value enhancement and project supply chain performance will be less than optimal. For example, in the case of highly technical projects, it is not atypical to have fifth or even sixth tier suppliers upstream in the project supply chain (Pinto and Rouhiainen, 2001). Managing the dynamic interrelationships and interactions that exist among these suppliers is considerably more complex and requires the effective integration of these supplier and their project activities into the larger framework of supply chain management.

## Design and Operations

Design and operations play several critical roles in a supply chain. New product designs often seek new solutions to immensely challenging technical problems. In the face of uncertain customer demand, changes to the existing supply chain may have to be made to take advantage of these new designs. They require consideration of trade-offs between higher logistics- or inventory-related costs and shorter manufacturing lead times. The operations function creates value by converting the raw materials and components to a finished product. This function is present in every phase of the supply chain and is responsible for ensuring quality, reducing waste, and shortening process lead times.

## Logistics

Logistics involves the transfer, storage, and handling of materials within a facility and of incoming and outgoing shipments of goods and materials. By ensuring that the right amounts of material arrive at the right place and at the right time, the logistics function makes a significant contribution to effective supply chain management. In project management, the logistics function requires a thorough understanding of customer requirements, reduces waste throughout the supply chain in order to reduce costs, and ensures timely completion and delivery of projects.

## Inventory

Inventory control is an essential aspect of effective supply chain management for three reasons. First, inventories represent a substantial portion of the supply chain costs for many companies. Second, the level of inventories at various points in the supply chain will have a significant impact on customer service levels. Third, cost trade-off decisions in logistics, such as choosing a mode of transportation, depend on inventory levels and related costs. In project-based organizations, inventory-related costs can be substantial. It is obvious that effective inventory management can only be achieved through the joint collaboration of all members of the supply chain.

# SCM Issues in Project Management

The benefits of utilizing the total supply chain management approach in the traditional make-to-stock manufacturing and retail environments have been well documented. Increasingly, project organizations and project managers are realizing that the integration of the total supply chain in managing projects, as opposed to a firm-focused approach, has the potential of reducing project schedule and cost overruns and the chances of project failure. However, as shown in Figure 10.1B, the typical chain for a project is considerably more complex. Problems associated with scope changes, resource constraints, technology, and numerous suppliers that may require global sourcing makes the total integration of the project supply chain risky and challenging. Consider, for example, the $200 billion Joint Fight Striker program, a mammoth and one of the most complex project management undertaking in history. "The principals of this project supply chain include

1. a consortium comprised of Lockheed Martin, Northrop Grumman and BAE Systems, overseeing design, engineering, construction, delivery and maintenance,
2. a matrix of partners, including Boeing, engine-makers Pratt & Whitney and Rolls-Royce and a handful of other subcontractors, all of which will lean on their own myriad suppliers for hundreds of thousands of components,
3. a multifaceted customer, the Pentagon, which is representing the U.S. Air Force, Navy and Marines, as well as the British Royal Navy and Air Force" (Preston, 2001).

Integrating and managing the total supply chain for this project is a Herculean task that will involve careful balancing of different vested interests and collaboration among all these partners to meet the stringent cost, quality, and delivery criteria set by the customer, the Pentagon. If the project's goal is focused only at the department or at the individual company level, instead of the total project supply chain, value optimization for the project cannot be achieved.

Projects in the construction industry are notorious for ill-managed supply chains. A recent research study of the UK construction sector found that fundamental mistrust and skepticism among subcontractors and other supply chain relationships was quite prevalent in this industry (Dainty et al., 2001). Such a lack of trust among supply chain partners will have a detrimental effect on the project delivery process. The key issue here is how to foster the necessary attitudinal changes throughout the project supply chain network to improve project performance.

Effective inventory management is yet another important supply chain issue in projects. In the airline industry, for example, enormous inventory inefficiencies such as duplication of distribution channels and excessive parts in storage have led to increasing costs for the total supply chain. In addition, a significant portion of every dollar invested in spare parts inventory constitutes holding and material management costs. Clearly, efficient inventory management throughout the total supply chain for projects in this industry has the potential for significant reduction in project life cycle costs.

Value optimization in projects cannot occur without the joint coordination of activities and communication among the various project participants. Consider, for example, a development project for an aluminum part to be delivered to an airline customer that was originally designed with a certain anodize process specified in the drawing. When the part designed is ready for production, the supply chain department of the project development group will then typically choose from a list of its favorite suppliers to get the lowest possible price. Often, these companies are rarely the ones that worked on the development hardware, and not surprisingly, they all will have different design changes that they would like to enforce on engineering to efficiently produce the part that will fit their particular set of processes. This can often lead to substantial increase in costs by way of engineering modification and requalification efforts. Furthermore, in the interest of price reduction, if the supply chain department later changes the design, without communicating or coordinating with the engineering department, to allow a supplier organization to use a different anodize process to fit its capabilities, then the part that will be delivered to the customer will be different from what the customer wanted, with colors that may be aesthetically displeasing when installed in the aircraft. Much time, money, and effort may have to be expended to rectify the situation with the irate customer. Project managers should be aware that without the joint collaboration of all project stakeholders working toward a common goal for the overall project, suboptimization will occur and the project is likely to fail.

Accurate, timely, and quality information on supply-chain-related issues is often not available to project managers and, as a result, causes them to make suboptimal decisions. Effective project supply chain management requires an infrastructure that can accelerate the velocity of information and will enable all project participants to collaborate throughout the project life cycle. For example, in a chemical plant construction project, the Global Project and Procurement Network uses the Internet to streamline and accelerate information flow that enables all supply chain participants to collaborate from plant design through operation and maintenance (Cottrill, 2001).

The terrorist attack of September 11, 2001, has heightened interest in security matters in the management of the total supply chain for many organizations. The challenge for many project organizations may range from designing facilities that are secure against outside intrusion to ensuring that the product can be protected from tampering till it reaches the end consumer. Ensuring security throughout the total supply chain is an enormous problem that will require project managers to provide unique and innovative solutions.

# Value Drivers in Project Supply Chain Management

Value drivers in a project supply chain are those strategic factors that significantly add or enhance value and provide a distinct competitive advantage to the chain. The typical value drivers for a project supply chain are listed in Table 10.1.

The customer is the most important value driver in project supply chain management. In the context of project supply chains, the project client is the final recipient of the completed project. It is this customer's definition or perception that determines what constitutes value in a project. All other upstream supply chain activities in a project are triggered by this concept of customer value. If the customer values price, then all supply-chain-related activities of the project should focus on efficiency and eliminating waste throughout the total supply chain. On the other hand, if the customer values completion of the project on time or ahead of schedule, then all of the project supply chain activities should be geared toward achieving this goal. Thinking in terms of customer value requires project managers to have a clear understanding of customer preferences and needs, profit and revenue growth potential of the customer, and the type of supply chain required to serve the customer, and they must make sure the inevitable trade-offs that need to be made are indeed the correct ones (Simchi-Levi et. al, 2003).

The need to significantly lower or control project costs will also drive changes and improvements in the supply chain. In the retail industry, for example, the policy of everyday low prices required Wal-Mart to adopt the cross-docking strategy in its warehouses and distribution centers and strategic partnering with its suppliers. In the personal computer industry, Dell Computer Corporation uses the strategy of postponement (i.e., delaying final product assembly until after the receipt of the customer order) to lower its supply chain costs.

### TABLE 10.1. PROJECT SUPPLY CHAIN VALUE DRIVERS.

| Value Drivers | Definition |
|---|---|
| Customer | The final customer at the end of the project supply chain |
| Cost | Total cost incurred at the end of the project supply chain |
| Flexibility | The ability of the project supply chain to quickly recognize and respond to changing customer needs |
| Time | Refers to on-time delivery or delivery speed of completed projects to the end customer |
| Quality | The ability to deliver a completed project that meets or exceeds end customer expectations |

Flexibility, the ability to respond quickly to changes in customer needs or project scope, is yet another important value driver in project supply chains. For example, the willingness of the project organization to provide the client the freedom to make significant design changes through development with the help of a strong and supportive engineering staff will enhance the value of the project supply chain (Pinto and Rouhiainen, 2001). Dell Computer Corporation is a classic example of a company that used flexibility to enhance customer perception of value. By allowing the customers to configure their own personal computer systems, Dell gained a significant competitive advantage in its industry.

The dimension of time has always been an important success factor in project management. Time in the form of project scheduling, in conjunction with cost and quality, represents the three most important constraints in projects. In event project management such as the Olympic Games, for instance, the dimension of time is of overriding importance, as the whole world is watching and the games must start on time. In other project-oriented situations, however, cost or quality can be the more important value drivers, and trade-offs in terms of time may have to be made in such projects. In any event, the ability to complete a project on time or ahead of schedule will certainly contribute to value in project supply chain management. In the retail industry, for instance, several time-based supply chain strategies such as continuous replenishment systems, quick response systems and efficient consumer response evolved as a direct result of the value-adding nature of time.

Quality, in a project context, is defined as achieving the project objectives that are "fit for purpose." "Fit for purpose means that the facility, when commissioned, produces a product which solves the problem, or exploits the opportunity intended, or better. It works for the purpose for which it was intended" (Turner, 1999). Project quality, simply defined, is that the project's product meets or exceeds customer expectations (Turner, 1999). Quality has several dimensions. For example, a person wanting to buy a Steinway grand piano for a price of $25,000 is more likely interested in the performance dimension of quality, whereas a person who wants to buy a Baldwin vertical piano for $5,000 is probably looking for a piano of consistent quality. Understanding the level of quality a customer wants in a project, ensuring the functionality of the project's product at that level of quality, and delivering the project at a reasonable price and time that will delight the customer should be the ultimate goal of every project manager. Meeting or exceeding the quality expectations of the customer adds value by fostering and sustaining customer loyalty and goodwill for years—long after the project is completed. Achieving this level of quality in projects, however, is easier said than done. It requires the total commitment to quality by each and every member of the total project supply chain and the integration of all their quality management activities.

# Optimizing Value in Project Supply Chains

## Choosing the Right Supply Chain

A fundamental prerequisite for value optimization in projects is the choice of the right supply chain for the project. More often than not, less-than-stellar supply chain performance is due to the mismatch between the nature of the product and the type of supply chain chosen to produce that product (Fischer, 1997). In the context of a project environment, this implies that first and foremost, the nature of the project, whether it is primarily functional or primarily innovative, should be clearly delineated. The next step is to choose the right supply chain for this project that will directly contribute to its core competencies and provide a distinct competitive advantage. Without having the right supply chain that is best for a particular project, value optimization in projects cannot be achieved.

## Total Quality Management (TQM)

Quality, as defined by the customer, is the primary value driver in project supply chain management. Therefore, optimizing value in projects requires an unyielding commitment to total quality by all members of the project supply chain. A way to achieve this commitment is to adopt and integrate Total Quality Management (TQM) in project supply chains. "TQM is a holistic approach to continuously meeting customer needs and aims at continual increase in customer satisfaction at continually lower real cost. Total Quality is a total systems approach (not a separate area or program), and an integral part of high-level strategy. It works horizontally across functions and departments, involving all employees, top to bottom, and extends backwards and forwards to include the supply chain and customer chain" (Rampey and Roberts, 1992). The integration of the quality management activities of the project supply chain members through TQM is vital to complete a project in such a way that the multiple objectives of the customer in terms cost, quality, time, and safety can be met. The construction industry, for instance, is increasingly embracing TQM to solve its quality problems and ensure customer satisfaction. Quality assurance has always been difficult in this industry, as the products are one-off, the production processes are nonstandardized, and project design changes are frequent. Furthermore, the general contractor for a construction project is totally dependent on the goods and services of suppliers and other subcontractors to meet the quality requirements of the customer. The integration of quality management activities in a construction project supply chain through the application of TQM will enable the general contractors, suppliers, and other subcontractors to improve their own quality performance and will contribute toward optimizing customer value. For example, Shui On Construction Company in Hong Kong had successfully adopted TQM in 1993 and since then has been known for its good performance in building housing projects and has won the "Contractor of the Year" award three years in a row from 1995 to1997 (Wong and Fung, 1999).

# Project Supply Chain Process Framework

The rest of the discussion in this section will be based on a simple framework of the project supply chain process that is presented in Figure 10.2. In this figure, the square box represents the procurement component of the chain. The oval-shaped box represents the conversion or fabrication phase of the project. It is in this component of the chain where the project's product is created. The rectangular box is the front-end portion of the project supply chain, delivery of the completed project to the customer.

## FIGURE 10.2. PROJECT SUPPLY CHAIN PROCESS FRAMEWORK.

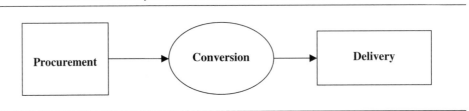

## Procurement

The procurement portion of the project supply chain is typically long, and it is not uncommon to find fifth or even sixth tier suppliers upstream. The greatest opportunities for enhancing value of the total project supply chain exist in this area. Procurement involves all activities that are vital in acquiring goods or services that will enable an organization to produce the product or complete a project for its customer. The decision to buy from an outside vendor should be made only after a thorough "make or buy" analysis. In general, an organization should produce a product or component if it directly contributes to its set of core competencies. Otherwise, the product or components should be purchased from outside suppliers.

Procurement involves identifying and analyzing user requirements and type of purchase, selecting suppliers, negotiating contracts, acting as liaison between the supplier and the user, and evaluating and forging strategic alliances with suppliers. For many organizations, materials and components purchased from outside vendors represent a substantial portion of the cost of the end product, and hence effective procurement can significantly enhance the competitive advantage of an organization. Managing these suppliers and ensuring that parts and components of appropriate quality are delivered on time is a truly daunting challenge. In 1997, for example, Boeing, in its desire to respond to an unprecedented demand for new airplanes, attempted to double its production overnight without realizing the impact such a move would have on its supply chain. Parts and worker shortages at the assembly stage forced Boeing to close its 747 and 737 assembly lines and the company was hit with a $1.6 billion loss. Four years later, Boeing, through the use of lean manufacturing techniques, began to revamp its supply chain process and now requires tighter integration with suppliers and just-in-time delivery of their parts (Holmes, 2001). In the aerospace industry, companies such as Boeing and Rolls-Royce typically incur 60 percent of their project cost and 70 percent of their lead time because of purchased materials.

Hence, effective procurement strategies such as international sourcing, long-term supplier contracts, partnering with suppliers in project design, and risk and information sharing can maximize these companies' purchasing power, contribute to their business success, and significantly enhance the value of their supply chain. In 1993, Sikorsky Aircraft adopted the method of supplier Kaizen and realized the benefits of supplier long-term commitment and partnering for its future growth, declining prices and shorter lead times (Foreman and Vargas, 1999). "Supplier Kaizen is a method of bringing the suppliers to the same level of operations as the parent company, through training and improvement projects, to ensure superior performance and nurture the trust that is required for strong partnerships" (Foreman and Vargas, 1999).

The construction industry, as a whole, is characterized by significant distrust and antagonism within existing supply chain relationships. The key to future performance improvements in this industry is through the adoption of effective procurement strategies such as supplier selection and partnering, e-procurement, and supplier Kaizen. A recent Hong Kong-based study of factors affecting the performance of the construction industry has shown that the methods used for selecting the overall procurement system, contractors, and subcontractors are critical and the use of information technology/information systems can facilitate appropriate selection through all stages of the construction supply chain (Kumaraswamy et al., 2000). In the context of supplier selection, the series of international standards on quality management and quality assurance called ISO 9000 developed by the International Organization for Standardization (ISO) can be highly useful. For instance, companies that are ISO 9001 certified have demonstrated to an independent auditor that their systems and operations have met the rigorous international standards for quality and therefore can be included in the list of potential suppliers.

*Supply Chain Relationships.* Value optimization in projects cannot be achieved in the absence of close and trusting relationships among the project participants. Building trust and integrating the information systems among the supply chain members can lead to the elimination of certain redundant processes and simplification of sourcing, negotiating, and contracting procedures. Planning efficiency and project performance would improve, as suppliers are in a better position to provide valuable inputs to project planning because of the availability of timely and accurate information (Yeo and Ning, 2002). A recent study of two construction projects in the UK has shown that significant supply chain benefits and improvements can be realized through close partnerships and involvement of suppliers and subcontractors very early on in the project (Ballard and Cuckoo, 2001). Through partnering, all members of the supply chain are involved in translating the design concept into reality and ensuring that the appropriate cost criteria are met. The suppliers, along with other partners can be more innovative; problems can be resolved early, as there are more open channels of communication; and the end result will be a project that is completed on time, of higher quality, at a lower cost, and that gets the clients better value for their money. It is estimated that in the construction industry, supply chain partnering alone would lead to a 10 percent reduction in cost and time, similar increases in productivity and quality, and a 20 percent reduction in defects and accidents (Watson, 2001).

*Supplier Development.* Supplier development is yet another strategy that can add value to the procurement phase of the project supply chain. General Electric Company, as part of their global sourcing initiative, has a program for supplier development that involves providing extensive training by GE personnel to vendors in improving their own operations. Vendors who have improved their operations to the level of quality and efficiency that GE requires are awarded long-term contracts. In the final analysis, procurement in a project context requires extensive planning and coordination of project activities with suppliers. Strategies such as supplier Kaizen, partnering based on trust, vendor development, information and risk sharing, long-term strategic alliances with suppliers, and integrating quality management activities of suppliers through TQM will significantly reduce procurement and inventory costs, shorten lead times, and improve quality of purchased materials, and the value of the total project supply chain will be enhanced.

## Conversion

The next phase of the project supply chain shown in Figure 10.2 that requires attention for value optimization is the conversion or fabrication phase of the chain. This is the project venue where the project's product is actually created, as in the case of new product development, creation of a new software package, or building an offshore oil-drilling vessel (Pinto and Rouhianen, 2001). The degree of success, in terms of value, that can be achieved in this area, to a large extent is dictated by the efficiency and effectiveness of the procurement phase of the project supply chain. As in the case procurement, the challenges encountered in this phase will depend on whether the project is relatively routine or highly complex. Regardless of the nature of the project, however, several strategies that have proven to be successful in the traditional manufacturing environment can be employed to enhance value in the conversion phase of the project supply chain. For instance, Boeing Corporation, in order to thwart the stiff competition from Airbus, is employing lean manufacturing practices to effect an innovative company-wide implementation of gigantic, moving assembly lines in its commercial aircraft division. For Boeing, such a technological advancement is reckoned to speed up production by 50 percent and increase its profit margins to double-digit levels on commercial aircraft sales (Holmes, 2001). The application of lean manufacturing techniques can add value in a project environment by eliminating waste and unnecessary inventories and by shortening process lead times.

## Delivery

The final phase of the project supply chain process in Figure 10.2 is the delivery of the completed project to the customer. Normally, the transfer of the completed project to the client is relatively straightforward. In recent years, however, the project delivery process has undergone some significant changes, particularly in the case of clients from foreign countries. For example, in construction projects for large plants, some foreign countries require the project organization to operate the plant jointly with the foreign client for some extended period of time to mitigate potential start-up problems and to reduce the risk to the foreign client (Pinto and Rouhianen, 2001). While this increases the risk to the project organization, it also has the potential to enhance customer value. Clients are becoming increasingly more risk-averse, and the willingness of the project organization to assume additional project risks is certain to add more value and provide a distinct competitive advantage to the total project supply chain.

## Integrating the Supply Chain

The obvious key to value optimization in projects is the total integration of the various components of the project supply chain. Several strategies can be implemented to achieve this goal. First, as shown by a recent study of two demonstration projects in the United Kingdom, development of "work clusters" and the application of concurrent engineering principles can lead to project supply chain integration, which in turn can improve value, eliminate inefficiencies, and reduce project costs (Nicolini et al., 2001). Second, project supply chain integration can be achieved through collaboration and standardization of business processes among the project supply chain partners. Such collaboration, however, requires understanding and managing the differences and interests of all the project supply chain members to create a common vision and work culture (Padhye, 2001). Third, accelerating information velocity by building an Internet-based supplier network for procurement purposes can further facilitate collaboration and integration in project supply chains (Cottrill, 2001). Building such a network also presupposes the presence of a viable IT/IS infrastructure among the project supply chain members. For example, the Joint Strike Fighter project discussed earlier in this chapter will require the various organizations involved in the design, engineering, manufacturing, logistics, finance and so on to collaborate over the internet to meet the stringent cost, quality, and time requirements set by their customer, the Pentagon.

Project supply chain integration and therefore value optimization requires the supply chain partners to change traditional thinking and practices. Effecting such a change requires the commitment and involvement of the people in each organization in the project supply chain. The impetus for achieving such commitment and involvement should come from the senior management of the project supply chain partners. Ultimately, it is the responsibility of the senior managers to prepare their organization for change, to overcome the cultural and organizational barriers to change, and to achieve cross-functional and cross-business unit cooperation (Burnell, 1999). Without the senior managers assuming the role of project champions, project supply chain integration and, hence, value optimization cannot be achieved.

In addition to the strategic initiatives discussed, the following specific steps can be undertaken to add value to a project (Hutchins, 2002):

1. *Flowchart the project supply chain processes before the project is initiated.* Such a flowchart will show the various links or steps involved in completing the project. Each step will potentially have a customer and a supplier. The flowchart can identify potential areas of redundancies, waste, or other non-value-adding activities in the chain and thus facilitate the use of lean management initiatives to eliminate them.
2. *Standardize processes.* Standardization of processes throughout the project supply chain by the use of methods such as simultaneous design, concurrent engineering, lean manufacturing, mistake proofing, total productivity maintenance, and collaborative teamwork will ensure consistency in the chain.
3. *Control process variation.* It is essential that processes across the total project supply chain are monitored and controlled for variation. For example, variability in lead times or

quality in materials and production processes should be controlled. Once the supply chain processes are stabilized, they can be improved.

4. *Prequalify suppliers through supplier certification.* Ensure that suppliers in each link of the project supply chain process are QS-9000 or ISO 9001 certified. Such certification guarantees a pool of quality suppliers.

5. *Audit the project supply chain processes and take corrective and preventive actions.* Processes should be audited periodically for improvement and risk identification. Corrective action should be taken to eliminate the root causes of nonconformances and deficiencies that were uncovered through the audit. Preventive action ensures the recurrence of such problems.

6. *Measure project supply chain performance.* Measure project supply chain performance through the development and use of performance metrics and competitive benchmarking. Without the availability of specific quantifiable performance metrics, project supply chain performance in terms of both efficiency and customer satisfaction cannot be gauged. Such metrics will convey immediately how the project supply chain has been performing over time or in comparison with the best-in-class competition.

## Performance Metrics in Project Supply Chain Management

Measuring project supply chain performance is a complex and challenging endeavor. Appropriate metrics should be carefully developed at the planning stage of the design of the total project supply chain. Involvement of all members of the project supply chain is critical to ensure that meaningful metrics are developed and will be used to monitor the performance of the total project supply chain. This will require reconciling differences and reaching consensus among the supply chain members on which metrics are appropriate for comparison with those of the best-in-class competition to measure success.

While there are a number of approaches to classify performance metrics, we will use the project process value drivers—time, cost, quality, and flexibility—to examine project supply chain performance. These performance metrics categories are presented in Table 10.2 (Coyle et al., 2003).

Time, in particular, project completion time, has always been considered an important measure of project performance. However, in addition to project completion time, this metric should capture other elements of time such as operational and start-up times, and procurement and manufacturing lead times. Furthermore, for routine projects the potential variability in these times should also be measured to track consistency and reliability of the project supply chain. For example, assume that historically the estimated completion time for routine construction projects has been 36 weeks. How frequently the project supply chain achieves this completion time is an indicator of consistency and reliability and can provide important insights for future improvements to the supply chain.

Some of the cost metrics noted in Table 10.2 are fairly straightforward. The important caveat here is that the emphasis should be on the cost incurred for the total project supply chain and not on just the cost incurred by the project organization. The total project supply chain cost is multidimensional and includes several elements, such as procurement and manufacturing cost of materials and goods, inventory costs, and so forth. Focusing on the

### TABLE 10.2. PROJECT SUPPLY CHAIN PERFORMANCE METRICS CATEGORIES.

| Performance Category | Performance Issues |
|---|---|
| Time | 1. Was the project completed and delivered on time?<br>2. What is the potential variability in project completion times?<br>3. Was the completed project operationalized on time to the satisfaction of the customer?<br>4. Were the purchased materials and manufactured components delivered on time by upstream suppliers?<br>5. What is the potential variability in procurement lead times? |
| Cost | 1. Was the completed project within budget for each of the project supply chain member?<br>2. What was the total project supply chain cost?<br>  • Procurement cost of purchased materials<br>  • Manufacturing cost<br>  • Inventory-related cost<br>  • Transportation cost<br>  • Project acceleration costs<br>  • Cost of liquidated damages<br>  • Other relevant costs: administrative, etc. |
| Quality | 1. Did the project meet the technical specifications and does it provide the functionality desired by the customer?<br>2. Was the customer satisfied with the service provided during start-up, implementation, and final project transfer?<br>3. Were the purchased raw materials and manufactured components defect-free?<br>4. Was the completed project's product reliable and durable during its life cycle? |
| Flexibility | 1. Was the customer accorded reasonable freedom within reasonable a time frame to make changes to the project scope, design, or specifications?<br>2. Were the upstream suppliers responsive to the reasonable needs of their downstream partners in terms of delivery time and quality issues? |

total cost incurred will enable project participants to identify inefficiencies in the supply chain and facilitate coordination to devise ways to eliminate them. The ultimate goal is to optimize value by reducing waste and unnecessary cost throughout the supply chain.

Like cost, the quality metric also has several dimensions. In a project context, the most obvious ones are the dimensions of performance—that is, the functionality of the project's product and conformance to design or technical specifications. In addition to these dimensions, the level of service provided to the customer during the start-up and implementation phase and throughout the project's life cycle are also important quality measures. Ultimately, it is the customer perception of quality that matters, and the response of the project supply chain to meet this value perception should be the focus of this metric.

The last project supply chain performance metric category is flexibility. This metric measures the willingness and ability of the project supply chain to respond to reasonable changes in scope or design requested by the customer. Building an effective configuration and change control system that spans the total project supply chain can help achieve such flexibility and provide a distinct competitive edge to the value chain.

## Project Supply Chain Metrics and the Supply Chain Operations Reference (SCOR) Model

Project supply chain metrics span the entire supply chain, with specific focus on common processes, and they should capture all aspects of supply chain performance. The SCOR model developed by the Supply Chain Council (SCC) provides the framework to track such performance and has been the basis for supply chain improvement for both global as well as site-specific projects (Yeo and Ning, 2002). By integrating the well-known concepts of business process reengineering, benchmarking, and process measurement, the model provides a cross-functional framework for improving supply chain performance. It spans all aspects and interactions of the supply chain, from the customer's customer all the way back to the supplier's supplier. "The SCOR model provides standard descriptions of relevant management processes; a framework of the relationships among the standard processes; standard process performance metrics; and standard alignment to features and functionality. The ultimate aim is to produce best-in-class supply chain performance" (Coyle et al., 2003). The model uses five key aspects of a supply chain—plan, source, make, deliver, and return— as building blocks to describe any supply chain. The SCOR model can be adapted to describe a project supply chain, as shown in Figure 10.3.

In a project supply chain context, the planning process in Figure 10.3 encompasses all aspects of planning, including the integration of the individual plans of all supply chain members, into an overall project supply chain plan. The planning phase essentially involves understanding customer needs and project scope; the best course of action to meet the sourcing, producing, and delivery requirements of the project; and developing the criteria to evaluate the total project supply chain performance. The sourcing phase focuses on all processes related to procurement, such as identifying, selecting and qualifying suppliers, contract negotiation, and inventory management, and so on.

The "make" process encompasses all aspects of creating the project's product, such as design and testing, and building and completing the project. It also includes systems and processes for quality and change control and performance reporting. The delivery process covers all aspects related to the final transfer of the completed project to the customer including installation, start-up, and so forth to the satisfaction of the customer. The return phase encompasses all activities that may range from addressing problems associated with the completed project's functionality at the customer site to return of raw materials to the vendor.

The SCOR model for a project supply chain is composed of three levels. At the top level, the scope and content of the model is defined and performance targets based on best-in-class competition are established. The next level focuses on the configuration of project's supply chain. The last level includes process elements such as performance metrics, systems

FIGURE 10.3. THE SCOR MODEL ADAPTED FOR A PROJECT SUPPLY CHAIN.

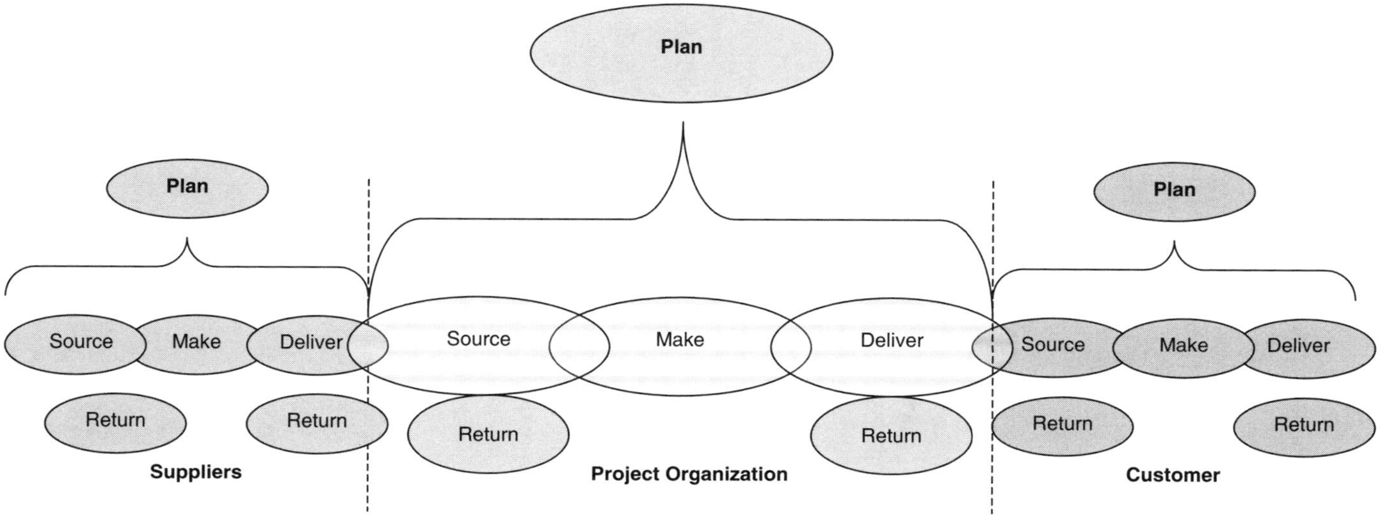

Source: Supply Chain Council Inc. www.supply-chain.org.

and tools, best practices, and the system capabilities to support them. As the SCOR model is based on standard processes and standard language, meaningful performance metrics for the project supply chain can be developed.

# Future Issues in Project Supply Chain Management

Project supply chains in the twenty-first century will encounter a number of challenges, as well as opportunities for improvement. First, the availability and power of information technology will dramatically transform project supply chains. It will facilitate the virtual integration of project supply chains and thus provide the benefits that accrue from tight coordination, partnering, quick and efficient communication, focus, and specialization. The Internet and related e-commerce technologies can be exploited to overcome major systemic constraints. The challenge is to create and build a boundary-spanning information infrastructure that enables quick and efficient information sharing and communication.

Second, the trend toward globalization in project supply chain management will accelerate, as it has the potential to provide significant cost advantages. Boeing Corporation, as part of an initiative to reduce the number parts handled in its production lines, has partnered with European suppliers to procure higher-level assemblies (Sutton and Cook, 2001). Finally, businesses are increasingly concerned about the environment and are undertaking environmental projects to reduce costs, to reduce pollution and hazardous materials, to improve manufacturing performance and quality, to improve relationship with external stakeholders, and to proactively deal with environmental regulation. This trend toward environmental friendliness will require the supply chain for such projects to address issues such as recycling, reuse, asset recovery, minimization of waste, and handling and disposal of hazardous materials (Carter and Dressner, 2001).

To effectively respond to those challenges and exploit the opportunities, project supply chains need to adopt a comprehensive and integrated supply chain perspective. Such integrated supply chains will significantly enhance their value and enjoy a distinct competitive advantage.

# Summary

The retail and the traditional manufacturing industries have enjoyed great success by adopting the principles and strategies of supply chain management. More recently, project-based organizations have also realized that the use of SCM-related strategies can significantly enhance value in projects.

Supply chain management is a set of approaches that can be used to integrate the network of all organizations and their activities in producing and delivering a product or undertaking and completing a project so that systemwide costs are minimized while meeting or exceeding customer requirements.

Given intense global competition, businesses have to embrace effective supply chain management approaches to remain viable and provide value to their customers. Significant benefits accrue to companies that effectively manage their supply chains. The benefits include lower costs, shorter lead times, increased productivity, greater customer satisfaction, and higher profits.

Customers, suppliers, design and operations, logistics, and inventory are critical areas of a supply chain. Value optimization in supply chains requires that these critical areas be effectively managed. While project supply chains encounter many of the same issues and challenges faced by supply chains in other industries—such as effective inventory management, supplier partnering, coordination of activities and effective communication among supply chain members, availability and sharing of information, security, and so on—the complexity of project supply chains makes management more difficult and challenging.

The important value drivers in project supply chains are customers, cost, flexibility, time, and quality. Strategic management of these factors by choosing the right suppliers, adopting TQM and lean management approaches, supplier partnering, and, above all, having a customer focus, will ensure value optimization.

Specific steps that can be taken for value optimization in project supply chains include process flowcharting, process standardization, process control, prequalifying suppliers, periodic supply chain audits to take preventive and corrective action, and measuring project supply chain performance. The four process value drivers of time, cost, quality, and flexibility constitute the major performance metrics categories for project supply chains and the Supply Chain Operations Reference (SCOR) model can provide an excellent framework for measuring and tracking performance in projects.

Increasing globalization, advances in information technology, and environmental concerns are some of the challenges that project supply chains will face in the future. Strategic thinking and an integrated approach to managing project supply chains can overcome these challenges and lead to success and value optimization.

# References

Ballard, R., and H. J. Cuckow. 2001. Logistics in the UK construction industry. *Logistics and Transportation Focus.* 3(3):43–50.

Betts, M. 2001. Kinks in the chain. *Computerworld* 35(51):34–35.

Burnell, J. 1999. Change management is the key to supply chain management success. *Automatic I. D. News* 15(4):40–41.

Carter, C. R., and M. Dresner. 2001. Purchasing's role in environmental management: Cross-functional deployment of grounded theory *Journal of Supply Chain Management* 37(3): 12–26.

Cottrill, K. 1997. Reforging the supply chain. *Journal of Business Strategy,* 18(6):35–39.

———. 2001. Engineering a value chain. *Traffic World.* 265(9):21–22.

Coyle, J. J., E. J. Bardi, and C. J. Langley. 2003. *The management of business logistics: A supply chain perspective. 7th ed.* Cincinatti: South-Western.

Dainty, A. R. J., G. H. Brisco, and S. J. Millet. 2001. Subcontractor perspectives on supply chain alliances. *Construction Management and Economics* 19(8):841–848.

Fisher, M. L. 1997. What is the right supply chain for your product? *Harvard Business Review* 75(2): 105–116.

Foreman, C. R., and D. H. Vargas. 1999. Affecting the value chain through supplier Kaizen. *Hospital Materiel Management Quarterly* 20(3):21–27.

Holmes, S., (2001). Boeing Goes Lean. *Business Week*. (June 4): 94B–94F.

Hutchins, G. 2002. Supply chain management: A new opportunity. *Quality Progress* 35(4):111–113.

Kumaraswamy, M., E. Palaneeswaran, and P. Humphreys. 2000. Selection matters: In construction supply chain optimization. *International Journal of Physical Distribution and Logistics Management* 30(7/8): 661–669.

Mohamed, S. 1996. Options for applying BPR in the Australian construction industry. *International Journal of Project Management* 14(6):379–385.

Nicolini, D., R. Holti, and M. Smalley. 2001. Integrating project activities: The theory and practice of managing the supply chain through clusters. *Construction Management and Economics* 19(1):37–47.

Padhye, A. 2001. Apply leverage to ensure business process integration in your supply chain. *EBN* 1282:PGL38.

Pinto, J. K., and P. J. Rouhiainen. 2001. *Building customer-based project organizations*. New York: Wiley.

Preston, R. 2001. A glimpse into the future of supply chains. *Internet Week* 885:9–10.

Rampey, J., and H. V. Roberts. 1992. Perspectives on total quality. *Proceedings of Total Quality Forum IV*. Cincinnati.

Simchi-Levi, D., P. Kaminsky, and E. Simchi-Levi. 2003. *Designing and managing the supply chain: Concepts, strategies and case studies*. 2nd ed., p. 11. New York: McGraw-Hill/Irwin.

———. 2003. *Designing and managing the supply chain: Concepts, strategies and case studies*. 2nd ed. New York: McGraw-Hill/Irwin.

Stevenson, W. J. 2002. *Operations management*. 7th ed. New York: McGraw-Hill/Irwin.

Sutton, O., and N. Cook. 2001. Quest for the ideal supply chain. *Interavia*, 56(657):24–27.

Turner, J. R. 1999. *The handbook of project-based management*, p. 150. Marlow, UK: McGraw-Hill.

Watson, K. 2001. Building on shaky foundations. *Supply Management* 6(17):22–26.

Wong, A., and P. Fung. 1999. Total quality management in the construction industry in Hong Kong: A supply chain management perspective. *Total Quality Management* 10(2):199–208.

Yeo, K. T., and J. H. Ning. 2002. Integrating supply chain and critical chain concepts in engineer-procure-construct (EPC) projects. *International Journal of Project Management* 20:253–262.

CHAPTER ELEVEN

# PROCUREMENT: PROCESS OVERVIEW AND EMERGING PROJECT MANAGEMENT TECHNIQUES

Mark E. Nissen

Effective procurement is critical for effective project management. Depending upon the specific type of project being managed, over 50 percent of the total project cost can be attributed to parts, supplies, and services procured, and for many high-technology projects, this procurement fraction can approach 90 percent. Further, because of long lead times, procured items nearly always define the critical path through the project schedule network. And dependence upon markets for procured items can create project management difficulties in terms of agency (e.g., information asymmetries, incentives) and coordination (e.g., redundant project management organizations, interorganizational communications). In short, if a project manager is not managing procurement, then he or she is only managing 50 percent or less of the project as a whole.

Despite this critical role, however, the term "procurement" remains broadly defined and is used to describe a variety of entities (e.g., functions, organizations, systems, processes). The term is also evolving through time, as the activities associated with procurement have become increasingly important to enterprise success. For instance, procurement was once descriptive of the simple clerical activities associated with purchasing well-specified items, but it has evolved in some organizations to describe instead strategic partnering efforts made by senior executives.

In the former case, all that is required is buying an item that has already been specified, from a vendor that has already been selected, for a purpose that has already been determined (e.g., a part to be installed on an assembly line). But procurement also involves the activities associated with deciding whether an item will be made in-house or purchased from outside vendors (i.e., the make/buy analysis), and deciding from which vendor—or collection of vendors—to purchase an item is commonly included within the responsibilities assigned to procurement organizations. Procurement further represents a central activity in terms of supply chain management, which seeks to integrate the processes and activities of vendors, suppliers, producers, and customers, and procurement executives are routinely relied upon to shape enterprise strategy based on opportunities to form partnerships, alliances, and joint ventures with "vendors."

As implied, the project management professional has a number of different lenses through which to view procurement. For instance, procurement has long been referred to in functional terms, which depict a division of labor (e.g., buying items vs. making them), specific job tasks (e.g., market research, obtaining vendor quotations), and worker skills (e.g., contract interpretation, negotiation). As another instance, procurement is also referred to in organizational terms, which depict a specific department or other organizational entity in the enterprise, complete with its own managerial hierarchy, worker roles, and organizational responsibilities.

Procurement is further referred to in terms of a system, which involves examination of its inputs (e.g., requirements, information), outputs (e.g., purchase orders, received vendor items), transfer function (e.g., vendor selection, vendor management), and environment (e.g., corporation, industry). But I find it particularly useful to describe procurement in terms of a process—actually, a set of processes interlinking vendors, producers, and customers along the supply chain—with its attendant work activities (e.g., requirements determination, source selection), actors (e.g., market researchers, buyers), organizations (e.g., purchasing, contract management), and technologies (e.g., electronic catalogues, communication networks). Although every lens offers a unique perspective, the process view is inherently cross-functional, interorganizational, and systemic, and it enables one to focus on those aspects that are most important in terms of project management, even supporting analytical efforts that can effect dramatic performance improvement.

In this chapter, I adopt a process perspective to provide a focused overview of procurement, and I illustrate how it interacts with and interrelates to other key enterprise processes of importance to the project manager. Because this chapter is followed by more detailed treatments of supply chain management, procurement practice, bidding/tender management, and contract management, I refrain from delving into these areas in any depth. Instead, I leverage the process perspective and show how it can be used to improve procurement performance on the project. This performance improvement focus presumes you understand the basic elements of project management (e.g., project planning, staffing, coordination, monitoring, intervention) and discusses some emerging project management techniques from current research. In addition to providing an overview of procurement in the project management context, I hope to make a contribution in terms of these new techniques for the project manager's toolbox, and we critically review the current state of procurement practice to provide some guidance for effective management. The chapter closes with key conclusions pertaining to the process view and emerging techniques for the management of procurement in the context of project management.

## Procurement Process

Nissen (2001) describes procurement in terms of complementary processes, which provides a useful perspective for understanding and management.

## Complementary Processes

Two complementary processes are involved with procurement: (1) customer buying and (2) vendor selling. Although these customer and vendor processes can be viewed as separate intraorganizational activities within each of the respective buying and selling enterprises, a strong case can be made for viewing such activities *together*, as an integrated interorganizational procurement process. This reflects the integrated focus of prominent procurement models, such as the one used by Gebauer et al. (1998) to discuss the revolutionary potential of Internet- and Web-based procurement. Other models, such as the one proposed in Kambil (1997), include variations on a similar process flow and set of activities.

I find the General Commerce Model (Nissen, 2001) to be particularly useful for the task at hand, because it is specifically designed to highlight those aspects of project procurement that lend themselves to process integration. Process integration (i.e., managing all supply chain participants and activities as single coherent whole) represents an emerging project management technique for use in the modern enterprise. In particular, this model makes explicit the activities and media associated with *exchanges*, which tie together the various participants and activities.

Exchanges are those activities that constitute procurement proper, and they demarcate interorganizational process integration points; such integration points highlight avenues for a project manager to influence procurement process performance. The General Commerce Model diagram presented in Figure 11.1 depicts the process flow (from left to right) associated with a procurement transaction or relationship. A transaction generally occurs over a relatively short period of time, whereas a relationship is more enduring. Whether ephemeral or enduring, most transactions and relationships can be seen to progress through the steps along the process flow depicted in the model. Clearly, these steps represent procurement at a very high level.

Referring to the figure, from the buyer's perspective, note that the process begins with the identification of some need (B1) and proceeds through sourcing (B2) and purchasing

## FIGURE 11.1. GENERAL COMMERCE MODEL.

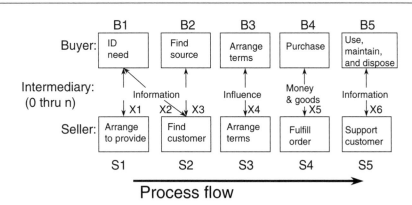

*Source:* Adapted from Nissen, 2001.

(B4) to the use, maintenance, and ultimate disposal of whatever product, service, or information is purchased (B5). The simple purchase by a price-taker (e.g., of commercial off-the-shelf, or COTS, products) will not generally include negotiation (B3) or involve significant terms other than price, whereas the exchange of influence through negotiation can become very involved in more complex procurements. The seller's process begins with some arrangement to provide (S1) a product, service, or information (e.g., through internal research and development, service process design, information acquisition) and proceeds through customer search (e.g., marketing and advertising; S2), pricing (S3), and order fulfillment (S4) to customer support (S5). The seller of COTS products may similarly not engage in negotiation.

The arrows connecting these high-level process steps are used to represent key items of exchange between buyer and seller. For instance, information is exchanged at several points along the process flow, as are money and goods (or services, information) and even "influence," as delineated at the negotiation stage. As depicted in the figure, zero or more levels of intermediaries (e.g., brokers, dealers, agents) can also participate in the process. Additional exchanges may also take place between such intermediaries and the buyers and sellers. In either case, by examining the activities through which information and other exchanges are made, one can acquire insight into those offering good potential for process integration, which can improve project performance dramatically.

## Enterprise Supply Chain Illustration

Here, I draw from recent field research (Nissen, 2001) to describe an enterprise supply chain for illustration. As delineated in the General Commerce Model, we find this enterprise supply chain is actually composed of two complementary process instances: customer buying and vendor selling. Specifically, the buying part of the process pertains to work done by the supply department at a medium-size government facility in the United States. As a government institution, this facility is subject to a full complement of procurement policies and procedures, not unlike those that govern the purchasing activities of most large enterprises, in both the public and private sectors. The selling part of the process pertains to work done by a leading-edge U.S. technology development company. This firm is a leader in its COTS product market and maintains an active research and development activity that drives frequent product introductions, updates, and releases. This kind of rapid product evolution has been noted as problematic for procurement in major enterprises such as the large corporation and government agency (see Nissen et al., 1998).

Referring to Figure 11.2, note that this enterprise procurement process is used to instantiate the General Commerce Model. Notice that the process includes a supply department intermediary, which performs specialized purchasing activities on behalf of diverse users (e.g., engineering, manufacturing, or sales personnel) in the organization. The enterprise process depicted here includes a more detailed delineation of activities than the General Commerce Model in Figure 11.1 does. For instance, the "ID need" (B1) and "find source" activities (B2) from Figure 11.1 are decomposed and expanded to account for a number of activities and exchanges that take place between the user, contractor, and supply department—for instance, conduct market survey (X2), complete purchase request (PR) form (X2'), research sources, and issue requests for quotation (X3). Exchanges internal to the buyer

## FIGURE 11.2. Enterprise Supply Chain Process.

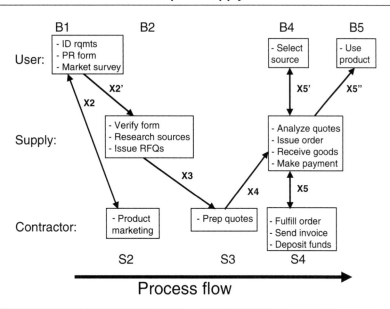

*Source:* Adapted from Nissen 2001.

organization are differentiated from their interorganizational counterparts by the prime symbol (e.g., X2', X5'). As another instance, the "purchase" activity (B4) from Figure 11.1 is similarly decomposed and expanded to account for more detailed activities and exchanges—for instance, analyze quotations, select source (X5'), issue order, receive goods, and make payment (X5).

As is appropriate for a general model, not all of the General Commerce Model activities are relevant to this particular enterprise supply chain process. For example, the buyer does not arrange terms (B3) for simple COTS purchases. With more complex procurements, arrangement of terms by the government buyer can be extensive. Alternatively, notice the seller effectively performs this activity (S3) for both parties when it prepares quotations. Such quotations will generally include information pertaining to product specifications, warranty, delivery, payment, and other terms set by the seller in addition to price. In one sense, this enterprise procurement process has evolved and specialized so that only the seller arranges terms. The other activities are relatively straightforward and map neatly to the General Commerce Model. From start to finish, we note a total of seven exchange points in the process between buyer, seller, and intermediary.

## Exchange-Point Analysis

In this section, I build upon the procurement process perspective mentioned previously to discuss exchange-point analysis and outline how it can be used by the project manager to improve procurement process performance. As noted previously, exchange points highlight opportunities for process integration, which is noted for performance improvement opportunities (Nissen, 2001). The key is, exchanges between different organizations along the supply chain (e.g., buyer, seller, intermediary) are associated with process *friction*, a phenomenon that is known well for increasing project cost and schedule (Nissen, 1998a).

Friction occurs whenever information or other exchange items must pass through separate organizational hierarchies (e.g., for management approval) or cross-functional specialties (e.g., from engineering to procurement), or involve market-based contracting mechanisms (e.g., legal documents). In terms of procurement process innovation (Nissen 2000), such exchanges are expressly labeled as "handoffs" to highlight their associated performance degradation. To the extent that the project manager can decrease friction, project coordination costs can be decreased and time can be saved in terms of project schedule. The key to process integration is to convert key elements of procurement and the supply chain process from one based on laws and commercial norms pertaining to markets (e.g., treating one's vendor as an *external* corporate entity) to one that relies upon trust and goal alignment through the hierarchy (e.g., treating one's vendor as an *internal* project group). Williamson (1975) provides a thorough treatment of markets and hierarchies.

Even though a customer and vendor may be organized as distinct legal entities (e.g., corporation, governmental agency, military unit), when they are mutually engaged in a common project effort, project-focused subsets of both customer and vendor organizations can align their processes with the common goal of project performance (e.g., satisfying project requirements, meeting cost and schedule estimates). This requires a different mindset for the project manager, one that diverges from "us versus them" thinking and envisions the customer and all supplier tiers as one integrated virtual organization (Davidow and Malone, 1992). Consistently, current best practices in industry and government project management include trust-based operations between customers and vendors (STSC, 2000).

However, in procurement as with shoes, one size does not fit all, and trust-based procurement is not for every organization and situation. For instance, where items to be purchased represent fungible commodities (e.g., wheat, standard grade steel, computer cables) that are offered through standard terms, price generally represents the single decision criterion, and there may be little to gain through developing close relations with any single vendor. Alternatively, where criteria include factors other than price alone, trust-based relationships can be critical. For instance, if you are using just-in-time or similar techniques to minimize inventory costs, you are likely to depend upon vendors to keep your line running, and you may need them to make short-term production adjustments and rush deliveries to meet your unanticipated demand changes. Consider this heuristic: Where a vendor becomes critical to your success, it may be worthwhile to invest in trust-based relations.

## RFP/RFQ Example

To help make these ideas concrete, consider, for example, the kinds of information exchanges that routinely occur between customers and prospective vendors associated with a request for proposal (RFP) or quotation (RFQ). Such exchanges are delineated as "X3" in Figures 11.1 and 11.2. A typical buying organization will spend considerable project time and effort developing specifications for an item to be procured, conducting market research to identify the kinds of product capabilities possessed by industry vendors and preparing often-lengthy quasi-contractual documents to request formal proposals (i.e., offers) or binding price quotations.

In turn, a typical selling organization will spend considerable project time and effort analyzing specifications, preparing cost and schedule estimates, and preparing often-lengthy contractual documents in the form of formal proposals or firm quotations. Because of their quasi-contractual and often binding nature, such documents generally must pass through one or more levels of management review—including legal analysis—before being officially transmitted through informational exchange. Although RFQs and price quotations can sometimes be prepared in a matter of minutes or hours for simple commercial and catalogue items, RFPs and proposals associated with large and/or complex projects often require weeks or months to prepare. This RFP/proposal information flow depicts a prime example of an exchange point that is ripe for process improvement through process integration.

For instance, the project manager can ask him or herself: How would such information be exchanged *within* our own organization? Would the same number and levels of management review be required? Would the lawyers be required to review all of the documents before transmittal? Would the documents require several days' processing time to wend their way through two organizations' internal mail systems and some third-party delivery enterprise (e.g., the post office)? Would the specifications require exact definition before being shared with product scheduling and manufacturing managers? If the answer is yes, then one's internal organization represents a rich area for process redesign. But if the answer is no, then exchange point analysis has identified several promising opportunities for process improvement, and such improvement can be made, in many cases, simply by treating the customer or vendor on a project as an "insider" as opposed to an external legal organization.

Why not empower project managers of both buyer and seller organizations to approve all project transactions, for example, with a caveat that lawyers from both sides must eventually (e.g., long after the items have been procured and used for the project) agree on a common set of terms and conditions? Some aspects of this suggestion are included under the rubric *alpha contracting* (Nissen, 1998b), but exchange point analysis enables the project manager to reach more deeply into the fundamental elements that join buyer and seller processes.

This specific approach (e.g., project manager empowerment, deferred legal review) is not so important as its ability to illustrate the kinds of procurement process changes that can be envisioned through exchange point analysis. Indeed, notice that this particular approach involves no technological innovation or even organizational change; only the respective attitudes (e.g., trust) between project customer and vendor require change, along with some intraorganizational rule changes (e.g., project manager authority) and process-work flow modifications (e.g., legal review).

Clearly, other changes such as employing network technology to electronically link customers and vendors, establishing electronic catalogues and virtual malls, and implementing intelligent software agents can further innovate the procurement process associated with a project, and organizational changes (e.g., forming project-oriented joint ventures) offer even more dramatic opportunities for buyer/seller process integration. Additionally, the other exchange points noted in the preceding figures (e.g., X1, X2, X4, X5) offer further opportunities for process integration and the associated performance improvement. But this example should convey the key ideas and illustrate how opportunities for procurement process improvement can be envisioned through exchange point analysis in the context of project management. The process view of procurement is central to envisioning such opportunities.

## Procurement Guidance

The current state of procurement practice reflects dynamic interaction between two opposing forces, and as such it remains in considerable flux. On the one hand, strong drivers both internal (e.g., increased corporate accountability) and external (e.g., electronic commerce) to the project organization press for change. On the other hand, procurement has been practiced by organizations for hundreds if not thousands of years, and through a Darwinian process, the systems and techniques in practice today are proven in terms of efficacy. Indeed, most organizations have volumes of procedures describing the procurement process in considerable detail, and resources such as the PMBOK lay out all the basic process steps. So what should the project manager think about the state of procurement today? What advice can we give for keeping the best of the old without missing opportunities of the new? Based on continuing research into advancing the practice of procurement, I offer guidance in the following seven project management heuristics or rules of thumb:

1. *Don't tinker with procurement systems.* The colloquialism "if it ain't broke, don't fix it" is not an exemplar of good grammar, but it captures an important consideration in terms of project procurement: Most systems and procedures serve a useful purpose and are understood by the people in your organization. Unless there is compelling reason for change, the project will likely be better off if you refrain from tinkering with the procurement process. Many managers seem compelled to involve themselves in every detail of a project, and such detailed involvement is clearly warranted in many areas (e.g., maintaining political support for the project, assessing complex technical/financial trade-offs). But system change is costly, as people's performance levels decrease reliably when they are required to learn new systems and procedures. Particularly in mature organizations with stable environments and established procedures, procurement systems is likely to be the last area requiring detailed project management attention.

2. *Do manage the critical path.* Notwithstanding the guidance in the preceding heuristic, I noted at the beginning of the chapter that procurement often lies on the critical path of a project, and we discussed exchange point analysis as an approach to systematically identifying opportunities for process enhancement. As such, any significant improvement in procurement efficiency or efficacy can effect direct improvement to the project as a whole, and the project manager sits in a prime position to view potential improvements. However, it's important for the project manager to balance the potential for performance gains through procurement enhancement with project risk. Many techniques for decreasing procurement cycle time (e.g., concurrent vendor development, close

customer-vendor interaction, increased trust) require additional coordination and increase project risk. Reducing the *most likely* project duration (e.g., by accelerating procurement schedules) only makes sense so long as management is also willing to accept the *worst-case* scenario that may emerge if coordination fails and vendor projects spin out of control.

3. *Question the matrix.* Most modern projects are organized using matrix management; that is, most people working on modern projects belong to two organizations: a functional group and a program or product group. In the case of procurement, the question is whether and how to integrate procurement people into the project organization. On the one hand, where standard commodities are required for a project, and specifications and schedules can be developed well in advance of their need, there is little advantage to having procurement specialists join the project team. Indeed, excluding procurement specialists from the project team reduces the number of people requiring direct supervision by and attention of the project manager. On the other hand, many project organizations depend critically upon certain key vendors, and understanding specific vendors' capabilities in detail can be central to project success. In such cases, procurement specialists on the project team may be indispensable.

4. *Balance efficiency with flexibility.* Efficiency is key to controlling costs and earning profits. But so is developing a product that sells and adapting to marketplace shifts. Efficiency is often obtained through standardization of procedures, specialization of labor, and buying in quantity with long lead time. However, standardized procedures are not generally flexible to change; specialized personnel require time and money to train for different tasks; and large-quantity, long-lead time contracts can be expensive to change or break. The project manager must maintain a balance between efficiency and flexibility, and such balance is likely to shift over the life cycle of a project, from one project to the next, and certainly across different technologies and industries.

5. *Benchmark.* Procurement systems and processes tend to be quite visible outside of organizations, and the astute project manager can periodically scan the environment to ascertain how various other organizations (e.g., competitors, partners, suppliers, customers) conduct their procurement processes (e.g., in terms of the four previous heuristics). Through such periodic looks, one's own procurement processes can be compared to those of other organizations to assess where one stands. Such assessment is commonly referred to as *benchmarking*, and it represents a relatively simple and inexpensive source of ideas for potential improvement—as well as confirmation that one's own processes are performing well (or at least adequately). Specialists and managers within the procurement organization are likely to perform benchmarking on a continuous basis, so procurement benchmarking information is often as close as a telephone call.

6. *Time carefully advancing information technology.* As noted, procurement is an exchange-oriented process, and information represents by far the principal object of exchange. Thus, advancing information technology offers omnipresent potential to enhance the procurement process, and because information technology advances so rapidly (e.g., annual doubling in terms of performance/price), last year's infeasible approach may very well develop into this year's competitive advantage and next year's crisis. Unfortunately, information technology improvements to a procurement process generally take

substantial time for implementation, so one must anticipate promising new technologies well in advance of their integration into the project organization. Further, information technology implementations often represent complex projects themselves, so the project manager can find him- or herself managing *two projects* (i.e., one concerning information technology and one concerning the organization's products) simultaneously. Careful timing is required here, as a project manager cannot generally afford to get too far ahead or too far behind in terms of information technology.

7. *Manage software procurements closely.* For over 50 years, software projects have been consistently underestimated in terms of cost, schedule, and complexity, and the success rate of large software projects in particular is dismally low. All projects clearly share many similarities, but it is important to note those aspects of software projects that make them unique (e.g., software is intangible, quality is difficult to evaluate, requirements are hard to specify and keep static) and manage them separately. Further, organizations rely increasingly upon vendors for software expertise, so the project manager depends upon the procurement system to select a capable vendor and establish means for effective coordination. As software becomes increasingly complex and products become increasingly software-intensive, managing the procurement of software can only become more critical to project success. There is no substitute for people with software (procurement) experience, and your project may benefit from one or more specialists if it meets the criteria above.

## Summary

Effective procurement is critical for effective project management; if a project manager is not managing procurement, then he or she is only managing 50 percent or less of the project as a whole. The project management professional has a number of different lenses through which to view procurement, but I find it particularly useful to describe procurement in terms of a process, as it enables one to focus on those aspects that are most important in terms of project management, even supporting analytical efforts that can effect dramatic performance improvement.

In this chapter, I adopted a process perspective to provide a focused overview of procurement and illustrated how it interacts with and interrelates to other key enterprise processes of importance to the project manager. But, presuming you understand the basic elements of project management, I introduced the emerging project management technique from current research called exchange point analysis. In addition to providing an overview of procurement in the project management context, I hope to make a contribution in terms of this new technique for the project manager's toolbox, and we critically reviewed the current state of procurement practice to provide some guidance for effective management. In terms of project guidance, we developed and discussed seven heuristics for the project manager: (1) Don't tinker with procurement systems; (2) Do manage the critical path; (3) Question the matrix; (4) Balance efficiency with flexibility; (5) Benchmark; (6) Time carefully advancing information technology; and (7) Manage software procurements closely.

In closing, research to develop the kinds of emerging project management techniques and heuristics outlined in this chapter seeks to extend the state of the art in terms of both the theory and practice associated with project management. As such, techniques and heuristics have now emerged from the drawing board and laboratory to help inform practice. I feel they are appropriate for inclusion in a book such as this. Additionally, I remain very interested in tracking projects in which emerging techniques and heuristics are employed. Project tracking along these lines will help to assess and refine further the associated project management knowledge, and I hope to incorporate important lessons learned through practice into other techniques and heuristics that are still being conceived and developed. This represents the kind of partnership between the academic and practitioner that can decrease the friction often associated with exchanges between theory and practice, and such partnership is essential to enable the flow of project management knowledge from the laboratory to the field. I welcome the opportunity to continue our existing partnerships and engage in new ones.

# References

Davidow, W. H., and M. S. Malone. 1992. *The virtual corporation.* New York: Harper Business Press.

Gebauer, J., C. Beam, and A. Segev, A. 1998. Impact of the Internet on procurement. *Acquisition Review Quarterly*. Special Issue on Managing Radical Change. 5(2):167–184.

Kambil, A., 1997. Doing business in the wired world. *Computer* 30 (5, May): 56–61.

Nissen, M. E. 2002. "An Extended Model of Knowledge-Flow Dynamics," *Communications of the Association for Information Systems* 8, pp. 251–266.

———. 2001. Agent-based supply chain integration. *Journal of Information Technology Management*. Special Issue, Electronic Commerce in Procurement and the Supply Chain. 2(3):289–312.

———. 2000. *Contracting process innovation.* Vienna, VA: National Contract Management Association.

———. 1998a. Redesigning Reengineering through Measurement-Driven Inference. *MIS Quarterly* 22(4):509–534.

———. 1998b. Alpha contracting JSOW style. *National Contract Management Journal* 29(1):15–32.

Nissen, M. E., and J. Espino. 2000. Knowledge process and system design for the Coast Guard. *Knowledge and Process Management Journal*. Special Issue: Into the "E" Era. 7(3):165–176.

Nissen, M. E., and E. Oxendine. 2001. Knowledge process and system design for the naval battlegroup. *Knowledge & Innovation: Journal of the KMCI* 1(3):89–109.

Nissen, M. E., K. F. Snider, and D. V. Lamm. 1998. Managing radical change in acquisition. *Acquisition Review Quarterly*. Special Issue on Managing Radical Change. 5(2):89–106.

STSC. 2000. *Guidelines for successful acquisition and management of software-intensive systems.* Version 3.0). Software Technology Support Center, Hill AFB.

Williamson, O. 1975. *Markets and hierarchies: Analysis and antitrust implications.* New York: Macmillan.

CHAPTER TWELVE

# PROCUREMENT SYSTEMS

David Langford, Mike Murray

No single issue has dominated project management in certain industries more than procurement. Yet what is it? The *Oxford English Dictionary* defines procurement as the "act of obtaining by care or effort, acquiring or bringing about." The Association for Project Management (APM, 2000) describes procurement as "the process of acquiring new services or products. It covers the financial appraisal of the options available, development of the procurement or acquisition of suppliers, pricing, purchasing, and administration of contracts. It may also extend to storage, logistics, inspection, expediting, transportation, and handling of materials and supplies." Within this generic definition we can see that procurement in a project management sense has a wider definition; it is really about management on behalf of a client or user of a product that is delivered using a project process.

The APM definition of procurement suggests that this process is undertaken in a linear and sequential mode. However, much anecdotal evidence would suggest that the process is often less rational, conducted in an iterative mode, and often influenced by political game playing, groupthink, and even unethical or illegal decision making. Risk management procedures are rather ineffective at combating such behavior. Indeed, the generic definition of the term requires the service of a shoehorn to make it fit different industrial sectors and nationalities. This is evident if we consider the description of the procurement process given below. The definition is taken from a U.S. Web page (www.mgmtconcepts) advertising instructional courses for procurement managers. It clearly considers procurement to be undertaken by an in-house department involved in a buying-selling relationship with external suppliers. The course syllabus is broken down into seven areas that represent six key procurement processes: procurement planning, solicitation planning, solicitation, source selection, contract administration, and contract closeout.

1. *Procurement planning.* Factors in the decision, make-or-buy analysis, and contract type selection
2. *Outsourcing and partnering.* Reasons to outsource, global outsourcing market, buyer/seller relationships, the partnering process
3. *Solicitation planning.* Tools and techniques, specifications, procurement document contents, evaluation criteria
4. *Solicitation.* Developing qualified sellers lists, contacting prospective sellers, conducting a bidders conference
5. *Source selection.* Screening and weighting systems, proposal scoring, contract negotiation strategies, making the decision, elements of a contract.
6. *Contract administration.* Roles, responsibilities, and coordination; kickoff meetings, project performance responsibilities, change control procedures, contract administration, payment system
7. *Contract closeout.* Contract documentation, steps in the claims process, termination of contracts, lessons learned

This syllabus is particularly suited to organizations in the manufacturing and service industries that have procurement departments. However, although the generic principles of buying-selling are appropriate to all, the six key processes noted are less effective in describing the key milestone in procurement within the construction and engineering industries. Owners in this sector often rely entirely on external procurement advisers (public clients are likely to have internal procurement departments/procurement champions) who may be the lead consultant, and the importance of the brief-design-manufacture interface is critical.

Procurement involves selecting from a range of acquisition options. Buying is not necessarily the only, or even necessarily the preferred, one: Making, renting, or leasing may be equally valid options alongside buying. Such options can, however, lead to confusion in project-based industries. Manufacturers and purchasers of products may talk of acquisition, whereas in construction the term "procurement" is most often used. However, even within this sector, the different professionals use preferred terminology. An estimator working within a contracting organization may talk of buying or acquisition, whereas the construction manager will consider these activities to be defined within the boundary of procurement. Thus, as the APM definition makes clear, the term "procurement" constitutes the system/process in which subactivities such as acquisition take place.

This chapter focuses on the strategic issues surrounding the decision to procure anew. The concept of procurement can, however, be seen to differ within various industrial sectors. The procurement structures used by the construction, aerospace, and motor vehicle industries can be considered broadly similar in that an end product is produced via a distinctive design and construction process. Nevertheless, they can also be considered a service industry, and within these three sectors, significant differences in practice exist—the high levels of subcontracting (also referred to as outsourcing) used in construction, for example.

Other industrial sectors may commission special in-house projects as a means to launch a new product, but procurement may be less concerned with design and construction than on "procuring" (often termed "solicitation") the material resources that are used to manufacture their products.

While many of the illustrations used in this chapter are taken from UK construction practice, the broad sweep of the movement from traditional methods to performance-based procurement systems has been observed in many project-based industries around the world. As such, the illustrations are local but the concepts are universal. Indeed, the increasing interest in company benchmarking has led to much cross-fertilization of procurement practice between industry sectors.

## Strategic Issues in Procurement: Examples from UK Construction

In an era of the "cult of the customer" where business relationships are perceived to be the dominant driver of business performance, there is increasing interest in the role of client satisfaction in respect of the project process. Government reports (in the United States and the United Kingdom) since the 1940s have complained that many project-based industries from construction and provision of IT services to defense have not served clients well (Morris, 1997; Murray and Langford, 2003) and that overruns on time, budget, and poor quality beset the project process.

## Client Satisfaction

Naturally one of the most important client roles in the procurement system is the provision of a brief. This "statement of need" has to flow from the client's expectation in terms of the function of what is to be delivered, and performance in terms of time, cost, and quality. (See the chapter by Davis et al. on requirements management.) Clients will typically want an appropriate level of involvement in the whole procurement process; this involvement may be minimal, but client satisfaction needs to be defined early on, as it will have powerful implications for the selection of the procurement route. The issue of risk sharing between the client and the project delivery team is at the heart of such early decisions about procurement, and as Rowlinson (1999) says, the client's role is a strategic one—it sets the project objectives and the project contractor's role is to turn such objectives into a finished product that satisfies the client, the users, and society at large.

Procurement—and to a large extent therefore project management practice—varies between industries. But even within industries, practice (of both) varies.

In the construction industry, for example, procurement strategies of today still follow methods of procurement that were available to the medieval builder. The master mason designed and built the structure; he (for it invariably was) designed and managed the works by engaging with other craft guilds to undertake such packages such as woodwork, roofing, glazing, and so on. Since that time professional roles have emerged that have fragmented the design and construction process. (See the chapters by Morris on construction, and Cooper on process modeling.) Most government reports into project-based industries since World War II have complained of this fragmentation, and since the late 1960s, attempts have been made to integrate the design and of many of the professional roles used to deliver major projects. (This has paralleled other initiatives aimed at improving project management practice.) Let's look in a little detail at the various procurement options taking construction as an example.

# Sequential Models

## Traditional Approach

Here the key feature is the separation of design from construction/manufacture with the sequence of events being design-tender-build/manufacture, and the design phase being divided up into roles or specialist consultants. In this system the lead designer has a key role in leading the design function and managing the project. The project delivery company is likely to be selected by competitive tender, and the work of delivering the project will be subcontracted to specialist trades. The structure is as shown in Figure 12.1.

## Accelerated Traditional

Because of dissatisfaction with the traditional elongated and often disputatious process, this model is often replaced with a fast-track model that overlaps design and manufacture/construction. With this model, project work starts as soon as sufficient design is available. The process is shown in Figure 12.2.

The design work is carried out with the same range of consultants, but the contractor will be selected by a negotiation or a pricing of a schedule of likely combinations of materials and processes. The roles and relationships will be as those in the traditional model.

## Design-Build

The disadvantages of the traditional methods are that the organizations charged with design on the one hand and with realizing the design on the other are separated. Consequently there is a lack of single-point integration and responsibility, and fragmentation increases. To

## FIGURE 12.1. TRADITIONAL PROCUREMENT.

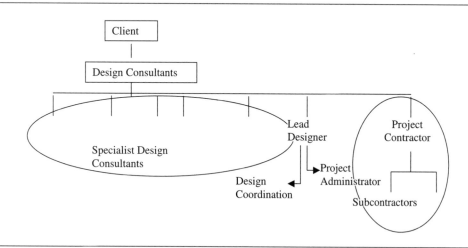

### FIGURE 12.2. TRADITIONAL (ACCELERATED) PROCUREMENT.

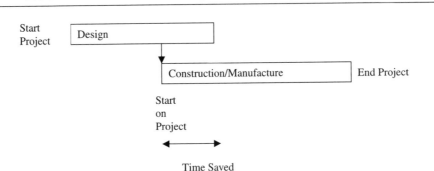

counter this, most project-based industries have a form of design-build model providing greater unification of responsibility (though in construction the traditional system still retains a strong presence throughout the world). In construction this method of procurement has evolved such that it has many variants. Broadly speaking, the design-build organization tenders/bids or negotiates with the client prior to commencing design work. Part of the tendering or negotiations will be schemas that demonstrate how the design-build organization is going to respond to the client's needs. In such circumstances, different costs will be associated with design solutions, and thus bid evaluation can be complex. Three main organizational configurations can be seen.

- *Integrated design-build.* This occurs when the design, costing, and implementation expertise lie within one organization and the contract for work is between the design and deliver company and the client.
- *Separated design-build.* This arrangement puts together a temporary organization comprising the necessary design and construction expertise. The project team is a consortium of independent practices for design and a construction or manufacturing firms, put together for a specific bid. It may well be that this arrangement has some semi-permanence in that every time a design-build job is imminent, a broadly similar set of firms configure themselves to present a design-build solution. Alternatively, bespoke arrangements are made to suit a particular kind of project.
- *Novated design-build.* Third, a more recent innovation is novated design build. The word "novated" is taken from the meaning of *novation*, which means substitution of a new obligation for the existing one. In this case the "new obligation" is for the project contractor to take over "existing" designs drawn by a lead designer. The system works by the client commissioning a design firm, often selected through an invited competition, to create the concept, key features of the design, and outline budget. When the conceptual work is agreed, then the designer's work is handed over to the project contractor who then "owns" the design and so develops it to a condition that enables the design to be realized. The method has the benefit of delivering solutions that have a schema design

developed without production issues dominating the designers' thinking, thus allowing design flair to flourish as the contractors develop the design with an eye to retaining the design concept while producing production drawings. This brings benefits in terms of assembly, function, and value. In short, the method seeks to maximize the design benefits of the traditional system and the production benefits of the design-build model.

The contractual relationships of all three variants are shown in Figure 12.3.

## "Management" Systems

This model is characterized by the separation of the management system and the operational system required to deliver projects. The technical system—the design and manufacture/construction—are undertaken by specialist organizations, while a construction manager (or management contractor), or project management organization, provides the integrating management system.

The management organization is appointed early on in the project process, its role being to coordinate the design, procurement, and logistics to be used in common by all contractors and to provide the project control (overall scheduling and cost management). In the United Kingdom there are two principal variants to the construction form of this system: management contracting (MC) and construction management (CM), the differences being shaped by the relationship between the parties to the contract. The two forms are summarized in Figure 12.4.

As Tookey et al. (2001) have pointed out, such arrangements are often tailored for a particular project with payment systems, legal arrangements, and specialist contractor selection methods all being bespoke. This has an influence upon professional roles and relationships. Walker and Newcombe (2000), for example, see that the social system of a project is influential with the personality and power bases of the leaders of the various organizations engaged in the project shaping performance in practice. This theme will be explored in greater depth later.

## FIGURE 12.3. VARIANTS OF DESIGN-BUILD.

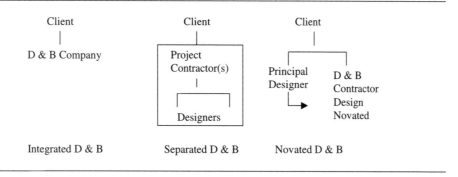

## FIGURE 12.4. MANAGEMENT SYSTEMS.

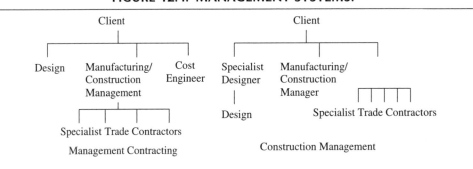

The advantages of the system are really drawn from Adam Smith's (1838) classical economic theory proposed in the *Wealth of Nations*, that of specialization producing economic benefits because of lower costs and more efficient production. The use of specialist trade contractors enables designs and installations to be rolled out as the project progresses, thus ensuring a fast-track approach. Moreover, the specialist packages can be tendered and so have the dubious benefit of work being let for the lowest cost. (This philosophy is waning in the face of "best value" rather than lowest-priced procurement strategies).

In construction, "project management"—as a procurement form rather than as a discipline—is similar to construction management, the principal difference being that the scope and authority of the project manager is likely to be greater that the CM. The PM organization will be appointed early and acts as a quasi-client. The PM may be asked to undertake on behalf of the client functions such as a site acquisitions, arranging the funding, obtaining necessary permissions—in effect undertaking all of the client-led feasibility studies. The project manager will commission the designs and constructors. In short, the PM acts in lieu of the client organization.

## Explaining the Changes

The drive toward a more professional role-based procurement model has surpassed the more contractual arrangements of the past. Curtis et al. (1991) see the spectrum as shown in Figure 12.5.

These changes in construction procurement need to be understood in the context of organizational theories and social relations between the parties. In the following, the framework used draws upon the construction industry, although the argument applies in other project based industries.

## Organizational Theories and Procurement

Of key importance is how the players in the project process make sense of the world of procurement. Hence, procurement is an "issue" that carries a definition of something that

## FIGURE 12.5. ROLE-BASED PROCUREMENT.

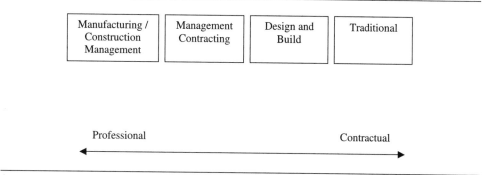

*Source:* After Curtis et al. (1991). Copyright CIRIA Publication SP81.

is "problematic and requiring action." In the context of procurement of projects, major clients, including government, have complained of poor performance of the project-based industries since World War II. Murray and Langford (2003) reviewing postwar government reports into the UK construction industry conclude that procurement is one of the "issues" for the industry to address. Similar evidence may be found in defense, shipbuilding, and other capital project industries.

The earlier models of traditional procurement have been found wanting. The emphasis upon "contractual" rather than "trust" relationships used to link the parties together has all too often, as the Latham report (1994) complained, ended in court. Such contractual arrangements relied upon a machine metaphor (Morgan, 1986) to explain how the procurement system worked. Each role in the process was carefully defined and well understood and was underpinned by contracts that spelled out the legal obligations and responsibilities of each party. Codes of practice for tendering and plans of work proceduralized the process. The parties in the contract have a role that may be replaced by others performing a similar role in the same way that a defective clutch on a car is replaced. The organization behaves in a hopefully predictable way.

As projects become more complex in technical and organizational terms, the machine metaphor begins to be no longer capable of containing the aspirations of different groups in the process. Clients become unsettled by delays and cost overruns and seek to assert authority over the project process. Dealing with a multiorganization and multicontractual environment expressed in the traditional paradigm becomes irksome. Hence, in a bid to work with "single-point responsibility," organizations encouraged the growth of design-build contracts. In Morgan's terms this would be the "organic" metaphor at work. The contractors who formerly were engaged as the principal builders of a project and had little to do with design organically evolve to fit the needs of the new environment. The client buys construction services from one organization, and the contractor, by integrating design with construction, reduces the physical and financial risks associated with converting someone else's design into reality. The shift in client expectations created preferences for changed procurement systems.

A third epoch may be detected in the development of different procurement routes: the move from design-build to management forms of procurement. Powerful clients and contractors looking at project based work in the United States in the early 1980s saw that practice there seemed able to deliver projects faster and cheaper than in the United Kingdom. The U.S. Construction Management form came into the UK environment as an alternate to management contracting and stimulated other forms of innovation procurement, not least project management. Clients and constructors no longer just polished the mirror to "tell it like it is." The challenge for such innovative organizations was to "tell it like it might become" and to unseat conventional assumptions about roles and relationships in the procurement process.

The metaphor here is of the organization as a "political" entity. Different interest groups have conflicting ideas of how to move forward. The contours of the professional roles start to change: Designers, for example, saw a diminished responsibility with fewer managerial duties. Cost advisors to the client saw a new future in project management as traditional ways of pricing projects began to fade away and new consultants such as value engineers, program managers, health and safety managers, and so on began to emerge. Inevitably in such paradigm shifts in roles and relationships, bitter arguments break out. The idea that "quality" refers to the product over long periods of time, it is said by some, is replaced by the concept that quality lies in the project process. This further diminishes the power of the design professionals.

In breakdown situations such as this, the old political order begins to fragment. Disputes then become more prevalent. Clients and government consequently have to seek new order. In the UK construction industry, for example, legislation has been introduced to restore order by making arbitration mandatory, ordering prompt payment to trade and specialist contractors. The phase of the political metaphor is complete.

This shift from traditional to management contract can be read as a reflection of the shifting nature of social hierarchies in society. In the traditional system the designer was "commissioned" as an independent artist and the contractor was "contracted" because legal constraints were needed to enforce performance. In this model the designer with extensive power and authority represents the ruling class and the contractor the working class. As society progresses, the class barriers are perceived to fall and political and social plurality is encouraged. The power exercised by the designer in relation to other professionals diminishes, and power and authority becomes more equally spread amongst the participants in the project process. In the new situation, the power of the client is emphasized and brought into greater relief. In the UK construction industry, the Egan report, "Rethinking Construction" (1998), legitimized the new authority of the client.

The new epoch is governed by the cultural metaphor. Here strong uniform norms of behavior are expected from all players in the project team. Epithets such as "teamwork," "singing from the same songsheet," and "pulling together" become watchwords of the cultural phase. In the UK construction sector, the trigger for this was the Egan report, which identified key issues in the procurement process that commanded reform. Central to these was the one that the procurement process should concentrate on customer focus with a strong emphasis upon measurement of a range of outcomes that could lead to improvement for all. In the United Kingdom the transformation of the industry culture is well under way and has shaped the procurement practices by the introduction of supply chain management to not only align the technical aspects of procurement but also to create project cultures that harmonize or at least make less diverse the cultural differences between participants in a project. (See the chapter by Venkataraman.)

The UK government sought to reshape government procurement by installing "best value" rather than "price" as the leading agent of contractor selection. Again, cultural values were seen as an important part of the "best value." Government agencies such as the UK Ministry of Defence and the Health Service refashioned their procurement practices to engage "prime contractors" who manage projects based upon functional and performance driven criteria. For example, the army may wish to procure a facility for keeping soldiers fit, a prime contractor is selected, and an early role will be to define whether the requirement is best delivered by the provision of a gym or an outdoor assault course. In this environment it is important for all those seeking to work with major clients to be compliant to this new culture. This means integration of software systems and capacity to engage in e-procurement. The journey is depicted by Kumarasawamy et al. (2003) in Figure 12.6, and we shall turn now to consideration of this new procurement paradigm.

## Performance-Based Methods of Procurement

Performance-based procurement systems (PBPS) are being implemented within a wide range of project-based industries as a means to overcome problems associated with performance. The culture of soliciting suppliers on the basis of lowest bid price has been common in many industries, and particularly international construction. The low barrier to entry in this sector has exacerbated such conditions, and the dissatisfaction often experienced by the end user of the product/service is well documented. However, experienced owners who regularly commission projects are now seeking "best value" rather than simple capital cost reduction in a product/service.

Within project-based industries, best value procurement incorporates the following principles:

- Integrating value management and risk management techniques within normal project management
- Defining the project carefully to meet the user needs
- Taking account of whole life costing
- Adopting change control procedures
- Use of partnering arrangements
- Not appointing suppliers on lowest cost

The end to a reliance on formal contacts is also a core philosophy behind PBPS, whereby the relationship between buyer and sellers emphasizes partnership rather than distrust. Competition is based on clear targets for improvement, in terms of quality, timeliness, and cost.

The terminology used by different organizations to describe this relatively new model of procurement is evident if we consider NASA. Goldin (1999) refers to the implementation

## FIGURE 12.6. EXAMPLE OF A FORCE FIELD "AGAINST" RELATIONAL INTEGRATION IN A CLIENT-CONTRACTOR RELATIONSHIP.

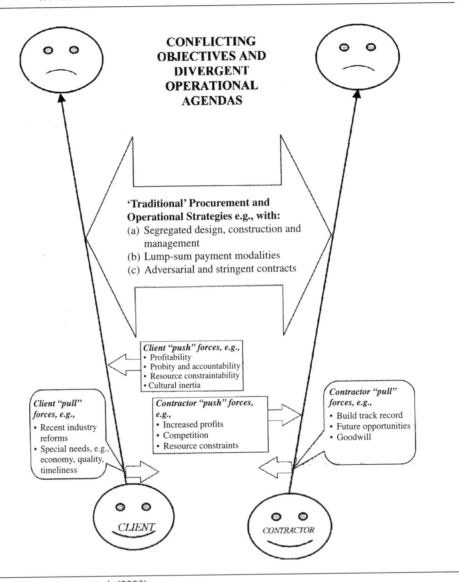

*Source:* Kumarasawamy et al. (2003).

of performance-based contracting (PBC), whereby all aspects of acquisition are structured around the purpose of the work to be performed as opposed to how the work is to be performed or broad and imprecise statements of work.

Kashiwagi (1997) notes that the success of PBPS rests in the supply of performance information and suggests that a facility owner use the supplier information in the list that follows. Although referring to the U.S. construction industry, any purchaser could use this performance information checklist.

- Expertise and experience
- Price
- Contractor margins, financial stability, and payment of subcontractors
- Previous size of jobs
- Previous types of contracts
- Completion rates on time and below budget
- Performance of previously constructed facilities of facility systems
- Personnel proposed for construction management.

The list should not be seen as exhaustive. Prequalification procedures are subject to evolution, and the current need to demonstrate both sustainability and ethical compliance are but two examples; both are increasingly added to such a list.

Kashiwagi notes the importance of information in facilitating a culture of trust between parties. The more information, the less distrust. It allows all parties to understand who and what all other parties bring to the partnership, including positive and negative characteristics. Figure 12.7 shows the PBPS proposed by Kashiwagi based on input from, inter alia, Motorola, Honeywell, IBM, McDonnell Douglas, Phelps Dodge, the State of Wyoming, and the U.S. Army command.

The transparency noted by Kashiwagi is also a core feature in the procurement of automotive parts at the Nissan UK plant in Sunderland. The "open-book" accounting allows Nissan to guarantee a fixed profit margin based on open-book accounting. If the price of raw materials rises, Nissan pays the supplier extra; if the raw material prices fall, Nissan expects to pay less. The Nissan factory also uses performance data to assess its suppliers. The benchmarking between the automotive and construction industry can be seen to exist (see Figure 12.8).

# Partnering

The concept of partnership within business transactions has longevity within many cultures. Conventional wisdom suggests that Japan may offer excellent benchmarking potential. The work of W. Edwards Deming in the 1950s and his contribution to the managerial revolution—Total Quality Management—can be viewed as a catalyst for the interest in partnering today. Deming's help in rebuilding Japan's post-WW II economy centered around continuous quality improvement, and today we consider this to be achievable through the use of

## FIGURE 12.7. PERFORMANCE-BASED PROCUREMENT PROCESS.

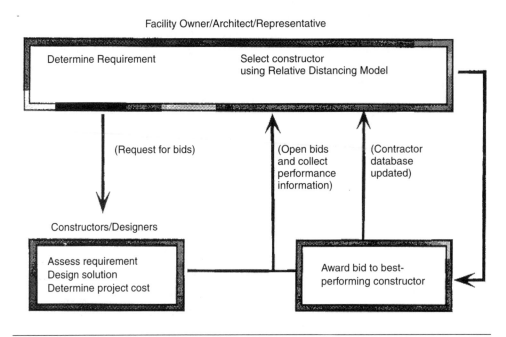

Source: Kashiwagi (1999).

integrated project teams who share mutual objectives and who can resolve disputes with a win-win rather than win-lose outcome.

Although the concept of partnering proliferated throughout much of Japan's industry, particularly car manufacturing, and although it has been subsequently adopted in Western manufacturing, process engineering, and defense-aerospace industries, it is not free from critical appraisal. Townsend (1996), for example, has doubts over the conventional belief that Japanese construction firms are more productive than Western counterparts.

Despite the potential for problems in partnering relationships, two UK construction industry reports, "Constructing the Team" (Latham, 1994) and "Rethinking Construction" (Egan, 1998) have both advocated the need for project teams to work together on serial contracts where a culture of continuous learning can assist in promoting project success. However, despite these laudable recommendations, the implementation of partnering in a project process is often confusing for all involved. Bresnen and Marshall (2000) have argued that partnering is an imprecise and inclusive concept capturing within it a wide range of behavior, attitudes, values, practices, tools, and techniques.

A recent survey of the benefits derived from partnering (Galliford, 1998) polled the views of over 500 managers and directors from organizations that directly appoint construction companies (see Table 12.1). The results show that clients regard partnering as providing

## FIGURE 12.8. REPLACING CONTRACTS WITH PERFORMANCE MEASUREMENT.

**Replacing Contracts with Performance Measurement**
Nissan UK and Tallent Engineering Ltd have no formal contract beyond an annual negotiation of the cost and quality of rear axles that Tallent produce for Nissan's cars, and rigorous targets for improving performance. Each morning Tallent receives an order from Nissan detailing the precise mix of axles required by Nissan and five times a day Tallent delivers to Nissan's Sunderland Plant. If a problem were to occur with quality Tallent would send engineers to Nissan to fix it on the car production line. If a problem resulted in a significant loss of production, Nissan would expect to be compensated by Tallent for lost business or vice versa, but this has never happened and both sides work hard to ensure it cannot. Both Nissan and Tallent use similar non-contracts agreements with the firms delivering their construction projects.

Nissan's QCDDM supply chain management system is acknowledged to be among the most effective in the world. It measures all suppliers on Quality, Cost, Delivery, Design, and Management against negotiated continuous improvement targets. For each element, the supplier is marked on a range of product and process items which are aggregated on a weighted basis to give performance percentage for the element. Competition is created across the supply chain by collating the performance information every month and informing each supplier of its performance in relation to the others.

*Source:* Egan (1998).

## TABLE 12.1. BENEFITS OF PARTNERING.

| Which of the following statements do you think best describe the benefits of partnering? | % |
|---|---|
| 1. Creates a sense of team-working on-site | 56 |
| 2. Identifies mutual objectives | 52 |
| 3. Solves problems faster | 49 |
| 4. Provides the best rout to a successful outcome | 39 |
| 5. Encourages the completion of projects within budget | 30 |
| 6. Encourages a higher quality of job | 23 |
| 7. Reduces lead in time | 23 |
| 8. Saves time | 19 |
| 9. Encourages completion of projects on time | 17 |
| 10. Reduces program time | 9 |
| 11. Saves money | 7 |

*Source:* Galliford (1998).

an atmosphere where a team can develop trust and cooperation and focus on achieving a common goal. And there is evidence that such conditions have the potential of reducing waste, thus offering better value for money while contributing to greater client and end user satisfaction (Chartered Institute of Building, 2002).

Partnering can be "project-specific" or "strategic," where the partners work together on a series of projects that are consecutive, if not continuous (CIB, 1997). Monaghan (2000) uses an interesting analogy to emphasize this difference when he talks of "partnering for comfort" (project) and "partnering for improvement" (strategic). Although project partnering is often considered to be less effective, Bennett and Jayes (1995) found that in a study of partnering in the United States, project partnering provided benefits, even where there was no possibility of the client providing further work. They do, however, recognize that strategic partnering provides more long-term benefits, since it allows for continuous improvement.

What are the major barriers to successful partnering in construction projects? Larson and Drexler (1997) note that partnering represents a paradigm shift in how one approaches construction projects. Their study revealed that a general level of mistrust existed between owners and contractors and that this was engineered by years of viewing and treating each other as potential adversaries. Of particular importance was the failure to build a true relationship of trust and the reliance of legal loopholes in documents. Other significant difficulties include the inability of the project management structures to synchronize goals of the numerous subcontractors, a misunderstanding of partnering concepts by upper management, and a failure to "walk the talk." Another investigation into the decline of relationships (Drexler and Larson, 2000) examined owner-contractor relationship over time to identify the factors that contributed to decline or improvement in the relationship. The research was based on response from 276 members of PMI. Four categories of owner-contractor relationship were identified:

- *Adversarial.* Parties perceive themselves as adversaries, with each party pursuing their own concerns at the other party's expense.
- *Guarded adversarial.* Participants cooperate within the boundaries of the contract.
- *Informed partners.* Participants attempt to sustain a cooperative relationship that goes beyond the boundaries of the contract.
- *Project partners.* Participants trust each other as equal partners with a common set of goals and objectives

Fifty-eight percent of the projects experienced some fundamental change in working relationship, either positive or negative. Projects that began as formal partnerships were most stable, with two-thirds ending as they had begun. Of the relationships that changed, half regressed to an adversarial relationship, while half progressed into some form of partnership. Several common themes were seen to develop as to why relationships improved or declined. Relationships that had deteriorated were characterized by unclear contracts and resulting litigation, changes in scope and schedule, personnel failing to perform, lack of trust, and underbidding contracts. Relationships that improved were characterized by trust and positive relationships, shared goals, teamwork and communication, personnel changes, and the presence of a clear contract.

In the United Kingdom the use of pioneering partnering contracts has formalized much of the ideology inherent in partnering. The Project Partnering Contract 2000 (PPC200) published by the Association of Consultant Architects (ACA) and the Engineering and Construction Contract (partnering option document) published by the Institution of Civil Engineers are two examples. The philosophy underpinning these contracts is built on a team-based multiparty approach within a fully integrated design/supply/construction process. The use of co-located offices and a common information system (project Web site) also help to promote a common objective, and the application of value engineering and risk management are considered prerequisites.

## Best-Practice Procurement

Three prominent owners who procure the services of the UK construction industry on a regular basis—British Airport Authorities (BAA), the Ministry of Defence (MoD), and the National Health Service (NHS)—have reformed their procurement practices and adopted a PBPS ideology.

### Framework Agreements (The Case of BAA)

In 1996 BAA implemented an initiative know as 21st Century Airports with the aim of establishing itself as the most successful airport operator in the world. This continues to be an ambitious challenge for a formerly public client privatized in 1987. The ambition to be "world class" had implications for BAA's development program, given that construction costs (around £500 million annually) have a direct impact on profitability. For those readers outside the UK, the relationship with the then chief executive and the current agenda to "rethink construction" should be made apparent. Sir John Egan, the chief executive of BAA between 1990 and 1998, had previously spent 11 years as chairman of Jaguar, the car manufacturer. In October 1997 he was commissioned by the deputy prime minister to lead a Construction Task Force, the remit being to examine the performance of the UK construction industry. Egan was considered to be a tough and demanding task master who had previously revolutionized the production process at Jaguar cars and was the driving force behind the changes in procurement practices taking place at BAA. Indeed, on comparing construction costs with other airport operators, he discovered that UK projects were comparable to those in Europe. In the United States, a different story was revealed: "In America my eyes were opened and I kept wondering: Is he talking in dollars—Christ that can't be possible, and they were saying: Yeah and we spent too much. They would be over budget and still be half the cost" (cited in *New Builder*, 1994).

BAA's framework process is based on contractors tendering to become framework partners, and the successful bidders are then allocated work on BAA's construction program for five to ten years. Under the agreements, the contractors are selected on various criteria that include price, quality of the company and its staff and products, attitude, and the ability to work with BAA. The successful designers and contractors are then designated into specific delivery teams grouped around different product ranges: baggage, fit-out, shell and core, and infrastructure. These prequalification "tests" were seen to strain many contractors, who considered them to be overly bureaucratic (European Union procurement legislation) and time-consuming and who were used to competing in a "lowest price wins" market. In contrast, the philosophy behind the framework agreements is based on continued improvements vis-à-vis longer-term and more open relationships where collaboration rather than confrontation is king (see Cox and Townsend, 1998, for an account of BAA's reengineering of the construction supply chain).

## MOD's Prime Contracting: Building Down Barriers

In the United Kingdom, the procurement of defense equipment such as tanks and aircraft has suffered major cost escalations and late delivery for many years, though recent moves to performance and partnering base procurement—"smart acquisition"—may now be reversing this trend (NAO, 2002a). "Building Down Barriers" (BDB) is a new procurement initiative pioneered by the MOD's Defence Estates (DE) department and supported by the government's Department of Environment Transport and Regions (DETR). Its purpose has been to create a learning mechanism for establishing the working principles of supply chain integration in construction. Two pilot projects established in 1997 involved the construction of recreation facilities and involved two "prime" contractors. In each project, the prime contractor was expected to integrate the supply chain into the building design, construction, and maintenance for a trial period of up to two years. The projects were monitored by the Tavistock Institute, who were also involved in establishing the project organization structures (see Holti, 1998; Holti et al., 2000).

The new organizational structure concept involved the concept of "work clusters." In each work cluster, the designers, subcontractors, and key suppliers were involved in a reasonably self-contained element of the building, undertaking a form of simultaneous engineering. Typical clusters included groundwork, frame, and envelope; swimming pools, mechanical, and electrical services; and internal finishes (see Figure 12.9). Within each work cluster the participants used techniques including value management, risk analysis, and risk management and contributed knowledge to anticipate the areas of interdependence that were likely to arise during detailed design and construction. The clusters were responsible not only for designing but also for delivering construction of their element of the building. Interdependencies that spanned the spheres of two or more clusters were resolved by taking them to an overall project team, where a cluster leader represented each cluster. Nicolini et al. (2001) note that the cluster process is in fact a collaborative, team-based approach to project decision making. The pilot projects are now completed and have shown positive benefits in labor productivity, material wastage, construction time, and whole-life costs.

## NHS Estates ProCure21

NHS Estates is the property arm of the UK's National Health Service (NHS) and has a capital works program worth about £3 billion a year. The aim of ProCure21 is to promote

## FIGURE 12.9. NEW INTER-ORGANIZATIONAL ARRANGEMENTS.

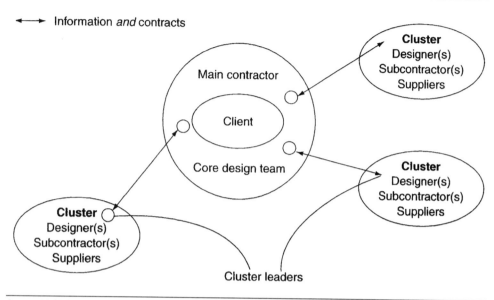

Source: Holti (1998). © Tavistock Institute of Human Relations.

better capital procurement in the NHS, the intention being to achieve this through delivering better-quality healthcare buildings and improved value for money by invoking a major cultural change. This reform involves the following:

- Establishing a partnering program for the NHS by developing long-term framework agreements with the private sector that will deliver better value for money and a better service for patients
- Enabling the NHS to be a best client
- Promoting the use of high-quality designs
- Monitoring performance through benchmarking and performance management

The need for such reform within NHS Estates is clear if we consider the disastrous procurement practices so prevalent in healthcare construction projects during the latter half of the twentieth century. Typically the gestation time to plan, design, and construct large healthcare projects lasted decades. During such a long process, the original design brief would "creep," with subsequent detrimental impacts on the original budget project program. Once completed, hospital projects were often found to be unsuitable for their end users, combined with excessive whole-life costs. Guys Hospital Phase 3 development in London is indicative of such conditions The National Audit Office (NAO 1998) investigated the project that was delivered three years late with an overspend of £68.7 million above its original budget. The key factors identified were as follows:

- Delay in putting design team in place
- Delay in resolving cost and funding problems
- Failure to freeze design and design changes
- Delay in designing the engineering services and producing drawings
- The insolvency of works package contractors
- Technical matters (defective copper pipework)
- Changes in statutory regulations (building regulations)
- Delays to the construction works
- Large number of claims associated with construction works
- Change to design team's fee rates
- Inflation

As with the Prime Contract initiative, ProCure21 means that NHS clients will work in partnership with substantially fewer suppliers. This will involve framework agreements with carefully selected companies from within the construction industry. The intention is to provide these companies with opportunities to undertake projects within an ongoing program of work at an appropriate level of profit. The current Procure21 pilot projects are said to be contributing to a saving of 4 percent in procurement costs. According to the initiative's program director, this can be contrasted with conventional procurement methods where an overspend of 8 percent is normal (Contract Journal, 2002d).

Principal Supply Chain Partners (PSCPs) who fulfill the selection criteria will be appointed onto a national framework. The PSCPs will take single-point responsibility for design and construction, and in the case of Private Finance Initiative (PFI) projects, facilities management and finance. Each PSCP has the responsibility for managing its integrated supply chain. Providers of key construction services appointed by the PSCPs are known as primary supply chain members (PSCMs). The PSCPs will also be responsible for appointing all other supply chain members. When a NHS client wants to construct a new scheme, it will select a PSCP, including its associated supply chain, from the framework agreement.

# Contemporary and Future Issues in Procurement

Technology and culture are two variables that have a large impact on the procurement process. Both are subject to constant and often unplanned pressure from both within and outside a project's boundary. The scientific management principles developed by the likes of Taylor and Galbraith (time and motion study) and adopted by Ford in their automotive plants were a product of this era. New project procurement enablers such as lean thinking and just-in-time are, however, steeped in process improvement. Thus, the procurement process could be said to be subject to management fads and fashions. This impacts on the various industrial sectors around the globe with varied degree and at different times.

The use of project intranets and electronic procurement has complemented the desire for integrated project teams and has permitted knowledge to be shared throughout project supply chains. Logistics management and its associated just-in-time philosophy are now a core ideology in some prestigious projects.

In addition, the recognition that adversarial "claims-ridden" procurement is destructive (notwithstanding the opportunities for construction lawyers!) has led to clients seeking less confrontational relationships where trust is the key motivator. This stakeholder view is becoming more apparent in society, and the topic of sustainability has encouraged clients to be more discerning about their supply chain, while contractors and consultants must demonstrate compliance with legislation. This has led to both external and internal forces moulding contemporary and future procurement strategies and to a post-reengineering world where strategic transparency and leanness are reflected in operational prefabrication, standardization, and mechanization However, one should not be deluded that risk and uncertainty have been removed from the procurement process. The sad toll of company insolvencies, accidents and deaths, litigation, and project failures continues to taint many project based businesses throughout the world.

## E-Commerce

E-business is the term used to describe activities that involve the sharing of information through electronic networks, including companies being able to sell or order and pay for goods online, check availability, and get further information on products. It can also include using IT for project collaboration (Construction Confederation, 2002). The Construction Confederation describes two types of e-business: "process" e-business that helps to manage the flow of information within industry supply chains and "transaction" e-commerce that includes selling products and services, the latter being on a business-to-business (B2B) or business-to-consumer basis (B2C).

The benefits of adopting e-business are commonly cited as a reduction in transaction costs and increased project personnel collaboration. The business case for the adoption of e-commerce is made explicit by Walker and Rowlinson (2000), who argue that clients who are dependent upon intranets, e-mail, and electronic transfer of information are unlikely to be impressed by contractors, suppliers, and design teams who have yet to grasp the technology of e-communication and are not using such tools effectively. This conclusion is supported by research undertaken by the Construction Industry Institute in the United States. Voeller (2002) found that owners are leading the implementation of e-procurement and that many of them were experimenting with or using e-marketplaces and reverse auctions. (However the attempts to initiate industry-wide e-procurement practices in the United Kingdom have been far from successful. A service termed Arrideo to include buying, tendering, procurement, project collaboration, contractual agreements, delivery, and invoice information and payment facilities [Contract Journal, 2000] failed disastrously [Construction News, 2002b].)

## Ethical Procurement

The concept of corporate social responsibility (CSR) and its associated drivers that include business ethics and transparency are becoming increasingly important in project-based industries. However, Loo (2002) notes that most books on project management fail to give this topic sufficient credibility and that this is repeated in both project management journals and conference proceedings. Such findings are perhaps worrying if we consider the findings of a survey conducted by the Ethics Resource Center in Washington. Nearly one-third of the 4,000 U.S. employees surveyed said that pressure had been put on them by their company to violate company policy in order to achieve business objectives (People Management, 1995).

Project-based industries, which are often characterized by interdisciplinary temporary teams, without a common objective, may perhaps provide an environment that facilitates enhanced conflict between business and society needs. The generic project procurement process has largely failed to account for unethical behavior or has perhaps unknowingly sanctioned such illegal activity! Risk management and value management are now considered a prerequisite in project-based industries, but how many organizations consider ethical management? One example that suggests improvements are being made is that of Lucas Aerospace UK. The dissemination of its ethics program throughout the company was seen to be crucial for its success following revelations that two of the company's North American divisions were found to have falsified tests on components supplied to the U.S. Navy (People Management 1995).

## Summary

Procurement is a constantly evolving phenomenon that accommodates the complex nature of project-based industries. This chapter has lent heavily on case study examples from the construction industry. No apology is offered for this, as it is perhaps the oldest known project-based industry. Its fragmented nature with multi-disciplinary and multi-organizational players present procurement managers with often unique challenges in fulfilling a project's completion.

It was also made explicit that the one-size-fits-all description of procurement is unhelpful in explaining the peculiarities within the various project-based industries. The frameworks used consist of much more than an explanation of a buyer-seller relationship and whether to outsource or not. However, the importance of PBPS may be illusive for far too many who procure projects, and impact of practices such as of organizational (project) learning, knowledge management, benchmarking, lean construction, and supply chain management remain as yet unknown. This is also the case for e-commerce despite its commercial advantage being clear to many.

If we are to move from twentieth- to a twenty-first-century project-based practices, then the recommendations suggested by Cain (2001) are appropriate for revitalizing the procurement process and the relationships within. He argues that best-practice procurement differs from all other forms of traditional procurement in two ways: It involves the abandonment of lowest capital cost as a value comparator, and it includes involving specialist contractors and suppliers in design from the outset. He concludes that these two key differences can be broken down into six primary goals that are essential for best-practice procurement:

1. Finished products delivers maximum functionality, which includes delighted end users.
2. End users benefit from the lowest optimum cost of ownership.
3. Inefficiency and waste in the utilization of labor and materials is eliminated.
4. Specialist suppliers are involved in design from the outset to achieve integration and buildability.
5. Design and construction of the building is achieved through a single point of contract for the most effective coordination and clarity of responsibility.
6. Current performance and improvement achievements are established by measurement.

The next epoch in procurement will be to build social improvement practices into procurement. Best value will hopefully be judged beyond the current mantra of price; performance and value will be extended to incorporate agreed goals that have a special purpose. Here issues of gender and ethnicity balance in the project team will need to be considered when procurement decisions are being made. New businesses set up by groups formerly excluded from the current business culture will lead to new procurement frameworks. Thus, the policies of procurement operated by project-based industries have some way to travel before it may finally realize its business and social potential.

# References

Association for Project Management. 2000. *Project Management: Body of Knowledge. 4th Ed.*, ed. Miles Dixon. High Wycombe, UK.

Bennet, J., and S. Jayes. 1995. *Trusting the team: The best practice guide to partnering in construction.* Centre for Strategic Studies in Construction. Reading, UK: Reading Construction Forum.

Bresnen, M., and N. Marshall. 2000. Partnering in construction: A critical review of issues, problems, and dilemmas. *Construction Management and Economics* 18:229–237

Cain, C. 2000. Cited in Was the Gain Worth the Pain? *Contract Journal* (June 1): 32–33.

Chartered Institute of Building. 2002. *Code of practice for project management: For construction and development.* 3rd ed. Oxford, UK: Blackwell Science.

Construction Confederation. 2002. An Introduction to E-Business in Construction. Construction House, London. www.theCC.org.uk.

Construction Industry Board. 1997. *Partnering in the team: A Report by Working Group 12.* London: Thomas Telford.

*Construction News.* 2002a. Fraud Probe at Navy Dock, p. 1. August 1.

*Construction News.* 2002b. Domain name firesale as the construction net dream dies. 1–2.

*Contract Journal.* 2000. Five contractors set to launch B2B venture. July 26.

———. 2002a. Up e-revolution. p. 9. July 19.

———. 2002b. Sharing e-efficiencies. 10–11. August 29.

———. 2002c. Investment takes off. 14–15. August 9.

———. 2002d. Procure21 shaving 4% off building costs. September 26.

Curtis, B., S. Ward, and C. Chapman. 1991. *Roles, responsibilities and risk in management contracting.* Special publication 81. London: CIRIA.

Chrichton, C. A. 1966. *Interdependence and uncertainty: A study of the building industry.* London: Tavistock Publications.

Confederation of Construction Clients. 2000. *The Clients' Charter Handbook*. London. www. clientssuccess.org.

Drexler, J., and E. W. Larson. 2000. Partnering: Why project owner-contractor relationships change. *Journal of Construction Engineering and Management* 126(4):293–297

Egan, J. 1998. Rethinking construction. The Report of the Construction Task Force to the Deputy Prime Minister, John Prescott, on the scope for improving the quality and efficiency of UK construction. London: The Stationery Office.

Galliford. 1998. *Partnering in the construction industry*. Leicestershire, UK: Galliford UK Limited.

Goldin, D. S. 1999. Performance based contracting. Taken from www.ksc.nasa.gov/procurements/ nls/perfbase.htm (accessed June 24, 2003.)

Holti, R. 1998. The lost world: Virtual organisation in the building industry. 44–49. Discussion paper in the *Tavistock Institute Review* 1996/97.

Holti, R., D. Nicolini, and M. Smalley. 2000. The handbook of supply chain management. CIRIA Report C546. London: CIRIA

Kashiwagi, D. T. 1997. The development of the performance-based procurement system. 275–284. *ASC Proceedings of the 33rd Annual Conference*. University of Washington, Seattle, April 2–5.

Kumarasawamy, M., M. Rahman E. Palaneeswaran, S. Ng, and O. Ugawa. 2003. Relationally integrated value networks. *CIIFE Conference*, Loughborough, UK.

Larson, E., and Drexter, J. A. 1997. Barriers to project partnering: Report from the firing line. *Project Management Journal* (March): 46–52.

Latham, M. 1994. *Constructing the team: Final report of the government/industry review of procurement and contractual arrangements in the UK construction industry*. London: The Stationery Office.

Loo, R. 2002. Tackling ethical dilemmas in project management using vignettes. *International Journal of Project Management* 20: 489–495.

Monaghan, K. 2000. Trust me, I'm a contractor. *Building* 26 (May): 36–37.

Morgan, G. 1986. *Images of organisations*. Beverly Hills, CA: Sage Publications.

Morris, P. W. G. 1997. *The management of projects*. London: Thomas Telford.

Murray, M., and D. Langford. 2003. *Construction reports 1944–98*. Oxford, UK: Blackwell Science.

National Audit Office. 1998. Cost over-runs, funding problems, and delays on Guys' Hospital phase 3 development. Report by the Comptroller & Auditor General HC 761 1997/98.

———. 2002a. Ministry of defence: Major project report 2001, HC 330.

———. 2002b. The construction of nuclear submarine facilities at Devonport. Report by the Comptroller and Auditor General, HC 90, Session 2002–2003. London: The Stationery Office.

New Builder. 1994. World view, April 8, pp. 21–22 Nicolini, D., R. Holti, and M. Smalley. 2001. Integrating project activities: The theory and practice of managing the supply chain through clusters. *Construction Management and Economics* 19:37–47.

Nicolini, D., C. Tomkins, R. Holti, A. Oldman, and M. Smalley. 2000. Can target costing and whole life costing be applied in the construction industry?: Evidence from two case studies. *British Journal of Management:* 303–324

*People Management*. 1995. Business Ethics. pp. 22–34.

Smith, A. 1838. *The wealth of nations*. London: Longman.

Tookey, J. E., M. D. Murray, C. Hardcastle, and D. A. Langford. 2001. Construction procurement: Redefining the contours of organisational structures in procurement. *Engineering Construction and Architectural Management* 8 (1, February): 20–30.

Townsend, M. 1996. Is the Japanese way working? *Contract Journal* 21 (November): 22–23.

Voeller, J. 2002. E-commerce applications in construction. Research summary. Austin, TX: Construction Industry Institute, University of Texas at Austin.

Walker, A., and R. Newcombe. 2000. The positive use of power to facilitate the completion of a major construction project: A case study. *Construction Management and Economics* 18(1):37–44.

Walker, D. T, and S. M. Rowlinson. 2000. A construction industry perspective on the use of the Web for establishing a marketing presence. *International Journal of Construction Information Technology* 8(1):93–112.

www.mgmconcepts.com/scripts/mcicoursepage.asp?MCICourse=6126 (accessed June 24, 2003).

CHAPTER THIRTEEN

# TENDER MANAGEMENT

George Steel

Most projects will, at one stage or another, need to go to the marketplace to procure the expertise, materials, equipment, and services required for development, implementation, or execution. The way this is done can have a profound impact on project success and profitability.

This chapter on tender management addresses the business imperative of the procurement function and the evolution of the way in which tenders for goods and services are structured and managed. It describes the tender management process in detail, starting at the early stages of contract development and finishing at contract award. By and large the chapter looks at the tender management function from the perspective of the owner organization awarding the contract. Tender management is also a critical function within supplier organizations, and this is also discussed.

## The Business Imperative

The economics of top-class procurement are compelling. With the growth of outsourcing, third-party goods and services can account for between 40 percent and 80 percent of a project or company activity. The difference in value between the routine "three bids and a buy" approach by administrative staff going through a procurement process with little imagination and entrepreneurial drive, and the application of a world-class procurement process by an experienced and motivated team, can be from 10 percent to 30 percent or more. The impact on the bottom line is significant for major operating companies, as shown in the following example.

**Case Study:** A European oil company spent €100 million per annum on the development and refurbishment of its gas stations throughout Europe. Implementing the kind of world-class tender management process described in this chapter resulted in 10 to 30 percent savings of third-party spend on materials equipment and construction services.

The impact on contracting companies can be even more dramatic. This is due to their relatively low profit margins and high levels of spend on subcontractors and other outsourced services. (See Figure 13.1.)

## Evolution of the Procurement Function

Because of the potential impact on the bottom line, it is hardly surprising that the procurement function has evolved from a fairly routine administrative function into a major strategic lever of the project, or indeed the enterprise (Lassister, 1998; Latham, 1994). Initially, effective procurement meant casting user requirements in stone and squeezing individual suppliers till the "pips squeaked." This evolved into the "aggressive sourcing" methodologies

### FIGURE 13.1 IMPACT OF PROCUREMENT SAVINGS ON THE BOTTOM LINE.

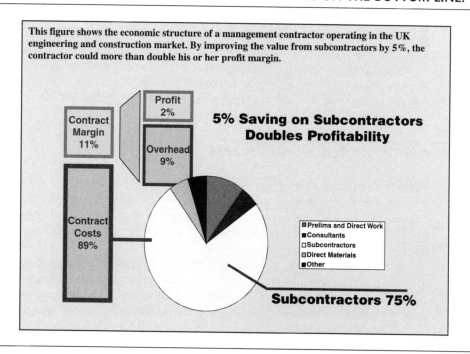

This figure shows the economic structure of a management contractor operating in the UK engineering and construction market. By improving the value from subcontractors by 5%, the contractor could more than double his or her profit margin.

pioneered in the automotive industry, with exhaustive analysis of the supplier market, aggregation of demand, reduction of suppliers, and very aggressive negotiation. In recognition of the importance of organizational linkages, aggressive sourcing soon merged with the "Business Process Re-Engineering (BPR)" movement to improve the way the organization translated business objectives into commercial reality.

BPR was supported by the systematic application of "value management techniques (VMT)" to optimize user requirements. As recognition grew of the mutual dependencies between owners, contractors, and subcontractors, so the whole concept of "supply chain management (SCM)" became the new mantra, as the vehicle for obtaining sustainable improvement from the marketplace. In the 1990s "alliance contracting" and "partnering" forms of contracting became very fashionable, particularly in the UK engineering and construction industry, where initiatives by Sir John Egan and Sir Michael Latham (Egan, 1998; Latham, 1994) stressed the importance of collaboration across the project supply chain.

Finally, the emergence of online tendering systems is streamlining and accelerating the tendering process, and reverse-auction capability is unleashing the ultimate in competitive pressure. This is potentially very powerful in the project environment, where there is often the need to balance the rigor, and duration, of the procurement process with project deadlines. Over the last year or so, the use of online reverse auctions has increased significantly, and many companies are bidding significant construction contracts and contracts for professional services online. A leading pharmaceutical group, for example, has started to use online auctions for nonspecialized buildings such as warehouses and for professional building services.

The way in which the project "industry" balances the use of the intense competition generated by online bidding, with the advantages of supply chain alignment, will be one of the major issues in project procurement over the next few years. The name of the game, however, is improving the bottom line of *your* project; incorporating the procurement and tendering methodologies that integrate every single value improvement technique referred to previously within your project management process. This chapter describes how leading companies are using the tendering process to improve the profitability of their projects.

## Overview of the Tendering Process

Figure 13.2 presents an overview of the procurement and tendering process covered in this chapter. The process starts with an evaluation of the project drivers and the statutory and corporate environment, along with the development of an execution strategy and contracting plan for the project. The process is complete when every package in the contracting plan has been awarded.

The process is generic and has been successfully applied to the procurement of commodities, complex equipment, and services of all kinds of projects—from IT systems to multimillion-dollar contracts for oil tankers and the engineering and construction of major process plants. Each of the steps indicated in the preceding figure is described in some detail in the following pages. The next section deals with strategic contracting issues at project

## FIGURE 13.2. PROJECT AND PROCUREMENT STRATEGY.

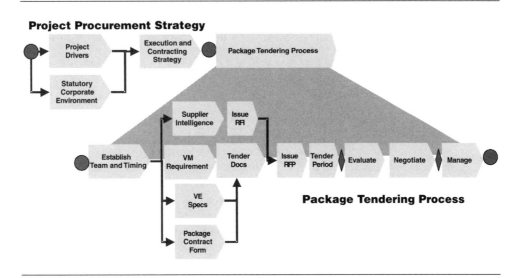

**Project Procurement Strategy**

**Package Tendering Process**

level. Subsequent sections deal with various aspects of procuring the specific work packages that form part of the overall project scope.

# Project Level Issues

## Understand Statutory and Corporate Compliance

It is absolutely essential that the tendering process reflects the statutory regulations and complies with corporate standards for the following reasons:

- Failure to comply with the statutory regulations that govern procurement in many countries, such as those in the European Union, may result in considerable embarrassment, abortion of the procurement process, and even expensive litigation.
- Similarly, failure to understand corporate standards for delegation of authority and contract approval, or the existence, may result in considerable delay prejudicial to the project.
- Many companies have framework contracts in place for certain materials, equipment, and categories of service. Framework contracts may save time and money for the project. The use of existing framework contracts may be an essential corporate policy.

The statutory and regulatory environment can have a profound impact on (a) the procurement process, (b) the time it takes to procure a given contract and (c) the composition of the team. This is especially true in most public sector procurement, where any minor non-compliance with process can result in elimination of the bidder or the need to retender the package. Most public institutions, and indeed many corporations, recognize the need for transparency in the procurement process and publish information on this on their Web sites.[1,2]

## Understand Project Drivers and Priorities

In many cases, project business drivers will have a significant impact on contracting strategy and process. For example, in a project with a high net present value (NPV), or where *time to market* is critical, the need to get resources on board and working or ensuring the delivery of critical equipment may outweigh the economies, which would result from a more rigorous procurement process. These factors will determine the project execution and contracting strategy and the priorities of Package Procurement.

Before embarking upon the development of a project contracting strategy or a package tendering initiative, the project team must have a very clear understanding of the project business case, the risks that have been identified, and particularly the impact of time on project benefits.

## Project Contracting Strategy

The development of an overall contracting strategy can be a complex subject in its own right, which goes beyond the intent of this chapter. The following points, however, will give some indication of the issues involved in developing and documenting an appropriate contracting strategy.

***Package Breakdown and Basic Contractual Philosophy.*** The contracting strategy adopted for a project will have a significant impact upon its eventual success and profitability. The contracting strategy consists of (a) Identification of all the *work packages*, that will be undertaken by third-party organizations, (b) the scope of each work package, and (c) the basic nature of commercial arrangements governing that work package, including, but not limited to, the transfer of risk.

***Competencies in Owner and Supplier Organizations.*** The optimal contracting strategy will be a balance between the competence of the owner organization to either undertake or to manage the work in question, and the competencies available of the supplier market. The point of departure is the definition of the minimum "core competencies" that the owner wishes to maintain within his or her own organization for strategic reasons. This will shape the services he or she wishes to buy from suppliers.

---

[1] Federal Acquisition Regulations 2001 S/N 922-006000008. provides a guide to the U.S. government procurement process.

[2] www.tendersdirect.co.uk provides a useful guide to EU procurement procedures.

**Case Study:** A European oil company requested suppliers to provide technical competence in electrical and instrumentation engineering as part of their offering. The company, however, regarded the electrical and instrumentation engineering as part of their core competence. This resulted in duplication of effort, confusion over roles and interfaces, and ultimately higher costs.

Similarly another oil company wished the engineering procurement and construction (EPC) contractor to provide overall project management of major refinery shutdowns. Unfortunately, the EPC contractor selected did not have this competence, and the oil company was required to take over the management of the turnaround project.

Major cost reductions were achieved in North Sea projects when owners stopped "man for man marking" and defined who was accountable for the result, and who was responsible for execution and reflecting this in their contracts.

*Appropriate Apportionment of Risk.* The contracting strategy will also define the way in which project risk is shared between the owner of the project and its suppliers and will also determine the shape of the owner project organization and the way the project is managed. The most appropriate apportionment of risk will depend on the following:

- Who has the financial resources to bear the consequences of the risk.
- Who has the best competencies to manage the risk.

Many owners will write contracts in which suppliers are asked to take risks that go way beyond their net asset value. More surprisingly, some suppliers will agree to accept such risks. Basically the owner is fooling him- or herself, because if the risk materializes, the contractor will be bankrupt and the owner will still suffer the consequences. Particularly in more complex contracts, it is probably in the interests of both the owner and contractor to ensure that risks are clearly defined and mutually understood as part of the contracting process and prior to contract award.

Risks are normally documented in a *risk register*, which lists the significant identified risks on the contract scope. In many cases the risk register is a key element in the contracting package. The risk register, shown in Figure 13.3, lists the risks that have been identified on the project, indicates the mitigating actions that can be taken, allocates responsibility for dealing with them, and specifies the contractual and commercial implications (Wideman, 1992).

## Contract Scope Statements Based on Work Breakdown Structure

Many, if not most, contracting problems stem from failure of either the owner or contractor to fully understand the scope of the work of a particular contract package. Failure to appreciate scope can lead to unrealistic expectations on the part of the owner, underestimation of time and cost by the contractor, and problems of interface between the various contractors working on the project.

## FIGURE 13.3. PROJECT RISK REGISTER.

| Ref | Risk | Impact | Actions | Cost Impact | Schedule Impact | Resp. | Contractual |
|---|---|---|---|---|---|---|---|
| **Loading Jetty** | | | | | | | |
| 10.1 | Sea Bed Conditions Uncertain | Could seriously impact the piling specification, hence schedule and cost | Try to convince sponsor to accept floating point mooring system instead of jetty (see detailed report on relative economics). More detailed survey required as part of design process | $1–3 Million | 3–6 weeks on activity but non critical | EPC Contractor | EPC contractor to be reimbursed on unit rate basis according to the piling schedule |

A very useful discipline is to base the *contract scope statement* on the project *work breakdown structure.* This technique ensures that all of the work necessary to complete the project is defined in one contract package or another. The technique also forces a very clear definition of the interfaces between contract packages. This is inevitably the area where problems will arise.

### Project Contracting Plan

Once the overall contracting strategy has been developed, the next step is to produce an overall *project contracting plan.* This plan should be an integral part of the overall project plan and ensures that the deliverables expected from each tender are delivered when they are needed to meet project requirements. The contracting plan reflects the deliverables and resources that can be made available, particularly from the supporting disciplines such as engineering, legal, and the like. The project and the procurement team must be extremely clear on the priorities of the different packages. In most projects the award of one or more contract package will almost certainly be on the critical path of the project.

Figure 13.4 illustrates the relationship between a series of design, procurement, and execution activities that are all part of an overall project plan.

# Procuring Project Packages

Once the overall contracting plan has been agreed at the appropriate level in the organization, the emphasis shifts from consideration of strategic issues to the systematic procurement of every package on the plan. The package tendering process is shown schematically in Figure 13.5.

## FIGURE 13.4. BASIC TIMING OF CONTRACTING PLAN.

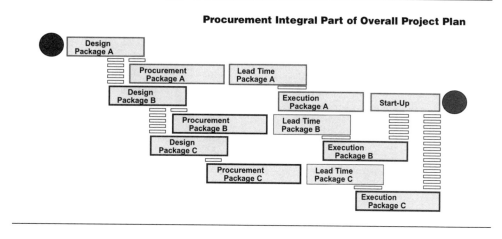

## FIGURE 13.5. PACKAGE TENDERING PROCESS.

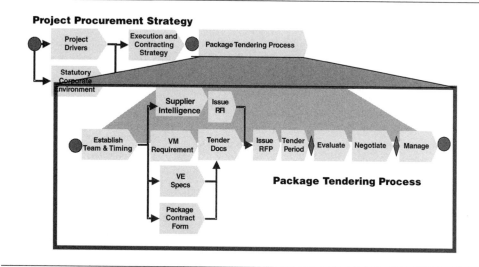

## Forming the Team and Defining Roles

Many traditional organizations still work in their functional "silos," where the end user specifies operating requirements, the technical department documents technical specifications, and the procurement department defines commercial conditions and procures the goods and services from the traditional list of approved and customary suppliers. Indeed, in some companies the engineering department will decide on the supplier and ask the procurement department to regularize their decision by placing the purchase order or contract. This does not tend to produce the best value.

> **Case Study:** Traditionally the technical department in a large organization would specify the complex equipment required for product analysis and ask the purchasing department to regularise the situation by placing a purchase order. By procuring the equipment through the tendering process described savings of 30 to 40 percent were achieved. Similar levels of saving on other complex products and services achieved through the application of more rigorous procurement processes.

Most companies now recognize that procurement is a team effort and that the structure and composition of the procurement team are critical success factors for anything other than the simplest of commodities. The following distinct functions and competencies may provide a useful framework for defining roles and responsibilities within the project team. However, it is worth noting that in smaller organizations, it would be perfectly acceptable and possibly desirable for one person to wear more than one hat.

- *Project sponsor.* The ultimate decision maker. Takes P&L responsibility for the contract in question.
- *Project management.* Provides overall leadership, pulling the team together and driving them through the most appropriate procurement and tendering process.
- *Procurement.* Provides detailed knowledge of supplier markets and trends in the industry.
- *Technical.* Provides detailed input on technical specifications either directly or as a conduit to the technical resources within the company or external experts in the field.
- *Operations.* Represents the ultimate users of the goods or services to be purchased and ensures operational requirements are respected.
- *HSEQ (Health, Safety, Environment, and Quality standards).* Supports functions such as financial, legal, and PR, providing expertise and counsel within their area of competence.

In fact, it is probably worth conducting a formal RACI workshop to ensure that everyone is absolutely clear on the roles and responsibilities of the various participants. (RACI stands for Responsibility, Accountability, Consulted, and Informed. In a RACI workshop all the individuals and functions responsible for undertaking the activities that make up a business process debate who is *accountable* (who carries ultimate responsibility for the result), *responsible* (who actually does the work), *consulted* (who must be consulted as part of the activity), and *informed* (who is informed of the decision once it has been taken). The consensus is documented in the kind of RACI Chart shown in Figure 13.6.

The specific accountabilities and responsibilities will, of course, vary from organization to organization, but this chart may serve as a useful basis of a workshop to debate and agree

## FIGURE 13.6. RACI CHART.

| Activity | Accountable | Responsible | Consulted | Informed |
|---|---|---|---|---|
| Gaining budget authority | BU | BU | Finance | Procurement |
| Preparing contract/purchase requisition | BU | BU | HSEQ, Procurement | Engineering |
| Determining contract/purchase scope | BU | BU/ Procurement | HSEQ, Engineering | |
| Determining technical specifications | BU | Engineering | HSEQ | Procurement |
| Determining HSEQ requirements | BU | HSEQ | Engineering | Procurement |
| Selecting appropriate terms and conditions | Procurement | Procurement | Legal, Engineering | BU |
| Selecting form of remuneration | Procurement | Procurement | BU, Finance | Engineering |
| Preparing tender evaluation criteria—commercial | Procurement | Procurement | BU, Finance, Engineering | |
| Preparing tender evaluation criteria—technical | BU | Engineering | HSEQ, Procurement | |
| Choosing suppliers/contractors to receive ITT | Delegated Authority* | Delegated Authority* | Procurement Engineering | |
| Preparing invitation to tender | Procurement | Procurement | BU | Finance, Engineering |
| Issuing invitation to tender | Procurement | Procurement | BU | Finance, Engineering |
| Receiving and safekeeping of tenders | Procurement | Procurement | | BU |
| Conducting tender opening | Procurement | Procurement | | BU |
| Conducting technical evaluation of tenders | BU | Engineering | HSEQ | Procurement |
| Conducting commercial evaluation of tenders | Procurement | Procurement | BU | Engineering |
| Clarification/negotiation with suppliers | BU | Procurement | Engineering, Legal, Finance | Legal, HSEQ |
| Preparing evaluation report and recommendation | BU | Procurement | Engineering, HSEQ | |
| Reviewing recommendation and approval | Delegated Authority* | Delegated Authority* | Procurement BU | Engineering, Legal, HSEQ |
| Awarding contract | Delegated Authority* | Procurement | Legal, Finance | All Functions |
| Maintaining contract file/audit trail | Procurement | Procurement | BU, Finance | |
| Post-award operational supervision | BU | BU | Procurement, Finance | Finance, Engineering |
| Contractor audit | Procurement | Procurement | Engineering, Finance | |

| | |
|---|---|
| RESPONSIBLE | Individuals are the "doers." |
| ACCOUNTABLE | Individuals are ultimately answerable to the board. |
| CONSULTED | Individuals must be consulted and involved before a final decision or action is taken. |
| INFORMED | Individuals need to be informed once a decision or action has been taken. |

on the specific responsibilities within your own organization or project. The ultimate effectiveness of the team will, however, ultimately depend on the level of domain expertise within each of the functions listed, along with the ability to work together creatively as a team and to communicate with the wider range of stakeholders who may be involved in or affected by the procurement initiative.

## Agree on the Tender Event Schedule

Once the team has decided on and documented the process, the next step is to establish the tender event schedule. The overall time required will depend on the nature of the contract and can be anything from a few weeks to many months. Figure 13.7 shows a typical tender event schedule of 16 weeks for a fairly complex alliance contract.

## Do Not Underestimate Resource Requirement

Working out a rigorous tendering process takes time. In the project environment there may be a large number of contracts to procure in parallel, and often the same people will be involved on different packages. Because the time frame for each package might appear quite relaxed, there is a tendency to overestimate the ability to handle a number of packages in parallel. This leads to an underestimation of the time, and hence resource, required. This

## FIGURE 13.7. TENDER EVENT SCHEDULE.

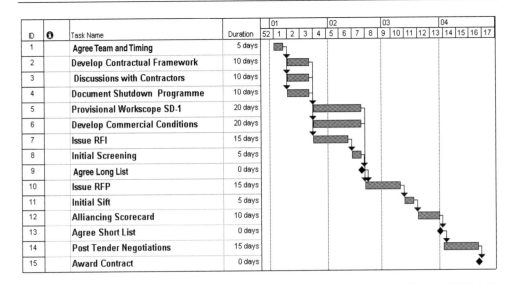

| ID | 🛈 | Task Name | Duration |
|----|---|-----------|----------|
| 1 | | Agree Team and Timing | 5 days |
| 2 | | Develop Contractual Framework | 10 days |
| 3 | | Discussions with Contractors | 10 days |
| 4 | | Document Shutdown Programme | 10 days |
| 5 | | Provisional Workscope SD-1 | 20 days |
| 6 | | Develop Commercial Conditions | 20 days |
| 7 | | Issue RFI | 15 days |
| 8 | | Initial Screening | 5 days |
| 9 | | Agree Long List | 0 days |
| 10 | | Issue RFP | 15 days |
| 11 | | Initial Sift | 5 days |
| 12 | | Alliancing Scorecard | 10 days |
| 13 | | Agree Short List | 0 days |
| 14 | | Post Tender Negotiations | 15 days |
| 15 | | Award Contract | 0 days |

is often particularly the case in the supporting functions without a dedicated project team member or in corporate functions such as the legal department who are offline and often "march to the beat of a different drum." Failure to assign adequate resources can often result in suboptimal procurement or schedule slippage or both.

One way around this problem is to insist on a fully resourced contracting schedule where the aggregate resource requirement is clearly demonstrated for all functions involved in the activity.

> **Case Study:** The critical path in the implementation of a petrochemical complex ran through the procurement of the main equipment. Selection of this was required to obtain the vendor design data required to undertake detailed design. This in turn paced the construction contracts. The contractor underestimated the level of effort required to meet the early procurement deadlines. A project delay of three months was incurred almost immediately which it was not possible to recover.

## Determine Potential Impact of Package on Project

The relative importance of the various packages will undoubtedly have been discussed when the overall project contracting strategy was being developed. One of the first tasks of the procurement team is to really think this through to the next level of detail:

- What percentage of the total project spend does the package represent?
- How critical is timing?
- What are the issues and specific risks inherent in the scope of work?
- Can we access experience within the company, or in the professional community?
- What were the lessons learned?
- What are the implications of all of the above on package procurement?

> **Case Study:** One of the packages on the construction of an air separation unit was the insulation of the cold box. The actual work did not take place till late in the construction schedule, so there was plenty of time to develop and negotiate the insulation contract. This activity had always been on the critical path on previous projects so time was of the essence and 24-hour working was required. This was built into the tender conditions and evaluation criteria from day one. On previous jobs this had involved expensive variations during execution. A saving of 15 percent and a reduction in time of five days over previous jobs was attained.

## Value Manage User Requirements

*Value management* is a set of techniques that can be used at project or package level to ensure that the equipment or service defined provides the best value for money for the owner. For example, by value managing the deliverables of the package, it may be possible to relax the functionality, timing, or operational constraints to reduce the price of the package without any prejudice to project profitability. Value management tends to apply to the scope, performance, and timing, rather than to optimization and technical specifications, which are covered under value engineering.

> **Case Study:** Many major projects are undertaken in remote locations. Typically the contractor will build a camp for his or her senior construction personnel that is demolished when the plant is completed. By advancing the construction of the operators' accommodation, the contractor could dispense with temporary accommodation for his or her project team and make a significant saving, which would be passed, in part, to the owner.

To encourage thinking, many companies invite participation outside of the project team in value management workshops, which are held at key points in the project life cycle to challenge the current definition. Many companies also use external facilitators with a particular experience in value management.

> **Case Study:** After a series of acquisitions a major bank required to restructure its branch network across Europe. Because of the costs involved, the board asked for an independent review. This review revealed that there was confusion over the bank strategy, which impacted on staffing levels. The policy on space allocation per full-time equivalent (FTE) was also ambiguous. In addition, architects were (a) allowing more space per FTE than necessary and (b) failing to design to need. By addressing these issues over a series of strategy and value management workshops, the scope, and hence potential cost, of the project was reduced by approximately 25 percent.

## Value Engineer Technical Specifications

Normally, drawings and technical specifications will be produced by the in-house engineering department or external consulting engineers working under the direction of the engineering representative of the project team. Indeed, many companies have a complete suite of technical specifications that form an integral part of their tender documentation.

In seeking to optimize the return from project expenditure, the project team must examine these specifications critically. Many companies have found that replacing prescriptive technical specifications by performance specifications, which define the output required rather than the inputs demanded, enables significant advantage to be taken of supplier experience and lower-cost, more standardized components.

Often technical specifications have evolved over many years. The process of subjecting engineering designs and standards to critical examination is called *value engineering* (VE).[3] Responsibility for driving this process must rest with the technical representative on the project team. Many companies build in VE workshops as a key aspect of their design and procurement processes. VE, however, does require time, and this must be reflected in the tender event schedule for the package.

---

[3] A value engineering professional association called VE Today is at www.vetoday.com.

**Case Study:** An international oil company was procuring the contract to paint their refinery tank farm. Technical specifications called for three coats on a particular painting system. By discussing the requirements with suppliers one was found who would guarantee a two-coat system. This reduced the overall cost of materials and labor by around 25%.

## Pick Supplier Brains

Before launching into the development of detailed package contract documentation, the owner organization must find out who the leading suppliers in the area are and pick their brains. Most suppliers are working with a wide range of clients, including competitors, and may be prepared to share some of the latest developments. It is advisable to try and do this before developing formal contract documentation. In this way it will be possible to incorporate these developments in the bidding scope and thus subject them to competitive pressure. If this is not done, the supplier may choose to offer them as an alternate. If these alternates are attractive, the owner may then be obliged to negotiate without the advantage of competitive pressure.

There is an ethical, and possibly legal, issue here about the incorporation of the ideas that emerge in this fashion into the competitive bidding process. In the international oil industry, this process is often formally structured as a design competition where bidders are paid to propose technical solutions. The owner thus obtains the intellectual property rights (IPR), which can be bid competitively. The winner of the design competition normally improves his chances of obtaining the contract.

## Determine Precise Contract Form for Package

***Contract Forms.*** While the basic contract form will have been established as part of the project contracting strategy, it has to be translated into specific and very precise contract documentation. Although it is always possible to develop a contract from first principles, in general it is advisable to use forms of contract in common use with an established body of case law.[4] Owners and contractors, and their lawyers, understand them. They can form more precise evaluation of their risk and will require less contingency as a result. Most major companies, however, adapt these basic contract forms to meet their own precise requirements based on their experience over the years.

***Creating Win-Win Conditions.*** In some forms of contracts the interests of the owner and contractor are directly opposed. For many years, in the UK construction industry, contractors were subjected to intense competition and were often forced to bid under their actual estimate of cost to win the work. As a response to this, many contractors developed a

---

[4]Most of the forms of contract in general use have been developed by the professional institutions, including Institute of Chemical Engineers (www.icheme.org), International Federation of Consulting Engineers (www.fidic.org), American Institute of Architects (www.aia.org), and Associated General Contractors of America (www.asce.org).

specialist function to examine tenders for weaknesses, in either the definition of scope or in contract conditions which could be exploited later as the basis of claims and variations. These claims and variations were then negotiated during the contract execution at very high profit margins when the owner had become totally dependent on the contractor for completing the project. As a result, both owners' and contractors' management devoted considerable attention to structuring and opposing variations and claims, often to the detriment of the actual project.

Better results can often be obtained if the commercial interests of the owner and contractor can be aligned. For example, in a high NPV contract where time to market is critical and the execution of the package is on the critical path, a bonus can be awarded for compressing schedule. The same principles can be applied to reducing the overall cost of the project. Figure 13.8 illustrates a typical incentivized commercial arrangement.

In the contract form shown schematically in the figure, the owner and the main engineering procurement and construction (EPC) contractor agree to work on the basis of a fixed fee to cover project management, engineering, procurement, and construction management costs, including overhead and profit. The direct cost of materials and equipment and construction contracts were estimated by the contractor and a target *total installed cost* (TIC) established. If the contractor manages to procure the equipment and complete construction for less than the target cost, the saving is shared with the owner on a pre-agreed basis and the contractor is paid a "bonus." If, however, this target cost is exceeded, the contractor and the owner share the overrun and the contractor suffers a *malus*. It is normal to limit the extent of both bonus and malus. It should be noted that the these incentivized

**FIGURE 13.8. INCENTIVISED CONTRACT ARRANGEMENT.**

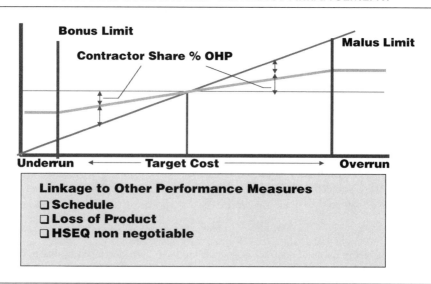

contracts are normally bid competitively where the contractors are required to specify both their fixed fee and the target they hope to achieve.

In the purest form of an "alliance open-book contract," only the fee for overhead and profit is fixed. All other costs are reimbursed "at cost" without any margin or profit. All costs incurred by the contractor are visible to the owner and subject to audit, hence the expression "open book." In theory these open-book contracts are volume-neutral so that all variations can be discussed on their technical merits.

> **Case Study:** An international oil company already had called for lump-sum turnkey (LSTK) bids for the engineering and construction of a major petrochemical complex. They were, however, concerned by the lack of flexibility to introduce changes to plant layout and specification to improve operating and maintenance. By moving to the incentivised target cost EPCm framework indicated previously, the owner could inject operating experience into the project design without incurring major variation charges. The EPCm scope was completed for 18 percent less than the original LSTK price, and the owner has a plant that is cost-effective to operate and exceeds performance targets.

While an incentivized or Win-Win contract framework can help promote alignment between the owner and contractor organizations, the real benefit comes from the ability of key individuals within the owner and contractor companies to collaborate professionally and fairly. The kind of incentivized and open-book contract forms referred to previously in fact may require a much more sophisticated management on the part of both the owner and the contractor to realize their theoretical advantages (ENR, 2001; Bennett and Baird).

> **Case Study:** A major refiner thought an alliance contract would be ideal for plant turnarounds, since the contractor could be engaged early to assist in detailed planning before scope was fully defined and then deal with changes on a commercially neutral basis. Neither the owner nor contractor management was used to working within such a framework, relative responsibilities were not well defined, and costs escalated out of control.

## Request for Information—Widening the Supplier Network

Many companies tend to work, by default or by design, with a limited number of suppliers who have provided good service in the past. These suppliers understand their business and have established relationships, often at various levels, within owner organizations. Indeed, it is conventional management wisdom to reduce the number of suppliers, and it is fashionable, and even sensible, to consider tighter linkages throughout the supply chain. Furthermore, it takes considerable time and effort to search out and qualify new suppliers, and in certain areas, particularly service contracts, there is an increased risk in the level of performance that will be actually delivered by untried organizations.

However, a recent study by McKinsey concluded that productivity gains over the last decades were essentially driven by a combination of innovation and competition. In such circumstances, the continual search for new and emerging suppliers is a major value improvement driver that must be balanced with the need to limit the overall number of suppliers and build better relationships with them.

> **Case Study:** A European petrochemical producer opened up the maintenance of its facilities and grounds to contractors with no specific process industry experience. This enabled a number of smaller, local contractors with a lower cost structure to compete for the work. The track record of these contractors and the competence and motivation of their management were carefully evaluated. There were some problems as the new contractors came to terms with a more industrialized environment but costs were reduced by 30 to 40 percent.

The identification of new suppliers is, however, a time-consuming task that needs to be approached in a very systematic and structured fashion. It helps if this identification and qualification of suppliers can be done off-line without the pressures of project deadlines. Figure 13.9 shows the approach used by a major oil company to identify contractors to work on refinery shutdowns in various regions across Europe. The following are details:

- Over 1,000 contractors were identified from various trade registers and supplier databases.
- Each was sent an e-mail describing the nature of the work and the basic prequalifications required and asking whether they wished to be considered.

## FIGURE 13.9. IDENTIFICATION AND EVALUATION PROCESS.

Example of Identification and Evaluation Process

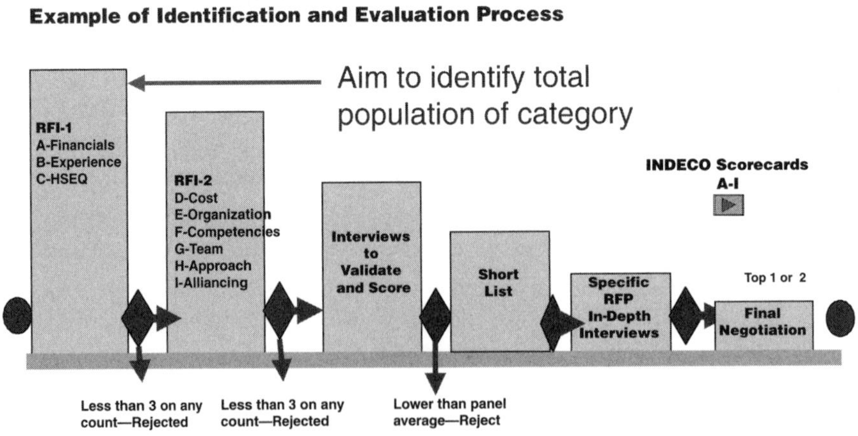

- The 300 who responded in the affirmative were sent a *request for information* (RFI) essentially asking for their financial statements, relevant experience, and HSEQ record.
- This information was used to identify about 50 potentially suitable contractors. These contractors were then sent a more detailed questionnaire soliciting information on their organization, competencies, and trade skills, including their approach to refinery shutdowns and alliancing.
- Short-listed contractors were then invited to interview to assess the quality of their management and validate their responses.
- Based on these interviews, a shortlist was created who were invited to bid for suitable projects.

A detailed set of information schedules and scorecards were established, as shown in Figure 13.9. The first RFI included Schedule A—Financial Structure, Schedule B—Relevant Experience, and Schedule C—HSEQ Credentials and Certification. RFI-2 sent to companies who passed the first cut included Schedule D—Cost Structure, Schedule E—Organization, Schedule F—Competencies, Schedule G—Project Team, Schedule H—Approach to the Specific Tender, and Schedule I—Experience and Approach to Alliancing. Companies who passed the next cut were invited to interview to validate the previous data and to present themselves and their ideas. Detailed interview guides were established that were agreed with and completed by the procurement team.

## Tender Documentation

***Clarity and Completeness of Tender Documentation is Critical.*** The structure and content of the tender documentation is at the core of the tender management process. The clarity and thoroughness of this documentation will determine the quality of the bids received and the prices tendered. It will also provide the basis for the execution and eventual cost of the contract. Any ambiguity will inevitably become the subject for debate, and possibly the source of contract variations and extra costs. The necessity of defining precisely what you want, what you are requesting the contractor to provide, and how you are paying for it cannot be overstated (Kennedy, 2000).

> **Case Study:** If you are in any doubt about the preceding, refer to the celebrated case of the British Museum where the client wanted a "Portland stone" masonry construction. The technical specification referred to the technical specification of Portland stone. The contractor constructed the building in French limestone, which apparently met the technical spec but which did not come from Portland. The resulting furore became a *cause celebre* in the British construction industry.

***Structure of Tender Documentation.*** Although a number of variants are possible, the classic structure of a tender documentation package is described in the following. Most companies will standardize on this or something similar to ensure completeness and consistency and to speed up the development of the tender package within their own organization.

- *Invitation to tender.* This is a standard "one-page" letter inviting the contractors to submit a bid for the package in question under the conditions defined and documented in the package.
- *Instructions to tenderers.* This includes instructions to contractors defining the form, place, and time of tender submission and the duration of validity. Instruction to tenderers will reference a series of sections to be completed by the tenderer. These sections will ultimately form an integral part of the contract between the owner and contractor.
- *Form of Agreement.* The basis of the tender defining the contracting parties and referencing the various supporting schedules (see Figure 13.10). This will be signed by the contractor and form the basis of his offer. This form of agreement will be underpinned by a series of sections precisely documenting contract scope and conditions. The tenderer will be required at this stage to identify any exclusions—that is, any conditions or any other aspects of the documentation that are unacceptable.

The following sections are typical on many project related contracts, although they may vary according to sector:

- *General conditions of contract.* Most companies accustomed to awarding contracts will have these as standard articles or legal "boilerplate". Figure 13.11 shows a typical heads of agreement. Articles requiring particular attention are those covering "limits of liability," "variations to contract," and the "termination."
- *Contract scope.* This schedule defines matters such as the following:
  - Services or end result to be provided by the contractor
  - Operating conditions and environment
  - Delivery dates
  - Testing and handover requirements and protocols
- *Technical specifications and standards.* This section defines any performance or technical specifications and standards to be met by the contractor. Any drawings or technical data will normally be included and referenced in this schedule.
- *Remuneration.* This section defines how the contractor will be paid for the services provided under scope. Payment events, incentive arrangements and payment conditions, fiscal responsibilities, and currency issues are normally defined in this section. In addition, this section generally contains a pricing schedule. The pricing schedule has vital commercial implications. Some considerations on structuring this are developed in the *Pricing Schedule* section, coming later in the chapter.
- *HSEQ.* This section documents compliance requirements with Health, Safety, Environment, and Quality standards (HSEQ).
- *Provided by owner.* This section defines the services, information, and facilities to be provided to the contractor by the owner. This could cover such things as free issue materials, office and workshop space, access to owner computer systems, and the like.
- *Contract administration.* This section defines how the contract is to be administered, including the names of designated individuals to represent each party and the method of serving official notices and when they are required.

## FIGURE 13.10. FORM OF AGREEMENT.

**FORM OF AGREEMENT**

This CONTRACT is made between the following parties:

xxxxxx  a company having its registered office at  xxxx, hereinafter called the COMPANY;

and

Contractor xxx  having its main or registered office at xxxxxx,  hereinafter called the CONT RACTOR.

**WHEREAS:**

1)              the COMPANY wishes that certain WORK shall be carried out, all as described in the CONTRACT; and

2)              the CONTRACTOR wishes to carry out the WORK in accordance with the terms of this CONTRACT.

**NOW:**

The parties hereby agree as follows    :

1)              in this CONTRACT all capitalised words and expressions shall have the meanings assigned to them in this FORM OF AGREEMENT or elsewhere in the CONTRACT.

2)              the following Sections shall be deemed to form and be read and construed as part of the CONTRACT:

|  |  |
|---|---|
| Section 1 | Form of Agreement |
| Section 2 | General Conditions of Contract |
| Section 3 | Scope of Work |
| Section 4 | Remuneration |
| Section 5 | Health, Safety, Environment and Quality |
| Section 6 | Items to be provided by the COMPANY |
| Section 7 | Contract Administration |

The Sections shall be read as one document, the contents of which, in the event of ambiguity or contradiction between Sections, shall be given precedence in the order listed.

3)              In accordance with the terms and conditions of the CONTRACT, the CONTRACTOR shall perform and complete the WORK and the COMPANY shall pay the CONTRACT PRICE.

4)              The terms and conditions of the CONTRACT shall apply from the date specified in Appendix 1 to this Section I - Form of Agreement which date shall be the EFFECTIVE DATE OF CO MMENCEMENT OF THE CONTRACT.

5)              The duration of the CONTRACT shall be as set out in Appendix 1.1 to this Section I - Form of Agreement.

The authorised representatives of the parties have executed the CONTRACT in duplicate upon the dates indicated below:

For:       COMPANY

Name:

Title:

Date:


For:       CONTRACTOR NAME

Name:

Title:

Date:

## FIGURE 13.11. GENERAL CONDITIONS OF CONTRACT.

| General Conditions Table of Contents | | | |
|---|---|---|---|
| 1. | Definitions | 20. | Patents And Other Proprietary Rights |
| 2 | Interpretation | 21. | Laws And Regulations |
| 3 | Company And Contractor Representatives | 22. | Indemnities |
| 4. | Contractor's General Obligations | 23. | Insurance By Contractor |
| 5. | Responsibility For Company Provided Items | 24. | Insurance By Company |
| 6. | Contractor To Inform Itself | 25. | Consequential Loss |
| 7. | Contractor To Inform Company | 26. | Confidentiality |
| 8. | Assignment And Subcontracting | 27. | Customs Procedures |
| 9. | Contractor Personnel | 28. | Completion |
| 10. | Co-Operation With Others | 29. | Defects Correction |
| 11. | Programme | 30. | Termination |
| 12. | Technical Information | 31. | Audit |
| 13. | Inspection And Testing | 32. | Liens |
| 14. | Variations | 33. | Business Ethics |
| 15. | Force Majeure | 34. | General Legal Provisions |
| 16. | Suspension | 35. | Liquidated Damages |
| 17. | Terms Of Payment | 36. | Limitations Of Liability |
| 18. | Taxes And Tax Exemption Certificates | 37. | Resolution Of Disputes |
| 19. | Ownership | | |

*Source:* INDECO.

Under the structure of tender documentation described in the preceding text, the submitted tender forms the basis of the contract binding on "contractor."

***Document Quality Assurance and Version Control Essential.*** Contract documentation is often assembled under considerable time pressure with input coming from many sources. Mistakes and omissions reflect badly upon the organization and may have extremely serious consequences on contract and project execution. It is therefore absolutely essential to operate a formal document management and quality assurance procedure. Ideally, each section should be checked and signed off by a competent person who is not the author.

## Pricing Schedule

The *pricing schedule*, which defines how and how much the contractors will be paid for his services, is a critical part of the documentation and is worthy of special note. The pricing schedule should be carefully constructed to provide the basis for the detailed comparison between the eventual tenderers. Ideally, the pricing schedule should provide the owner with enough information to compare the offerings of the tenderers on a *whole-life cost* basis. (See the section *Whole-Life Cost Analysis* coming up in the chapter for a detailed explanation and example of whole-life cost analysis).

> **Case Study:** A distribution company was soliciting bids for the maintenance of their national retail distribution system, which was split between a number of contractors. A unit rate contract form and schedule was developed and bid by over 20 contractors. These pricing schedules were inspected and any anomalous rates removed. A schedule of rates was then developed using the lowest rate from the competing contractors. In a second round of bidding, the short-listed bidders were requested to bid this schedule of rates at a discount. This reduced maintenance costs by over 20 percent. The contract has now been running for about five years and has been very successful for both parties.

The detailed breakdown of bid prices can be an invaluable aid in the evaluation of tenders, allowing the owner to identify areas of anomaly, where perhaps the tenderer has either misunderstood the scope of work or has developed a creative approach to reduce costs. A detailed pricing schedule can also be an invaluable aid during post tender negotiation, allowing the owner to use a *cherry-picking approach* during post tender negotiation. In this approach the prices of the competing bidders are compared and negotiated at elemental level. This evaluation, which can be done in Microsoft's Excel or in a database application, highlights where certain contractors may have overestimated or "padded" the element. Pressure can be put on these elements to reduce unit, and hence total, costs. An extreme example of this is described in the case study later in the chapter.

In more complex cases it can be very effective to produce a *pricing wizard* using Excel. In this way the bidders are obliged to enter their information in a very structured way. More importantly, the buyer can easily manipulate and compare the data. The example shown in Figure 13.12 shows the front sheet of a pricing wizard for a contract to provide technical maintenance on over 300 elevators in many different buildings in three countries.

## Manage the Tender Period

Control of information is absolutely critical during the tender period. During the tender period bidders actively seek to gain intelligence on their competitive situation. Even a chance

FIGURE 13.12. PRICING WIZARD.

**INSTRUCTIONS** **INDECO**

**Please read these instructions in conjunction with the following two documents :**

**"Preamble to the Pricing Table"**

**"Instructions to Tenderers"**

**NOTE : Failure to complete this document in accordance with the instructions may result in exclusion from the competition for failure to comply with requirememts.**

This pricing schedule contains two spreadsheets that you must complete.

The first spreadsheet, Full Details Schedule, contains information on each elevator. You must provide a price for each one. Where no price is given we will assume that you have included for its maintenance in the total at the bottom of the spreadsheet.

The second spreadsheet, Schedule of Unit Rates, contains information on Man-day / Man-hour rates, mark-ups and discounts. You must complete this chart.

In completing this pricing schedule you should follow these steps:

**1. Complete Full Details Schedule**

You can do this either by filling in the spreadsheet directly, or by using our Pricing Wizard. This Wizard will present the information for each lift and ask you to enter information on the number of visits you will require to maintain the lift and the unit rate for each visit. The wizard will populate the spreadsheet for you.

NOTE : If a data box in the elevator pricing schedule is not filled in, the item will be taken as included unless it is specifically stated as excluded by entering the words "not priced" in the table.

To access the Elevator Pricing Wizard (recommended) click here [Wizard]

**OR**

To complete the sheet entitled "Full Details Schedule" directly click here [Spreadsheet]

**2. Complete Schedule of Unit Rates sheet**

To complete the Schedule of Unit Rates click here [Schedule]

**3. Print and Check the schedules**

Print these schedules, check that you have included all the information requested, before returning it to us.

Click here to print your completed schedules [Print]

remark on apparently noncommercial matters may encourage contractors to maintain a higher pricing level. Also, inconsistency of information supplied to bidders may cause considerable confusion and even potentially embarrassing and costly litigation.

For all of these reasons, one single point of contact in both the owner and contractor organization is highly desirable (see Figure 13.13). All other contacts should be actively discouraged. All questions arising should be formally documented, and identical responses answers should be sent to each bidder. Some companies run bidder conferences where bidders have the opportunity to ask questions in an open forum. The answers given during this tender period form an integral part of the contract documentation and conditions and so have to be very carefully considered and documented. Failure to manage the tender period in a manner that is absolutely and transparently equitable can, particularly in the public arena, result in a legal challenge from unsuccessful bidders.

## Tender Evaluation

The object of the tender evaluation phase is to identify the proposal that offers the best value for money to the company. This can be more difficult than it sounds. Often the

## FIGURE 13.13. SINGLE POINT OF CONTACT.

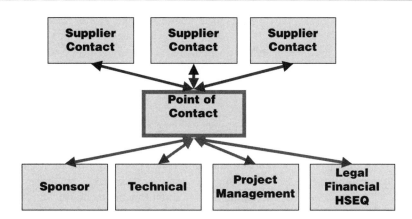

ultimate value for money to the company will depend on the quality of the contractor providing the service and the contractor's ability to meet performance targets and deadlines often over quite a long time frame. There may also be real or perceived different levels of risk associated with each tender. The more sophisticated tender evaluation systems described in the section *Whole-Life Cost Analysis* will take all of these factors into account.

***Tender Evaluation: The Formal Approach.*** In public institutions and many large organizations, the procedures governing tender evaluation are by necessity very rigorous and formal (see Figure 13.14). In such cases there is a formal opening of tenders often in front of

## FIGURE 13.14. FORMAL TENDER EVALUATION PROCESS.

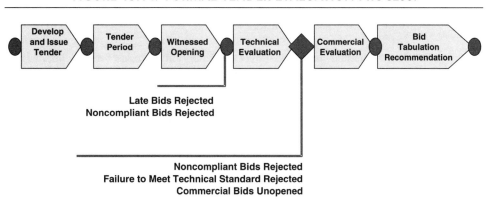

official witnesses. Each bid is divided into the technical and commercial package and the commercial package placed unopened in a secure locked filing cabinet or safe. The technical bids are then evaluated by a technical committee and graded according to the strength of the technical proposal. Bids that do not meet the technical requirements are rejected and the commercial offering left unopened. The commercial proposals of the bids still under consideration are then evaluated by a commercial committee and the overall cost to the owner determined. In many cases, particularly in the public sector, the lowest-cost bid of the proposals adjudges to be technically adequate is accepted.

This approach, while straightforward and transparent, does not, however, necessarily give the best value for money to the owner. Specifically:

- The offering of a technically adequate contractor may provide the lowest initial capital expenditure (capex), but this may be at the cost of higher operating expenditure (opex).
- There may well be a fairly significant spread in the technical and management quality of contractors who are all deemed to be technically adequate. Simply picking the technically adequate contractor with the lowest price may not necessarily provide the best value for money for the owner. There may also be a higher level of risk associated with delivering the required level of service between a contractor who is technically adequate and one who offers a higher level of management competence.

More sophisticated evaluation approaches, described in the following, can be used to take both these situations into account.

***Whole-Life Cost Analysis.*** The structure of commercial proposal should permit the evaluation of the *whole-life cost* of the service. For example, in the case of an equipment package, such as an air-conditioning unit, the whole-life cost normally expressed as net present value would include the following:

- Original capital cost
- Associated freight and handling costs
- Power consumption
- Cost of operation
- Cost of spares, service and insurance.

For many equipment packages, the cost of spares and service over the operating life can be many multiples of the original capital cost. Unless the tender has been very precisely framed, many of these costs will be at best estimates, with little hard commercial underpinning. It is also evident that basing the evaluation on capex alone could be grossly suboptimal for the company over the life cycle of the project or asset. Figure 13.15 shows the comparison between two equipment packages where the lowest capex cost is definitely not the best buy over a ten-year life cycle for the project. This is, of course, highly relevant in the case of lump-sum turnkey bids, where, unless it is very carefully specified, the contractor will inevitably select the equipment package with the lowest initial cost.

## FIGURE 13.15. WHOLE-LIFE COST.

### VENDOR-A

| Assumptions<br>Discount Rate | | 8% | | | | | | | | | |
|---|---|---|---|---|---|---|---|---|---|---|---|
| 000 Euros          Year | WLC | 1 | 2 | 3 | 4 | 5 | 6 | 7 | 8 | 9 | 10 |
| **Capex** | | | | | | | | | | | |
| Capital Cost | | 250 | | | | | | | | | |
| Insurance and Freight | | 10 | | | | | | | | | |
| Installation | | 25 | | | | | | | | | |
| Capex | **285** | 285 | | | | | | | | | |
| **Opex** | | | | | | | | | | | |
| Spares | 110 | | 10 | 10 | 15 | 10 | 15 | 10 | 15 | 10 | 15 |
| Service | 113 | | 10 | 10 | 12 | 12 | 12 | 12 | 15 | 15 | 15 |
| Consumables | 100 | 10 | 10 | 10 | 10 | 10 | 10 | 10 | 10 | 10 | 10 |
| Utilities | 120 | 12 | 12 | 12 | 12 | 12 | 12 | 12 | 12 | 12 | 12 |
| Opex | **443** | 22 | 42 | 42 | 49 | 44 | 49 | 44 | 52 | 47 | 52 |
| **Total Cost Ownership** | **728** | 307 | 42 | 42 | 49 | 44 | 49 | 44 | 52 | 47 | 52 |
| **NPV** | € 551.81 | | | | | | | | | | |

### VENDOR-B

| Assumptions<br>Discount Rate | | 8% | | | | | | | | | |
|---|---|---|---|---|---|---|---|---|---|---|---|
| 000 Euros          Year | WLC | 1 | 2 | 3 | 4 | 5 | 6 | 7 | 8 | 9 | 10 |
| **Capex** | | | | | | | | | | | |
| Capital Cost | | 225 | | | | | | | | | |
| Insurance and Freight | | 12 | | | | | | | | | |
| Installation | | 15 | | | | | | | | | |
| Capex | **252** | 252 | | | | | | | | | |
| **Opex** | | | | | | | | | | | |
| Spares | 180 | | 20 | 20 | 20 | 20 | 20 | 20 | 20 | 20 | 20 |
| Service | 113 | | 10 | 10 | 12 | 12 | 12 | 12 | 15 | 15 | 15 |
| Consumables | 150 | 15 | 15 | 15 | 15 | 15 | 15 | 15 | 15 | 15 | 15 |
| Utilities | 120 | 12 | 12 | 12 | 12 | 12 | 12 | 12 | 12 | 12 | 12 |
| Opex | **563** | 27 | 57 | 57 | 59 | 59 | 59 | 59 | 62 | 62 | 62 |
| **Total Cost Ownership** | **815** | 279 | 57 | 57 | 59 | 59 | 59 | 59 | 62 | 62 | 62 |
| **NPV** | € 600.81 | | | | | | | | | | |

### Commercial Evaluation

| Vendor-B | 12% | Lower Capex than Vendor-A |
|---|---|---|
| Vendor-A | 8.15% | Lower Whole Life Cost than Vendor B |

*Hard Money-Soft Money Evaluation.* In some projects the managerial and technically quality of the contractor can significantly affect the outcome and business benefit which the owner receivers from the service. "Hard money-soft money" evaluation is a technique that attempts to make this assessment on an analytical basis. ("hard money" refers to the actual costs defined in the tender; "soft money" refers to the estimates made on the basis of the perceptions of the team evaluating the bid on the basis of their professional judgment.)

Structured *scorecards* are used to quantify the perception of the less tangible parameters affecting the technical quality of the bid and hence the ability of the bidder to add value to the project. These scorecards are completed by each member of the review team. Scores are then debated until an overall consensus reached. Figure 13.16 shows an example of a scorecard used to evaluate the team proposed by a tenderer to undertake a significant project. Scorecards help focus the mind and provide the basis of communication for the evaluation team. These scorecard systems can be quite complex. Scorecards obviously enable us to express a perception of quality in hard numbers.

In the hard money-soft money approach, this perception of quality is translated into measurable benefits for the owner. For example, in the tender to undertake the maintenance turnaround of an oil refinery or power station, the quality of the management team proposed by a contractor is obviously a key success factor. How then to compare the offering of one contractor, who has a more experienced team, with another who has an adequate team and a lower price?

In hard money-soft money evaluation the owner team translates their perception of quality of the management team into the additional margin they can really generate by completing the project earlier. Ideally, these benefits should be linked to incentive arrangements in the contract form. Figure 13.17 shows the hard money-soft money evaluation of a contract for pipework refurbishment during the revamp of a process unit.

## Post Tender Negotiation

In virtually all public procurement, and in many companies, *post tender negotiation*, or negotiation of the price after tender submittal, is not permitted. Procurement professionals often debate whether post tender negotiation is in fact desirable. Theoretically, if bidders know there will be no further scope for negotiation, they will submit their best price in their original tender. In practice it doesn't seem to work like that. First of all, in most relatively complex proposals there is the need to debate the tender in great detail to ensure that both parties have a common understanding of the issues and the deliverables.

In fact, in the case of equipment procurement, the savings range is much higher because of the cost structure of the supplier organizations. In most multiround negotiating situations in a competitive environment, including the use of reverse auctions, prices tend to converge as expectations are progressively stripped away and suppliers cut to the absolute limit of their cost structure. Figure 13.18, taken from a real negotiation, shows how prices tend to converge in a typical multiround negotiation. Online reverse auctions also exhibit a similar pattern.

Negotiation is a very fine art where the invisible line between buyer and seller can be stretched to the limit but never broken. Experienced negotiators from both buyers and sellers

## FIGURE 13.16. EVALUATION SCORECARD.

| Scoring Guide Section G-Organisation and Project Team | | | | | | | | | | | |
|---|---|---|---|---|---|---|---|---|---|---|---|
| Non existent-totally unacceptable-no appreciation of issues | 0 | | | | | | | | | | |
| Poor by Industry standards-Disappointing | 1 | | | | | | | | | | |
| Below sector average-Below expectations | 2 | | | | | | | | | | |
| Average for the sector-What we expect | 3 | | | | | | | | | | |
| Quite good-Better than average | 4 | | | | | | | | | | |
| Quite exceptional-the best we have seen | 5 | | | | | | | | | | |

NOTE: Weightings of sections agreed by Team Prior to Evaluation Interviews.

| Ref: | Key Questions | Overall Weight | Weight Within section | Points Earned | Score | | | | | | Comment |
|---|---|---|---|---|---|---|---|---|---|---|---|
| | | | | | 0 | 1 | 2 | 3 | 4 | 5 | |
| **G-1 Organisation** | | | | | | | | | | | |
| | Linkage to Owner organisation appropriate? | | 20 | | | | | | | | |
| | All functions covered? | | 20 | | | | | | | | |
| | Structure coherent? | | 20 | | | | | | | | |
| | Resourcing adequate? | | 20 | | | | | | | | |
| | Delegations appropriate? | | 20 | | | | | | | | |
| | Other? | | | | | | | | | | |
| | **G-1 Sub-Total Organisation** | 20 | 100 | | | | | | | | |
| **G-2 Profiles of Key Staff** | | | | | | | | | | | |
| Appreciation of the experience motivation and flexibility of key staff | | | | | | | | | | | |
| | **Business Unit Director** | | | | | | | | | | |
| | Experience | | 50 | | | | | | | | |
| | Motivation & flexibility | | 50 | | | | | | | | |
| | **Overall Rating** | | 100 | | | | | | | | |
| | **Designated Project Manager** | | | | | | | | | | |
| | Experience | | 50 | | | | | | | | |
| | Motivation & flexibility | | 50 | | | | | | | | |
| | **Overall Rating** | | 100 | | | | | | | | |
| | **Designated Design Manager** | | | | | | | | | | |
| | Experience | | 50 | | | | | | | | |
| | Motivation & flexibility | | 50 | | | | | | | | |
| | **Overall Rating** | | 100 | | | | | | | | |
| | **Designated Procurement Manager** | | | | | | | | | | |
| | Experience | | 50 | | | | | | | | |
| | Motivation & flexibility | | 50 | | | | | | | | |
| | **Overall Rating** | | 100 | | | | | | | | |
| | **Designated Construction Manager** | | | | | | | | | | |
| | Experience | | 50 | | | | | | | | |
| | Motivation & flexibility | | 50 | | | | | | | | |
| | **Overall Rating** | | 100 | | | | | | | | |
| | **Designated Planning and Control Manager** | | | | | | | | | | |
| | Experience | | 50 | | | | | | | | |
| | Motivation & flexibility | | 50 | | | | | | | | |
| | **Overall Rating** | | 100 | | | | | | | | |
| | **G-2 Sub-Total Project Team** | 80 | | | | | | | | | |
| | *Overall Rating on Experience* | | | | | | | | | | |
| | *Overall Rating on Motivation & Flexibility* | | | | | | | | | | |
| **Overall Score Organisation and Project Team** | | 100 | | | | | | | | | |

### FIGURE 13.17. HARD MONEY-SOFT MONEY EVALUATION.

| Assumptions | | | | | | | | | | | |
|---|---|---|---|---|---|---|---|---|---|---|---|
| Cost/Day € | 50000 | | | Scope Creep Allowance | | | | Soft Cost | | | |
| COMPANY | Manhour Estimate Fixed Scope | Price ?/MH | Fixed Price | Day Rates ?/Hr. | Estimated Extra Hours | Estimated Cost Dayworks | TOTAL HARD COST | Over-Run Risk Days | Soft Cost | TOTAL COST | |
| Magyar Pipe | | | | | | | | | | | |
| Roma Pipe | | | | | | | | | | | |
| Gypipe | | | | | | | | | | | |

*From Manpower Planning*

will work throughout the pre-tender and tender period to develop the climate of the negotiation to their own advantage. From the owner's perspective, it is important to manage the environment during the tender period to heighten the desire of the bidder for the contract and lower the bidder's expectation on selling price. In fact, there is clear evidence that the harder someone expects the negotiation to be, the lower the price they will settle for (Field, 2003). The actual negotiations will be carefully planned as the use of time is a major negotiating lever. The various members of the team must clearly understand their roles and "script."

### FIGURE 13.18. PRICING CONVERGENCE.

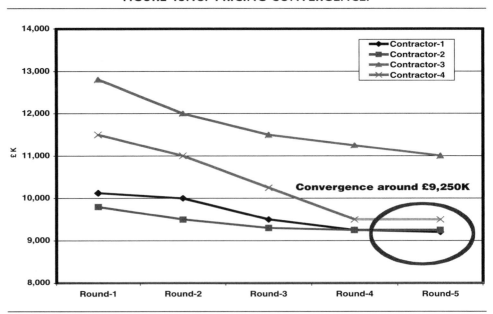

Remember that it is a vitally important time when often over a few hours or days, the ultimate profitability of the project or contract, involving years of work, can be made or broken. It is a time when the management teams from buyer and seller can develop a huge amount of mutual respect or loathing (P. Marsh).

## Transition to Execution

Once the contract has been awarded, the buyers and successful vendors celebrate, and in many cases depart, leaving the operating organizations with the task of making the contract work—which can often be quite difficult, particularly in more complex contractual arrangements. Many of the understandings developed within the negotiating team may not be fully understood by others and may result in difficulties throughout the life of the contract. It is therefore advisable to recognize the need to manage the transition from negotiation into operations. This is true in all project situations, but in large complex projects managed by an integrated owner-contractor project team where many members of the project team are new to each other, it can be a key success factor. Such companies will spend considerable time constructing a complete integration program, including the following:

- Social events and expeditions to promote "bonding" across the project team.
- Contract workshops where the owner and contractor team that negotiated the contract explain it to the operational staff and jointly answer questions of detail from the execution team.
- RACI workshops where the owner and contractor staff can work out how best to manage day-to-day operations and develop management alignment, eliminating any confusion over relative responsibilities.

All of these events improve communication and generate a common understanding and language across the project team. They may even result in mutual appreciation of professionalism and fairness. Such common understanding and language can only help resolve problems, remove constraints, and ultimately deliver a better result and business benefit for both parties

# Online Tendering Systems

The mechanistic aspect of the tendering process described previously lends itself to automation through the use of information systems. While the real value added in the tendering process is the development of creative contracting strategies and value managing requirements, the pure mechanics of issuing and producing tenders probably accounts for 80 percent of the level of effort and management time expended. This aspect can be really streamlined through the use of online tender management systems. Figure 13.19 outlines such a system.

The best of these systems support each step of the procurement process and provide online access as appropriate to everyone involved in the procurement process:

## FIGURE 13.19. TENDER MANAGEMENT SYSTEM.

**Web-Enabled Tender Management System**

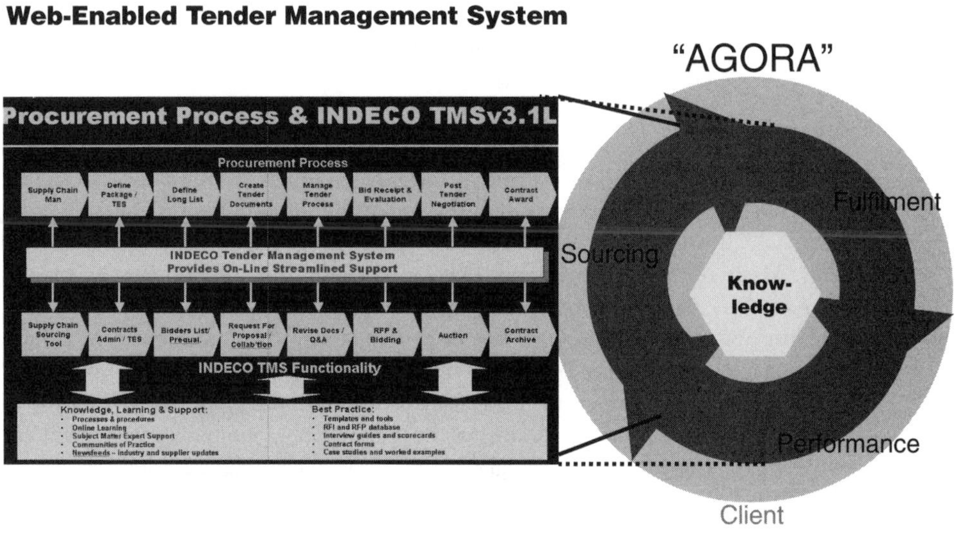

- Users can interrogate extensive supplier databases to identify potentially suitable suppliers.
- Request for Information can be published on the Web, completed, and processed online to produce a bidding list for a particular tender (see Figure 13.20).
- Questions and answers can be tabled and processed consistently online.
- Requests for proposals can be published and completed online, including all supporting documentation and drawings.
- Bids can be evaluated and a short list developed.
- Post tender negotiation can, where appropriate, be conducted through a reverse-auction process.

Tender management systems can readily be linked to a knowledge management database providing users access to case studies, templates, organizational documentation, and the like that is accessible at point of use. Such a system provides an extensive audit trail of every procurement transaction and can reduce the professional time to develop and conclude a typical tender by at least 25 percent.

## The Bidders' Perspective

By and large this section has been written from the perspective of the organization issuing the tender. If effective tender management is important to the "owner" organization, it can

## FIGURE 13.20. TRANSMITTING TENDER DOCUMENTS.

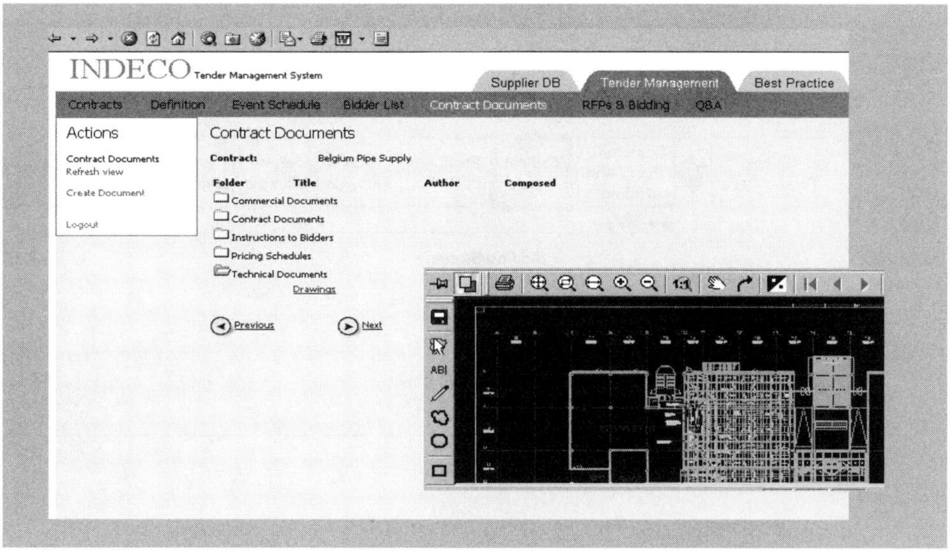

be even more so to the organizations responding to the tender. In many cases the risk a bidder assumes can be a very significant percentage of the bidder's net value. On the one hand a major contract can represent the opportunity for significant profit or strategic advantage. On the other, many companies have been bankrupted by contracts that went wrong. Figure 13.21 outlines the classic bidding process for a tendering company. Every company should have a formalized process for receiving, responding to, and submitting tenders. Tenders are the lifeblood of every contracting organization. They are mission-critical and must be treated as such.

## Decision to Bid

The decision to bid or not to bid is critical. Tenders should never be taken lightly. Producing a tender costs money, often a great deal of money, and will tie up the energies and time of the most creative people. Many contractors have been bankrupted by taking on uneconomic contracts. Every contractor should ask the following questions for every tender being considered:

- Is this opportunity in line with the strategic objectives of the company?
- Is this a client the contractor wishes to work with?
- Does the contractor have the competence and capacity to undertake this work?

## FIGURE 13.21. TENDER MANAGEMENT, BIDDERS' PERSPECTIVE.

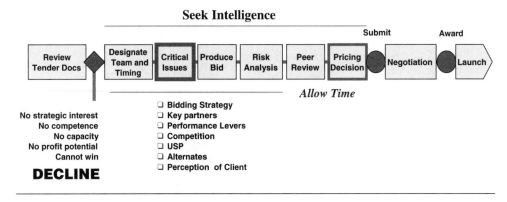

- Are there competitors the contractor cannot hope to beat?
- Does bidding this work make economic sense?

If the answer is no to *any* of these questions, the contractor should seriously consider whether it is advisable to bid. While it may be politically necessary to respond to a client with whom the contractor might wish to work on another project at another time, it is inadvisable to waste time and money on tenders the contractor does not really want to win. Figure 13.21 illustrates the typical tendering process we would expect to see within a typical reasonably sophisticated contracting organization.

Assuming the contractor wishes to tender, a bid manager is generally assigned with access to the appropriate in-house and external expertise and resources immediately. This bid manager is responsible for coordinating the company resources and producing a winning bid that will be profitable for the company.

The first key event in the bidding process is normally the *critical issues meeting*. This will be held as soon as possible and aims to table the collective wisdom of the contracting organization on the tender at hand. This meeting would seek to identify the following:

- The lessons learned from previous experience on similar work, or with the particular client.
- The performance levers in the scope of work that will generate real value for the owner organization.
- Any contacts, possibly within the client organization, who can provide intelligence on the tender at hand.
- Key suppliers who may provide a unique advantage and formulate ways to do this and deprive the competition of their services.

- The strengths and weaknesses of the competition and their likely response and pricing levels.
- The way we might differentiate ourselves from the competition and frame an offer the client cannot refuse.
- Any alternate approaches that might give a steep change in cost or any other client benefit.
- The major risks associated with the tender and how we can mitigate these risks.
- Any administrative, legal, or fiscal implications associated with the contract.

On the basis of this meeting a formal bidding strategy, budget, and schedule will generally be produced and responsibilities for the various aspects of the tender assigned. During the bidding period every attempt will be made to get as much intelligence on client thinking and the competitive position as possible. A formal risk analysis will be produced to support the pricing recommendation of the tender team. It is important to allow enough time for a formal peer review when the tender can be presented to senior staff, who have not been engaged on the tender, and subjected to challenge.

A formal memorandum will be tabled making the pricing recommendation to management. This memorandum will summarize tender scope, strategic implications, the key assumptions, the major risks, and mitigating strategies. The pricing of the tender should be recommended, including the basis of the negotiation strategy and the "walk away" price.

# Summary

Tender management is a vitally important management function for buyers and sellers alike. Done well, it can significantly improve the profitability of both project owners and contractors. Done badly, it can seriously jeopardize the profitability of owner organizations and the very survival of contractors. It is undoubtedly an exciting space but not one for the faint-hearted.

This chapter is based on many years' experience of project and tender management in many different sectors, in many different countries, and from the perspective of both an owner and contractor. Unfortunately, many of the lessons learned have been very expensive!

While details will vary from case to case, I absolutely believe that the approach to tender management described in this chapter is applicable in most sectors of industry and that the eventual success of any procurement and tendering initiative depends on the following three fundamental success factors:

- The composition and competence of the team and the quality of teamwork
- The rigor of the procurement and tendering process
- The creation of an innovative management culture that encourages the whole project team to strive for higher levels of performance

Tender management is a crucial period in the life cycle of most projects. It can have a fundamental impact on the success and profitability of the project. If this chapter provides a little enlightenment, or even reassurance, to those concerned, it will have served its purpose.

# References

Bennett, J., and A. Baird. *NEC and partnering: The guide to building winning teams*. London: Thomas Telford. ISBN: 0-7277-2955.

Egan, J. 1998. Rethinking construction. The Report of the Construction Task Force to the Deputy Prime Minister, John Prescott, on the scope for improving the quality and efficiency of UK construction. London: The Stationery Office.

ENR (2001) BP & Bovis boost relationship to remake oil giant facilities. ENR 2001.

Field, A. 2003. How to negotiate with a hard nosed adversary. March.

Kennedy, M. 2000. *Guardian*. December 6.

Lassiter, T. 1998. Balanced sourcing the Honda way. *Strategy & Business*. Booz Allen Hamilton, 4th quarter.

Latham, M. 1994. *Constructing the team: Final report of the government/industry review of procurement and contractual arrangements in the UK construction industry*. London: The Stationery Office.

McKinsey Report: Driving productivity and growth in the UK.

Marsh, P. Contract negotiation handbook: ASIN 0566-024039.

Value Engineering Professional Association. *www.vetoday.com*

Wideman R. R. M. 1992. Project and programme risk management. *PMI USA*.

CHAPTER FOURTEEN

# CONTRACT MANAGEMENT

## David Lowe

Contract management has been defined as ". . . the process which ensures that all parties to a contract fully understand their respective obligations enabling these to be fulfilled as efficiently and effectively as possible to provide even better value for money" (CUP, 1997). This process commences with the identification of the purchaser's needs and concludes with the completion of the contract. Further, the process has two dominant characteristics:

- *Risk identification, apportionment, and management.* Related to contract performance
- *Relationship management.* Between the purchaser and supplier

A prudent project manager and contract management team, therefore, will have a thorough understanding of the following:

- Procurement process and post-tender (bid) negotiation
- Assumptions made by the purchaser and the supplier
- Purchaser's expectations of the service relationship
- Contract terms and conditions; for example:
  - Purchaser's duties and responsibilities under the contract
  - Supplier's obligations under the contract
  - Main cost determinants, how they relate to the outputs and quality standards, and how they will be measured
  - Certification and payment mechanism
  - Purchaser's and supplier's rights if things go wrong
  - Legal implications of the contract for which they are responsible (CUP, 1997)

There is, however, a competing view: To facilitate successful projects, contracts should be "left in the drawer." Latham (1994) is sympathetic to this viewpoint; he considers the function of a contact is to serve the contract process, not vice versa. However, Hughes and Maeda (2003) contend that this view is incorrect; planning for future events in the contract process could be very problematical without knowledge of the contract.

The aim of this chapter is to develop a critical understanding of the factors that influence commercial contract practice. The intention is to provide the project manager with an overview of contract provisions and procedures. The chapter seeks to explain contract management in terms of generic contract provisions: the key components of contracts. However, to illustrate these principles, reference will be made, where appropriate, to the Fédération Internationale des Ingénieurs-Conseils (FIDIC) Conditions of Contract for Construction (first edition, 1999), also known as the FIDIC Construction Contract. Essentially, this contract is intended for building and engineering works designed by the employer; however, the contract does allow for the inclusion of some elements of contractor-designed civil, electrical, mechanical, and/or construction works. Members of FIDIC come from over 60 countries worldwide, and the FIDIC Construction Contract is an internationally recognized standard form.

# Contractual Issues

## Definitions

Goods and services necessary for the completion of projects are procured through the use of contracts with suppliers. Likewise, main suppliers use contracts to procure from subcontractors; while contracts in their own right, they are generally referred to as subcontracts (denoting a relationship to the main contract).

## Contracts

A *contract* is an agreement between two parties under which one party promises to do something for the other in return for a consideration, usually a payment. This places obligations on both parties to fulfil their part of the agreement. It is also the foundation for the relationship between the parties.

Elements of a valid contract vary depending on country of origin. Specifically:

- *Canadian law.* Intention to enter into the contract, consideration, capacity of the parties, offer and acceptance, and lawful object (Jergeas and Cooke, 1997).
- *English law.* Intention to be legally bound, consideration, capacity of the parties, and offer and acceptance.
- *French law.* Mutual assent, cause, capacity of the parties, and lawful object (French Civil Code, Article 1108).
- *German law.* Mutual assent, intent to confer a benefit, capacity of the parties, and lawful object (German Civil Code, §§ 518, 761, 780, 781).
- *United States law:* Mutual assent, consideration, capacity of the parties, and lawful object (American Law Institute, Restatement of the Law, Second: Contracts, §§ 3, 8, 12).

The parties to a contract are as follows:

- *The purchaser.* The party that acquires or obtains goods or services by payment or at some cost. Alternatively referred to as the buyer, client, customer, employer, owner, proposer, sponsor, user, and so on.
- *The supplier.* The provider of goods and services. Also referred to as the contractor, main supplier, main contractor, prime contractor, prime supplier, seller, vendor, and so on.

This text will refer to the purchaser and supplier unless reference is made to a specific form of contract.

Although not a party to the contract, most contracts make reference to the following:

- *A project manager.* The person who leads the purchaser's contract management team. Alternatively, the terms "architect," "contract manager," or "engineer" may be used. This text will refer to the project manager.
- *The subcontractor.* A supplier to the main supplier, main contractor, prime contractor, prime supplier, and so on.

While contracts can take the form of a single document, generally, commercial contracts comprise several documents. For example:

- *The contract agreement.* Itemizes the documents comprising the contract. It includes the identities of the parties and defines the scope of work, the contract price, and the schedule for its execution.
- *General specification and scope of work.* Describe the scope of work to be undertaken, the technical standards required, and the administrative procedures to control the implementation of the project.
- *General conditions of contract.* Normally a recognized standard form of contract. This details the obligations to produce and to pay, clarifies the offer and acceptance, allocates risks, describes the consequences of failure to pay or produce, and includes relevant issues, for example, insurances, bonds, safety, industrial relations, defects, and disputes, and so on.
- *Special conditions of contract.* These cover additions and amendments to the general conditions as required by the purchaser and specific circumstances of the project.
- *Administrative and coordination procedures.* Frequently the procedural aspects of a contract are covered separately as an appendix to the general conditions.

In practice, however, these documents will be interlinked, with some having greater importance than others. Because of the potential for conflicting information within these documents, an order of priority needs to be established prior to inviting bids.

## Letters of Intent

Occasionally, letters of intent are used as an interim arrangement to permit a successful bidder to start work in advance of signing a contract and in the knowledge that they will ultimately be awarded the contract. To be operative, a letter of intent must have the properties of a contract and refer to those elements of the bid where agreement has been reached. As a minimum, a valid letter of intent should always state that the purchaser

- intends to place a contract with the supplier;
- wishes the supplier to begin work in advance of the contract; and
- authorizes the supplier to begin work in advance of the contract.

A letter of intent may not be legally enforceable and may be revoked by the issuer without any redress to the courts where it is held to be an announcement of one party's wish to do something. A further disadvantage arises from the purchaser's declaration of intent: It will reduce the ability of the client to satisfactorily negotiate the outstanding terms. HM Treasury, Central Unit on Procurement (CUP, 1989) recommends the use of a "start-up contract" with appropriate controls, limits, and safeguards, rather than a letter of intent.

## Contracts and Orders

Legally there is no distinction between the terms "contract" and "order." Both refer to legally binding agreements for the supply of goods and/or services in return for some form of remuneration. Commonly, the term "contract" is used in relation to an agreement involving a longer time period and a greater outlay than a purchase order.

## Subcontracts

In parallel to the contract between the purchaser and the supplier (the main contract), the supplier employs subcontractors and suppliers of materials, plant, equipment and services, and so on. The supplier is generally held responsible for any subcontracted work, and the purchaser, within the main contract, retains the right to vet subcontractors and limit the extent to which the work is subcontracted. The supplier may be free to choose the terms of subcontracts, or alternatively the terms may be required to correspond, "back-to-back," with those of the main contract.

Purchasers may also reserve the right to nominate subcontractors, requiring the supplier to enter into a subcontract with a subcontractor chosen by the purchaser or the project manager, usually to carry out specialist work. Contracts, therefore, require specific clauses to manage specific risks imposed by nomination, for example, in relation to the default of the nominated subcontractor, and to ensure that the supplier pays the nominated subcontractor.

## Standard and Model Forms

Many industry sectors use standard or model conditions of contract as oppose to bespoke contracts. Wright (1994) contends that standard forms are used because they

- provide a recognized and predictable contractual basis;
- save time, both in writing and in negotiating the contract; and
- are familiar to the project/contract management teams, resulting in smoother-running projects, or at least in the avoidance of some mistakes that could disrupt progress.

Most organizations have their own standard conditions of contract such as standard sets of conditions of sale and purchase, for example, supplier's terms. Invariably, these contracts are biased, to a greater or lesser degree, in favor of the party that composed them. Alternatively, model conditions are used where the balance of power between the parties is approximately equal. Model conditions tend to be drawn up by an association including representatives of all parts of an industry and are, therefore, generally held, according to Wright, to represent a reasonable basis upon which organizations within that industry might be prepared to do business with each other.

## Legal Interpretation of Contracts

Contracts operate within the framework of law. Globally, European legal systems are dominant. Basically, three systems prevail: Those based on English, Roman, and Russian law, the latter being a fusion of Roman and English legal principles. As a broad assertion, however, the general principles of commercial contract law are the same the world over, but the detail will vary. Therefore, specific legal advice should be sought on the implications of the interpretation of contract clauses with regard to a particular legal system.

International contracts need to state which country's law or other jurisdiction will apply. The choice of law has a significant effect on the administration of a contract. It determines how the contract is formed and establishes the underlying terms of the contract. It also provides a structure within which the parties function, for example, laws concerning trading standards and practices, safety, tax, and so on. Further, it tends to establish where and how disputes are resolved. Where a choice of legal systems exists, it is important to contemplate the consequence of that choice before entering into the contract.

Where the contract is produced in more than one language, the contract needs to determine the ruling language. Likewise, the contract needs to state the language for communication purposes. This is important because, in the event of a dispute occurring, the exact words used in the contract will be carefully interpreted in order to determine the precise agreement made between the parties. Immense care, therefore, is required when selecting the words used in a contract as they can have different interpretations.

## Contract Terms

The terms of a contract are all the rights and obligations agreed between the parties, together with any terms implied by law.

### Express and Implied Terms, Incorporation by Reference

- *Express terms.* Those terms that are stated (written) within a contract
- *Implied terms.* For example, within English law, those terms that form part of a contract but are not expressed. Implied terms include the following:

- Conformity with statutes
- Supplier's responsibility for their subcontractors
- Fitness for purpose: provided the purpose has been communicated to the supplier and has not been overruled by an imposed specification
- Furtherance of purpose, where both parties endeavor to perform the contract as best as they can
- Duty to utilize competence and care
- Supplier's liability to execute the work at a reasonable pace
- *Incorporation by reference.* For example, where the contract makes reference to terms contained within other documents.

**Conditions and Warranties.** The terms of a contract under English law can comprise conditions and warranties. A *condition* is a key term within the contract: a promise of considerable magnitude. Failure to fulfil the promise entitles the injured party to terminate the contract and to claim damages for failure to comply. A *warranty* is a less serious term within the contract. Failure to comply with such a term entitles the injured party to claim damages but not to terminate the contract.

## Commercial Contracts

**Complexity of Contracts.** Commercial contracts are relatively complex documents; for example, construction projects generally require extensive contracts in order to express precisely the legal, financial, and technical facets of the project. As a result, one potential source of risk is the contract document. The contract conditions, therefore, according to Bubshait and Almohawis (1994), need to be assessed for clarity, conciseness, completeness, internal and external consistency, practicality, fairness, and effect on project performance—that is, on quality, cost, schedule, and safety. They present a simple and systematic instrument to evaluate these attributes.

**Commercial Manager.** The last 15 years has seen the emergence, primarily within large UK organizations, of the role of commercial management. A commercial manager has been defined as ". . . a person controlling or administering the financial transactions of an organisation with the primary aim of generating a profit generating whilst minimising associated risk" (Lowe et al., 1999). The function involves advising the organization on the use of contracts, formulating bespoke contracts, and in negotiating contracts.

# Contract Strategy and Type

The provisions of a contract should do the following:

- *Define the responsibilities of the parties.* For example, define the project's objectives and priorities; project finance, innovation, development, design, quality, standards, procurement, scheduling, implementation, installation; project management, safety, inspection, testing, commissioning, and managing operating decisions.

- *Allocate risk.* For example, financial investment in the project, project definition, design, performance specification, subcontractor selection, subcontractors' defaults, site productivity, delays, mistakes, and insurances.
- *Determine effective payment terms.* For example, for development, design, demolition, construction, fabrication, implementation, management, and others services (Smith and Wearne, 1993; Wearne, 1999).

Contracts are usually classified in terms of strategy (procurement methodology or organizational choice)—for example, traditional, design and build, turnkey, and management contracts—or by type (allocation of risk and payment terms)—for example, lump-sum, remeasurement, and target cost contracts. Contract strategy and type should be planned concurrently.

## Contract Strategy

For a discourse on the various procurement/contract strategies, see the chapter on procurement systems by Langford and Murray. The following are examples of contracts classified in terms of strategy.

### Design Combined with Production

- *Design and build contracts.* FIDIC Conditions of Contract for Plant and Design-Build; JCT Standard Form of Building Contract with Contractor's Design (1998); AGC 415 Standard Form of Design-Build Agreement and General Conditions Between Owner and Design-Builder (Where the Basis of Payment is a Lump Sum Based on an Owner's Program Including Schematic Design Documents) 1999 Edition.
- *Turnkey contracts.* FIDIC Conditions of Contract for EPC (Engineering, Procurement and Construction)/Turnkey Projects

### Design Separate from Production.
Two alternative organizational structures exist where design is separate from production:

- Sequential contracts
  - *Conventional or traditional contracts.* FIDIC Conditions of Contract for Construction; JCT Standard Form of Building Contract (1998).
- Parallel contracts
  - *Management contracting.* JCT Standard Form of Management Contract (1998), NEC Engineering and Construction Contract: Management Contract (1995).
  - *Construction management.* JCT Construction Management Documentation (2002); AIA A101™/Cma Standard Form of Agreement Between Owner and Contractor—Stipulated Sum—Construction Manager—Adviser Edition/A201™/Cma General Conditions of Contract for Construction (1992); AGC 230 Standard Form of Agree-

ment and General Conditions Between Owner and Contractor (Where the Basis of Payment is the Cost of the Work with an Option for Preconstruction Services), 2000 Edition.

Parallel contracts are advantageous where an early project completion date is crucial, the design requirements are uncertain at the outset, supplier involvement in the design process is advisable, there is a requirement to maintain the operation of existing installations, the segmented work is of a specialist nature, and/or suppliers have a limited capability.

### Alternative Organisational Arrangements

- *Term contracting.* Term contracting refers to a particular type of work to be executed over a given time period. It is commonly used for the provision of a service, for example, repair and maintenance work where the general nature of the work is known but the extent of it is not. Each individual order issued under the term contract becomes a discrete contract, and at this point the terms of the bid become binding. Example: JCT Standard Form of Measured Term Contract (1998).

**Strategic Cooperative Arrangements.** According to Smith et al. (1995), there are two major areas of operational difficulty in joint ventures, which have implications for both bid preparation and project implementation: conflict and culture. Likewise, Walker and Johannes (2001) found that equalization of power is crucial within joint venture partnerships, while the need to understand organizational cultural diversity was also seen to be pivotal.

Specific contract conditions are required, therefore, to ensure the establishment of a suitable organizational structure that will encourage the successful completion of the project and that will safeguard the purchaser in the event of the default or liquidation of one of the joint venture members. Example: AGC 299 Standard Form of Project Joint Venture Agreement Between Contractors, 2002 Edition.

## Contract Type

Essentially, there are two categories of project contract payment terms: price-based and cost-based.

### Price-Based

- *Fixed price.* Where the supplier is paid a fixed price or lump sum (a single tendered price) for the entire project. The terms "fixed" or "firm" usually indicate that the contract price will not be subject to escalation payments, whereas lump-sum contracts may. Additionally, fixed and firm contracts generally may not include variation clauses. However, the terms fixed and firm have no precise meaning. The payment terms included within a specific contract are the key factor.

Examples include AIA A101™ Standard Form of Agreement Between Owner and Contractor—Stipulated Sum/A201™ General Conditions of Contract for Construction

(1997); AGC 200 Standard Form of Agreement and General Conditions Between Owner and Contractor (Where the Contract Price is a Lump Sum), 2000 Edition; IChemE Form of Contract for Use in the Process Industries: Lump-Sum Contract (The Red Book), 4th Edition.

- *Measurement.* Where a list of the items and quantities of the work to be executed under the contract (bill of quantities) is incorporated into the bid/contract documentation: The purchaser pays a standard rate based on agreed productivity rates and unit rates.

     Examples include JCT Standard Form of Building Contract Private With Quantities (1998); NEC Engineering and Construction Contract: Priced contract with bill of quantities (1995).

- *Remeasurement.* Where the actual work carried out by the supplier is measured on completion, as implemented, based upon either

  - *an approximate bill of quantities.* A list of the items and approximate quantities of the work to be executed under the contract; where the purchaser pays a standard rate based on agreed productivity rates and unit rates.

       Example: JCT Standard Form of Building Contract Private with Approximate Quantities (1998).

  - *a schedule of rates.* A list of potential items to be executed under the contract; where the purchaser reimburses the supplier using agreed unit rates.

  - *a bill of materials.* A list of the materials expected to be used, together with a unit of measurement; where the purchaser pays a standard rate based on a pre-agreed composite unit of measure.

Price-based contracts incentivize the supplier; by working efficiently, cost can be controlled and profit maximized. Likewise, the supplier will generally only supply goods and services that meet the absolute minimum required by the specification. With regard to risk, price-based contracts require the supplier to bear a comparatively high level of risk: They are required to perform all the necessary work to meet the specification within a specified timescale. From the purchaser's perspective, the major limitation of a price-based contract is that it establishes a relatively inflexible contract structure.

### Cost-Based

- *Cost-plus.* Where the supplier is reimbursed all their entitled expenditure plus an agreed profit margin, which can either be a percentage of the final cost (cost plus percentage fee) or a fixed amount (cost plus fixed fee).

     Examples include NEC Engineering and Construction Contract: Cost reimbursable contract (1995); AGC 230 Standard Form of Agreement and General Conditions Between Owner and Contractor (*Where the Basis of Payment is the Cost of the Work*) 2000 Edition; IChemE Form of Contract for use in the process industries: Reimbursable Contract (The Green Book), 3rd Edition.

     Cost-based contracts have the benefit of being more collaborative, but they impose a much lower degree of control on the supplier, requiring more managerial effort by the purchaser. Compared with price-based contracts, the level of risk borne by the supplier will reduce, while that of the purchaser will rise; however, the contract will contend with high levels of change.

***Incentives and Contract Type.*** Incentive provisions can be incorporated within fixed-price and cost-reimbursable contracts. Herten and Peeters (1986) describe and illustrate three specific types: cost incentives, schedule incentives, and performance incentives. They also refer to multiple-incentive contracts, where two or more of these incentives are combined, either dependently or independently, in the same contract. Bubshait (2003) puts forward a fourth type, safety incentives, although he found only limited support for its value.

Incentive contracts are not as extensively used as they might be. According to Ward and Chapman (1994), this is perhaps due to a lack of appreciation of the limitations of conventional fixed-price contracts and/or of the ability of incentive contracts to motivate suppliers. However, within industrial projects, Bubshait (2003) highlights the variation in the perception of purchasers and suppliers concerning incentive/disincentive (I/D) contracting. While he found a general agreement on the effectiveness of I/D contracting in encouraging supplier performance, few organizations incorporate I/D principals into their contracts. Moreover, penalty systems were used, rather than incentive systems, to penalize the supplier for late completion.

Examples of incentive contracts include NEC Engineering and Construction Contract: Target contract with bill of quantities (1995); NEC Engineering and Construction Contract: Target contract with activity schedule (1995); AGC 250 Standard Form of Agreement and General Conditions Between Owner and Contractor (*Where the Basis of Payment is a Guaranteed Maximum Price*), 2000 Edition.

## Choice of Contract

The choice of contract type is one of the most significant strategic decisions, since it determines how the supplier is paid and how risk is allocated between the parties. As a general principle, contract type should aim to give the maximum likelihood of attaining the objectives of a project (Wang et al., 1996); they should be regarded as a means to an end.

Griffiths (1989) summarizes the advantages, problems, and resource requirements of the major contract type alternatives. In addition, based upon 93 R&D defense projects, Sadeh et al. (2000) found that contract type has a considerable impact on project success. Under increasing technological uncertainty, both parties to the contract benefit from cost-plus contracts, while fixed-price contracts generate more benefits when uncertainty is lower. They recommend two-stage projects. At the first stage, the preliminary design and feasibility study stage, where technological uncertainty is very high, they recommend the use of cost-plus contracts. At the second stage, the full-scale design and development stage, a fixed-price contract is preferable.

Likewise, Turner and Simister (2001) have demonstrated that, when using transaction cost analysis to indicate when alternative contract pricing terms should be adopted, it is uncertainty of the final product and not risk per se that determines the most appropriate type of contract. Further, they suggest that, if the purpose of a contract is to create a project organization based on a system of cooperation not conflict, then the requirement for goal alignment is more significant. This, they consider, requires that all parties to a contract should be properly incentivized, and that this is accomplished by incorporating contract pricing terms, as illustrated in Figure 14.1.

Turner and Simister (2001) conclude that the main criterion for selecting contract pricing terms is goal alignment, however, transaction costs are minimized *en passant*.

## Roles, Relationships, and Responsibilities

### Roles

A contract defines the roles of the two parties: the purchaser and the supplier. Additionally, the contract apportions roles: for example, project management, design execution and integrity, production supervision, and dispute determination.

***The Role of the Parties to the Contract.*** Generally, the purchaser will be involved in the following:

- Defining exactly what services are to be provided
- Setting service levels
- Providing relevant and timely information to the project manager and supplier
- Informing the project manager of under-performance

The degree of empowerment of a supplier is dependent upon the procurement approach adopted and the terms of the contract. Generally, the supplier will be responsible for the following:

- Deciding how to provide the service
- Delivering the service to specification;

### FIGURE 14.1. SELECTION OF CONTRACT TYPES.

|  |  | Uncertainty of the product | | |
|---|---|---|---|---|
|  |  | Low | High | |
| Uncertainty of the process | High | Fixed-Price Design and Build | Cost-Plus Design and Build Alliance | High Complexity |
|  | Low | Remeasurement Build Only | This situation was not researched | Low |
|  |  | Low | High | |
|  |  | Ability of the client to intervene | | |

*Source:* Turner and Cochrane (1993).

- Deciding priorities to achieve the service
- Meeting purchaser requirements within the contract terms and budget
- Monitoring the service delivery performance
- Development and implementation of agreed procedures
- Providing information as required by the contract (CUP, 1997)

The contract may also define the role of other parties (agents of the purchaser), such as project manager, engineer, architect, landscape architect, interior designer, quantity surveyor, superintending officer, clerk of work, and so on, as well as the relationships with other suppliers, including subcontractors, nominated subcontractors, and nominated suppliers, and so on.

## Relationships

*Rules and Procedures.* The establishment, from the outset, of transparent procedures will enhance contract management while reducing the disruption that problems may generate. While some will relate to the routine contract management activities, others will operate when needed.

Procedures will be required for the following:

- Performance/service management
- Risk assessment
- Contingency planning
- Payment submission, processing, and certification
- Budget review and control
- Change management—instigated by either the supplier or the purchaser
- Price adjustments
- Interrelationship of management and control
- Security
- Problem management
- Disputes resolution
- Compliance monitoring
- Termination requirements

Effective implementation of a project will be reliant upon the relationships between the parties, not necessarily on the contract or role definition. Further, it is crucial that these relationships be established at the outset, continually reviewed, and actively managed. Relationships need to be balanced between, on the one hand, flexibility and openness, and on the other, professionalism and businesslike behavior.

## Risk and Responsibilities

Contracts set down the rights and obligations of the parties to the contract and describe the responsibilities and procedural roles of those named within the contract. For any project, achievement of its objectives is the principal risk; this is borne by the purchaser. Likewise, the purchaser bears the key risks of any project, for example, in deciding to instigate the project, defining the project's scope and specification, selecting a contract strategy, and choosing a supplier.

Other risks relate to the design, implementation, and delivery of the project; contracts seek to allocate these risks to the parties. However, both parties may be at risk irrespective of the contract, for example, where forces outside their control frustrate the work. Kangari (1995) summarizes the attitude of U.S. construction contractors towards the allocation and importance of risks within contracts. He also reviews trends in these perceptions.

Ideally, the allocation of risk between the parties should be based upon the following:

- *Managerial principals.* A satisfactory completion of a project is more likely to be achieved through effective planning and supervision rather than requiring guarantees and imposing rights to damages for default.
- *Commercial principles*: A risk should be borne by the party best able, economically, to control, manage, or insure against its consequences.
- *Legal principles*: Unfair contract terms, for example, penalties may not be enforceable (Wearne, 1999).

Ultimately, it is not in the purchaser's financial interest to ask a supplier to absorb all risks. The purchaser's objectives are more likely to be attained through the use of contract terms that motivate the supplier to perform on time, economically, and so on, and where the risks transferred are not so great as to be detrimental to either party in the short or long term (Barnes, 1983).

**The Obligations and Entitlement of the Purchaser.** The purchaser has three main obligations: to enable the supplier to complete the works/product/service, to pay the agreed price, and to accept the works/product on completion. Contracts also include entitlements, such as the right to appoint a project manager or engineer and the right to employ and pay others to complete work, if the contractor fails to perform in accordance with the contract.

The purchaser discharges their contractual responsibility by paying the contractor the accepted contract amount, amended where required under the contract. For a discussion of the role and responsibility of the purchaser under the FIDIC suite of contracts, see Van Houtte (1999).

**The Obligations of the Supplier.** Contracts generally contain numerous clauses that command the supplier to either comply with an instruction or do something; the FIDIC Construction Contract includes over 80. For example, the contractor shall "design, execute and complete the Works in accordance with the contract; comply with instructions given by the Engineer; remedy any defects in the Works; and institute a quality assurance system."

The supplier discharges their contractual responsibility by fulfilling their obligations under the contract. For example, under the FIDIC Construction Contract:

The Contractor shall complete the whole of the Works . . . including achieving the passing of the Tests on Completion, and completing all work which is stated in the

Contract as being required for the Works or section to be considered to be completed for the purposes of taking over . . ." (Subclause 8.2).

***Management and Supervision.*** Purchasers often delegate the functions of contract administration, which Wearne (1992) refers to as a *concierge de contract*, and project supervision to third parties via contracts. These functions are often combined, for example, in contract strategies where design is separate from production; the initial designer, architect, or engineer usually undertakes both roles. More uncommonly, contracts separate these functions, for example, in the NEC the project manager is responsible for contract administration and the supervisor for ensuring the works are implemented in accordance with the contract. (NEC, 1995).

The supervision of a supplier, in terms of what, how, and when to supervise, is dependent upon the risks inherent in the project, the contract terms, and the inclusion of incentives to encourage satisfactory performance. Normally, the supervisor has no authority to amend the contract or to relieve either party of any duties, obligations, or responsibilities under the contract.

## Time, Payment, and Change Provisions

This section addresses the key areas of time issues, payment provisions, and change mechanisms within contracts.

### Time Issues

***Commencement.*** Contracts include a date for commencement of the project, usually determined by the purchaser or by negotiation between the parties. The commencement date should be set so that it enables the supplier to mobilize resources. Further, contracts should determine what happens in the event of the purchaser failing to provide access to a site or make available plant, service, or any other resources required under the contract, as such failure could frustrate the contract.

***Schedule.*** Generally, contracts include statements regarding the progress of the project. For example, the FIDIC Construction Contract requires the supplier to submit a detailed program (schedule) within a stipulated time frame and to submit revised schedules whenever the previous schedule is inconsistent with actual progress or with the contractor's obligations.

***Suspension.*** Contracts can include provisions that enable the purchaser to suspend the project. The consequences of suspension of the project, the entitlement to payment for plant and materials in event of suspension, the resumption of the project, and the possible termination of the contract as a result of prolonged suspension need also to be addressed within the contract.

***Completion.*** Usually purchasers specify a completion date or alternatively the number of calendar or working days authorized for executing the work. Failure to complete a project within this stipulated time limit can be grounds for a significant dispute between the parties. Numerous contracts include two completion targets: substantial completion and final acceptance. Generally, the contract will stipulate the procedures to be used to determine substantial completion, that is, where the supplier has achieved substantial performance.

## Payment Provisions

The contract states how, what, and when the supplier will be paid, for example, stage payments based on work completed at monthly intervals or milestone-based. Further, it will determine whether payment is incentivized and the level of retention. A purchaser, when planning a contract, should consider what payment terms are most likely to motivate the supplier to achieve the purchaser's objectives for the project.

***Fixed-Price Terms of Payment.*** Fixed-price terms of payment are appropriate for projects that are fully specified prior to inviting potential suppliers to bid and where the completion date of the project is more important to the purchaser than the need to make changes to the specification or any contract terms.

***Advance Payments.*** Advance payments, alternatively referred to as down payments or payment for preliminaries, are inducements to suppliers to commence work promptly; they also reduce the supplier's financing charges. A potential risk, however, is that the purchaser could lose the value of early payments if the supplier subsequently defaults. To avoid this, the supplier can be required to obtain a performance bond before receiving payment.

***Milestone and Planned Progress Payments.*** The supplier can receive payments "on account" in a series of payments for achieving defined stages of progress. Two examples are "milestone" payments, where payment is made based upon progress in completing defined segments of the project, and "planned payment systems," where payment is activated upon achieving defined percentages of a supplier's schedule. Early payment systems should reduce the supplier's risks and financing charges.

Stage payments provide the supplier with an incentive to complete the work promptly. Incorporating additional "bonus" payments for attaining a milestone ahead of schedule can increase this incentive. However, it has the disadvantages that the contract and its management are more complex, and disputes may arise if the milestones or equivalents have not been adequately defined or their achievement proved. Additionally, the contract should state what happens when a stage is achieved ahead of schedule, and what payment is due if a stage is missed but the subsequent one is attained.

***Payment Based upon Agreed Rates.*** In this provision, payment for work executed is based upon rates (unit prices) provided by the supplier when bidding for the contract, with the anticipated quantities of each item of work listed in a "bill of quantities" or "schedule of measured work." Unit rate terms of payment provide a basis for paying a supplier relative to the extent of work completed. The final contract price is calculated using fixed (pre-agreed) rates but is adjusted if the quantities change.

Alternatively, some contracts incorporate a "schedule of rates," where the rates are provided on the basis of indications of possible total quantities of each item of work in a defined period or within a limit of variation in quantity. Schedules of rates have the potential advantage of creating a basis for payment when the type of work is known but not the exact quantities. However, there is the potential disadvantage that supplier will incorporate un-economical rates.

***Contract Price Adjustment/Price Fluctuations/Variation of Price.*** Contracts can include terms for reimbursing suppliers for escalation of their costs as a result of inflation—for example, a clause that allows a contractor to raise prices in line with a pre-agreed index. Such terms have the potential advantage that suppliers have to attempt to forecast inflation rates, which may result in the submission of higher prices; however, a disadvantage for purchasers is that they generate uncertainty over the final project cost.

Where such terms are not incorporated into the contract, the supplier's prices are prone to be higher in periods of inflation; however, the final contract sum will be independent of inflation. Also, in periods of inflation the suppliers will have the incentive to complete their work more quickly to reduce the impact of increases in their costs.

***Cost-Reimbursable Terms of Payment.*** Cost-based terms of payment can be referred to as cost-reimbursable, prime cost, dayworks, time and materials, and so on. There are two versions:

- *Cost plus fixed fee.* Where the purchaser reimburses all the supplier's reasonable costs: employees on the contract, materials, equipment, and payments to subcontractors, plus usually a fixed sum for financing, overheads, and profit.
- *Cost plus percentage fee.* As above, but the fee is added as a fixed percentage.

Cost-plus contracts can be let competitively where the supplier is required to provide their rates per hour or per day for categories of personnel, equipment, and other services. While this is also a unit rate system, it varies from those mentioned previously, as payment is for cost not performance.

With regard to the supplier, under cost-plus contracts their risks are limited at the expense of potential profit. For the purchaser, cost-plus contracts are appropriate if all the categories' potential resources can be predicted although the exact extent of the work initially remains uncertain.

***Target-Incentive Contracts.*** Target-incentive contracts are a development of the reimbursable type of contract, where the purchaser and supplier agree at the beginning a probable cost for an as-yet-undefined project; however, they also agree to share any savings in cost relative to the target. However, if the target is exceeded, the supplier will be reimbursed less than cost-plus.

*Cost plus incentive fee.* Where the fee may fluctuate either up or down within set limits and in accordance with a formula linked to permissible actual costs. Veld and Peeters (1989) consider the most important aspect in cost incentives is the sharing factor. They note that the sharing formula can be nonlinear, it can vary between overrun and under run, and sometimes a neutral zone is introduced.

Al-Subhi Al-Harbi (1998) explains how purchasers and suppliers select the supplier's sharing fraction based upon a risk-averse, risk-taking, or risk-neutral perspective. Veld and Peeters (1989), Ward and Chapman (1994), Al-Subhi Al-Harbi (1998), and Berends (2000) provide examples of incentive fees expressed as equations.

Berends (2000) believes that in order to be effective, the incentive scheme must be aligned with the overall project objectives, not just the cost objective, of the owner. Further, it must provide a positive relationship between the supplier's performance and the supplier's profit margin. Subject to both parties having the ability to prepare realistic cost estimates, the contract negotiation process affords a means to deliver an effective incentive scheme.

**Convertible Terms of Payment.** A convertible contract incorporates an agreement that after any significant uncertainties have been decided it will be changed into a fixed-price or unit-rate-based contract. Potentially, such an agreement has the advantage of limiting the contract price once it is converted; however, there is little or no opportunity for competitive bidding.

**Periodic Payments.** Contracts may incorporate clauses that entitle the supplier to interim payments: payments on account. These payments are usually on a monthly basis, based on the estimated value of the project executed by the supplier in the preceding month and include any amount to be added or deducted under the terms of the contract, for example, retention.

**Retention.** The majority of large contracts incorporate a clause where a percentage of any payment due to the supplier, for example, as fixed price, milestone achievement, or value of work completed, is retained by the purchaser for a specified period (generally up to one year). The retained amount (retention) is paid (released) once the supplier has satisfactorily completed his or her obligations, for example, the rectification of any defective work.

Such a clause, for the purchaser, has the potential advantage of motivating the supplier to complete the project appropriately the first time, thereby, activating the release of the retention at the earliest opportunity.

## Incorporating Change

A purchaser's needs may alter during the period of a contract, for example, in quality, quantity, and even character. Where a supplier provides an extra service or additional work, they are entitled to request extra payment. Contracts, therefore, should incorporate terms that facilitate the effective management of change. Where, in the course of a contract, major variations are expected, suppliers should be granted greater empowerment to initiate change, possibly in the form of value management. This can be of significant advantage to the purchaser where there is the potential to incorporate advances in new technologies or new techniques.

Contracts should also provide a mechanism for pricing any changes in the project deliverables, although, with regard to the purchaser, these mechanisms are usually less competitive than those in the original contract. Any changes need to be managed by both the purchaser and the supplier; inevitably this adds to the cost of the contract

Change management requires the following stages: identification of a potential change requirement, contemplation of the full impact of the change, production of a formal change order, and notification to the parties of the agreed change. Additionally, each stage of the process should be documented and the decision maker should possess the appropriate authority to agree the change.

Latham (1994) identified variations as one of the main problems confronting the UK construction industry, regarding them as probably being a significant cause of disruption, disputes and claims. (See also the chapter by Cooper and Reichelt elsewhere in this book.)

For example, under the FIDIC Construction Contract the engineer can instigate variations at any time before issuing the taking-over certificate for the works, either by issuing an instruction or by requesting that the contractor submit a proposal. The contractor is required to comply with a variation, although the contractor can object to a variation where the goods required for the variation cannot readily be obtained. Further, the contract prescribes how a variation is to be measured and evaluated. Generally, appropriate rates or prices specified in the contract will be used or form the basis to derive new rates or prices to value the work. Where no appropriate rates or prices exist, the work is to be valued based on the reasonable cost of carrying out the work, plus reasonable profit.

## Remedies for Breach of Contract

Contracts need to incorporate provisions to manage the consequences of default by either party. This section introduces the concepts of performance indicators, liquidated damages, and termination.

### Performance Indicators (PI)

The contract should contain appropriate and achievable performance indicators. While it is essential to include the general principles of performance indicators within the bid documentation, the actual indicators are often determined during the contract negotiation stage with the preferred supplier—that is, before the contract price is agreed and the contract signed.

Determining an effective performance measurement system requires the identification of the following:

- The services to be provided
- The critical success factors for a particular contract
- The key performance indicators, targets, and measures
- The components to be measured or assessed, both in terms of the outputs and outcomes of the service (CUP, 1997).

## Liquidated Damages

Where it is possible to derive a preestimate of the loss to be suffered under certain circumstances, it is generally prudent to incorporate liquidated damages into the contract. Liquidated damages, however, should be a bona fide preestimate of the loss in the given situation.

"Liquidated damages" places a limited liability on the supplier to pay a specified sum for a defined breach in performance, for example, late delivery of a product or late completion of a project. The aim is to encourage suppliers to meet their contract obligations; however, their effectiveness may be limited, for example, where the cost of performing their obligations is more than the liquidated damages.

In addition to liquidated damages, contract terms can be used to motivate the supplier, for instance, by offering additional payments where the supplier completes on time or recovers time after a delay in meeting their contract obligations.

## Termination

Under certain circumstances a contract may need to be terminated. As this action results in a lose-lose situation, it is generally only used as a last resort, due, usually, to the unacceptable performance of the supplier. However, other exceptional circumstances include a change in government machinery, a change of policy, or a change in user needs.

Termination of a contract requires contingency plans to ensure continuity of supply, construction, or completion of a service or product. This is essential when the service is critical to the business and the supplier can terminate the contract. These contingency plans may need to be incorporated into the contract; for example, to effect transition from one supplier to another, the existing supplier could be required to cooperate with the new suppliers. Also, special contract provisions may be needed to deal with intellectual property rights in software developed or improved during the contract.

For example, under the FIDIC Construction Contract the employer is entitled to terminate the contract if the contractor abandons the works, subcontracts the whole of the works or assigns the contract without the required agreement, becomes bankrupt or insolvent, or gives or offers to give a bribe (This list is indicative, not exhaustive.)

Likewise, the contractor is entitled to terminate the contract: if the employer fails to provide reasonable evidence of their financial arrangements, or an amount due under an interim certificate; the engineer fails to issue a payment certificate; or where the employer substantially fails to perform their obligations under the contract. (Again, this list is indicative, not exhaustive, and most clauses contain a time frame for compliance.)

# Bonds, Guarantees, and Insurances

## The Principles of Bonds and Guarantees

Financial guarantees are issued by banks, insurance companies, surety companies, or a parent company so that funds will be available should the purchaser have a legitimate claim against the supplier. Chaney (1987) classifies the following types of guarantee:

- *Tender (bid) guarantee (bond)*. Typically required by purchasers to ensure that once a bid has been submitted, the supplier can be held to it, and, so far as possible, to exclude suppliers who lack the necessary financial resources to complete the contract. They will virtually always be conditional and have a time limit.
- *Repayment guarantee*. Generally incorporated where an advance payment is involved. They can either be "on-demand" or "conditional." An on-demand bond permits the purchaser to invoke the guarantee without having to establish default, while under a conditional guarantee, the purchaser can only invoke the guarantee once the supplier has admitted a breach of contract, or upon a ruling of a court or an arbitrator that the supplier is in default.
- *Performance guarantee*. Seeks to ensure that a purchaser can recover damages in the event of a supplier's failure to perform. Again, they may be classified as on-demand or conditional.
- *Retention guarantee*. An alternative to a retention clause, discussed earlier. While a purchaser may consent to this arrangement from the start of the contract, retention guarantees are more commonly used during the maintenance period, that is, upon attaining practical completion.
- *Surety bond*. The usual form of guarantee in North America. They vary significantly from the principle of indemnity: rather than merely paying the amount of the bond following a default, the emphasis is on the surety organizing the completion of project. A performance and payments bond is usually required on all publicly bid construction projects and may be required on some private projects (Carty, 1995).

The liability of the guarantor is therefore for costs incurred by the purchaser limited to the amount of the guarantee, while a surety has the added responsibility of arranging for the completion of the work by a third party. The guarantor does not insure the supplier; they merely provide a guarantee. Therefore, as a condition of issuing the guarantee, the supplier will be required to indemnify the guarantor to the extent that, if the guarantee is called in, the guarantor will only suffer financially if the supplier enters into liquidation. Further, the guarantor will generally require the supplier to provide details of their financial position and their capability and resources to undertake the contract. Depending upon who issues the guarantee, premiums can amount to 2 percent of the sum guaranteed per annum, a cost either directly or indirectly borne by the purchaser.

## The Principles of Insurances

Contracts state what types of insurance are required, determine who is to be responsible for obtaining the insurance, and specify the particular terms of the policies and limits of coverage.

# Claims

Smith and Wearne (1993) define a claim as a demand or request, usually for extra payment and/or time. Usually, a claim is an assertion of a right in a contract; but in others it is the converse: a submission outside the terms of a contract.

## Suppliers

Contracts include provisions for the submission of claims for time and money where the supplier's work is likely to be disrupted or delayed. In such circumstances the supplier is generally required to give notice to the project manager of such disruption or delay. Further, the supplier will be required to give specific details of what caused the delay and to quantify the likely delay or disruption.

In the first instance, contracts usually empower the project manager to agree or determine any matter, by consulting with each party so as to reach an agreement. If there is a failure to reach an agreement, then the project manager is usually authorized to make a fair determination. If the project manager concludes that completion is or will be delayed, the supplier will be entitled to either an extension of time, an extension of time and payment of any cost incurred, an extension of time and payment of any cost incurred plus reasonable profit, or payment of any cost incurred plus reasonable profit.

*Extension of Time.* Events that would lead to the supplier having an entitlement to an extension of time, under the FIDIC Construction Contract, include the following, where completion is or will be delayed due to the following:

- A substantial change in the quantity of an item of work included in the contract
- Exceptionally adverse climatic conditions
- Any delay, impediment, or prevention caused by or attributable to the employer, the employer's personnel, or the employer's other contractors on the site
- Or, if the contractor is prevented from performing any of their obligations under the contract by force majeure and suffers delay and/or incurs cost due to force majeure. (This list is indicative, not exhaustive.)

*Force majeure* is defined as an exceptional event or circumstance, which is beyond a party's control; could not have been reasonably provided against before entering the contract; having arisen, could not have been reasonably avoided or overcome; and that is not substantially attributed to the other party (Subclause 19.4).

***Extension of Time and Payment of Any Cost Incurred.*** Events that would lead to the supplier having an entitlement to an extension of time for completion and payment of any cost incurred, under the FIDIC Construction Contract, include the following:

- Encountering unforeseeable physical conditions
- Discovery of antiquities and the like found on the site
- Engineer's suspension of the works
- Suffering delay and/or incurring cost from rectifying any loss or damage to the works, goods, or contractor's documents due to the employer's risks. (This list is indicative, not exhaustive.)

***Extension of Time and Payment of Any Cost Incurred Plus Reasonable Profit.*** Events that would lead to the supplier having an entitlement to an extension of time for completion and payment of any cost incurred plus reasonable profit, under the FIDIC Construction Contract, include the following:

- Failure of the engineer to issue any necessary drawing or instruction within reasonable time.
- Failure by the employer to give the contractor right of access to, or possession of, the site (within a specified time limit).
- Executing work that was necessitated by an error contained in the contract or notified by the engineer. (This list is indicative, not exhaustive)

***Payment of Any Cost Incurred Plus Reasonable Profit.*** The supplier, under the FIDIC Construction Contract, is entitled to payment of any cost incurred plus reasonable profit if costs are incurred:

> . . . as a result of the employer taking over and/or using a part of the works, other than such use as is specified in the contract or agreed by the contractor (Subclause 10.2).

The use of words such as reasonable and unforeseen, which the FIDIC Construction Contract further defines as "not reasonably foreseeable by an experienced contractor by the date for submission of the tender," introduces ambiguity into contracts. The interpretation of words such as these has led to disputes. It is widely acknowledged that it does not mean what it says; rather, it is interpreted as what might have been foreseen and allowed for by an experienced contractor (Corbett, 2000).

## Purchaser's Claims

Likewise, the purchaser can claim against the supplier, that is, include an amount as a deduction in the contract price or payment certificate. Typical claims against the supplier, according to Bubshait and Manzanera (1990), relate to the following:

- Use of nonspecified materials
- Defective work
- Damage to property
- Late completion by the supplier

Purchaser's claims can be broadly categorized as liquidated damages, claims explicitly provided for by the contract, and claims for damages for breach of contract by the contractor (Corbett, 2000).

For a discussion of force majeure (Van Dunne, 2002) and claims under the FIDIC Construction Contract, see Seppala (2000) and Corbett (2000).

## Formal Procedures for Submitting a Claim

Generally, contracts require the project manager to assess and award extensions of time and/or expenses; however, prudently, most contracts place the onus for substantiation entirely on the claimant. Further, they state specific time frames for submission of details and assessment, and specify the course of action open if the claimant disagrees with the decision.

Kumarasawamy and Yogeswaran (2003) review the techniques and approaches available to substantiate claims for an extension of time and give recommendations on their use. These include global impact, net impact, time impact, snapshot, adjusted as-built critical path method, and isolated delay type techniques.

# Dispute Resolution

## Adversarial and Nonadversarial Dispute Resolution

Problems can arise when implementing contracts regardless of the relationship between the parties and the type of project. In addition to those mentioned earlier, problems can arise as a result of one or both of the parties having conflicting objectives; failing to anticipate significant risks or variations; failing to consult; making erroneous assumptions; and, at a basic level, making a mistake.

A major role of the project manager and, therefore, contract management is predicting potential difficulties in implementing the project to prevent problems turning into disputes. The project manager should, therefore, endeavor to reduce the effects of problems by instituting transparent procedures that are adhered to, by promoting collaboration, and by engendering a shared aspiration to resolve difficulties. The effect of problems can be reduced by the following:

- Establishing approved procedures
- Establishing and adhering to boundaries of delegated authority
- Making contingency plans
- Setting up regular reviews with both the purchaser and the supplier
- Instigating timely recognition and corrective action
- Implementing appropriate contract changes
- Escalating, where appropriate, the problem to senior management (CUP, 1997)

Traditionally, both in the United States and the United Kingdom, the preferred method of dispute resolution arising out of contracts was by litigation—a very slow and costly process. Contracts, therefore, generally incorporate methods of dispute resolution to avoid disrupting the implementation of the project and to resolve any dispute in a fair and timely manner. Examples include mediation, conciliation, dispute review boards, adjudication, and arbitration.

Fenn et al. (1997) distinguishes between conflict and disputes. Conflict, they assert, although an unavoidable fact of organizational life, has positive aspects to do with commercial risk-taking; alternatively, disputes afflict industry. They propose a conflict management/dispute resolution taxonomy (see Figure 14.2).

Elsewhere, Cheung (1999) presents a dispute resolution "stair-step" chart composed of the following steps:

- *Prevention.* Includes risk allocation, inceptive for cooperation, and partnering
- *Negotiation.* Includes direct and step negotiation
- *Standing neutral.* Includes dispute review board and dispute resolution adviser
- *Nonbinding resolution* Includes mediation, mini-trial, and adjudication
- *Binding resolution.* Arbitration
- *Litigation.* Judge

Antagonism and cost increase as one goes down the list (up the stairs). Moreover, resorting to arbitration or litigation is unlikely to improve the implementation of the project and should, therefore, only be used as a final measure. Furthermore, if they are used, the contract has effectively failed.

In the Hong Kong construction industry, Cheung found that when deciding upon a method of dispute resolution, the parties are mainly interested in the benefits, whether tangible or perceived, that may be gained. These benefits include prompt resolution, low cost, and preservation of the relationship between the parties.

Generally, governments are reluctant to legislate in regard to commercial contracts. However, in the case of construction contracts there are exceptions:

- In 1981 the State of California established mandatory arbitration for state agencies' construction contracts (Carty, 1995).
- In the United Kingdom, the Housing Grants, Construction and Regeneration Act 1996 established the right for a party to a construction contract to refer a dispute arising under the contract for adjudication, as outlined in the Scheme for Construction Contracts (England and Wales) Regulations 1998. The decision of the adjudicator will be binding until the dispute is finally decided by legal proceedings, by arbitration, or by agreement. For a detailed review of the implications of this act, see Paterson and Britton (2000).

## Dispute Resolution under the FIDIC Construction Contract

Initially, under the FIDIC Construction Contract the engineer will proceed to agree or determine any matter. In doing so:

**FIGURE 14.2. FENN ET AL.'S PROPOSED TAXONOMY.**

**Conflict Management**

*Non-binding*

Dispute review boards

Dispute review advisors

Negotiation

Quality matters:

    Total quality management

    Co-ordinated project information

    Quality assurance

Procurement systems

**Dispute resolution**

| *Non-binding* | *Binding* |
|---|---|
| Conciliation | Adjudication |
| Executive tribunal | Arbitration |
| Mediation | Expert determination |
| | Litigation |
| | Negotiation |

. . . the engineer shall consult each Party in an endeavour to reach agreement. If agreement is not achieved, the engineer shall make a fair determination, in accordance with the Contract, taking due regard of all relevant circumstances" (Subclause 3.5).

The contract also makes provision for the establishment of a Dispute Adjudication Board (Subclause 20.2), comprising either one or three suitably qualified persons, to adjudicate disputes that may arise among the parties. Where a dispute has been referred to the DAB and a decision, if any, has not become final and binding, the contract states that it shall ultimately be settled by international adjudication

For a discussion on the resolution of construction disputes by disputes review boards, see Shadbolt (1999) and specifically, under the FIDIC Construction Contract, see Seppala (2000).

## Flexibility, Clarity, and Simplicity

Effective contracts require flexibility, clarity, and simplicity, and should stimulate good management. Standard forms have been criticized for failing to provide these attributes. For example, standard forms of contract within the UK construction industry have been criticized for lacking clarity and simplicity; criticisms that could be leveled at standard forms of contract in many industries and countries. The possible reasons for this, according to Broome and Hayes (1997), relate to their origin, being derived from very old precedents; their age, the language and phrasing being derived from English contracts of the late nineteenth century; development by committee; partisanship; lack of direction; and amendment, where specific users heavily amend and supplement the standard form.

Likewise, a review of procurement and contractual arrangements in the UK construction industry (Latham, 1994) criticized the existing standard forms of contract. Contracts, it suggests, should be fair, comprise simple phrasing, set transparent management procedures, and encourage teamwork.

The New Engineering and Construction Contract (NEC, 1995), designed to be used internationally, includes virtually all Latham's recommendations. The NEC suite of contracts has been written to form a manual of project management procedures rather than an agenda for litigation. However, a survey by Hughes and Maeda (2003) found their respondents to be ambivalent about the concept of a spirit of mutual trust; moreover, they held that authoritative contract management would improve performance. This finding is divergent from the principles behind current steps toward innovative working.

Other recently revised standard forms of contract include the FIDIC Suite of Contracts (First Edition 1999), the American Institute of Architects, AIA Contract Document, for example, the A201-1997 General Conditions of the Contract for Construction, and the recent Associated General Contractors of America AGC 200 Series of General Contracting Documents.

# Summary

A contract, according to the Association for Project Management, Specific Interest Group on Contracts and Procurement (SIGCP, 1998), should be designed to be the basis for successful project management: being right in principle (contract strategy) and right in detail (contract terms). This chapter discussed the factors that influence commercial contract practice, providing the project manager with an overview of generic contract provisions and procedures.

To summarize, a good contract stipulates what, where, and when something is to be provided; identifies the supplier and purchaser; defines various roles, relationships, and responsibilities; sets standards; deals with issues of time, payment, and change; provides remedies for breach of contract; covers the issues of bonds, guarantees, and insurances; and contains mechanisms for the submission of claims and the resolution of disputes. Further, effective contracts require flexibility, clarity, and simplicity, and should stimulate good management.

Finally, the following best practice, in terms of planning the contract and contract management, is provided by the Association for Project Management, Specific Interest Group on Contracts and Procurement (SIGCP, 1998):

*Planning the Contract*

- Select suppliers that will best serve the interests of the project.
- Understand how a contract is formed and how it can be discharged.
- Choose the terms of a contract logically, taking into account the nature of the work, its certainty, its urgency, and the competence, objectives, and motivation of all parties.
- Consider how the contract will impact on other contracts and projects, and plan the coordination of the work carried out under the contract in relation to existing facilities and systems.
- Plan, before starting, how the contract will terminate.
- Envisage what can go wrong, utilize risk management, and allocate risks to best motivate their control.
- Define the obligations and rights of each party.
- Specify only what can be tested.
- Agree criteria for the assessment of satisfactory performance.
- Determine and incorporate effective payment terms.
- Ultimately, say what you mean and be clear about what you want.

*Contract Management*

- Control the contract through the appointment of a single manager, who has the authority to decide how to avoid problems and has experience of the potential conflicts of interest that can arise.
- Determine what power the suppliers' manager really has over resources.
- Scrutinize the contract and examine the obligations and rights of all parties.
- Recognize that, in the course of most contracts, objectives and priorities vary.

- Control variations and derive appropriate advantage from potential variations.
- Keep records and notes of reasons for decisions. Use routine headings.
- Distinguish between legal rights and project/commercial interests.
- Finally, realize that a contract should be a means to an end.

# References

Al-Subhi Al-Harbi, K. M. 1998. "Sharing fractions in cost-plus-incentive-fee contracts," *International Journal of Project Management* 5(4):231–236.

American Institute of Architects (AIA). 2002. AIA contract documents. www.aia.org/documents. Washington, D.C.: The American Institute of Architects.

Associated General Contractors of America (AGC). 2002. *AGC contract documents at a glance*. Alexandria, VA: The Associated General Contractors of America.

Barnes, M. 1983. How to allocate risks in construction contracts. *International Journal of Project Management* 1(1):24–28.

Berends, T. C. 2000. Cost plus incentive fee contracting: Experience and structuring. *International Journal of Project Management* 18:165–171.

Broome, J. C., and R. W. Hayes. 1997. A comparison of the clarity of traditional construction contracts and the new engineering contract. *International Journal of Project Management* 15(4):255–261.

Bubshait, A. A. 2003. Incentive/disincentive contracts and its effects on industrial projects. *International Journal of Project Management* 21:63–70.

Bubshait, A., and S. A. Almohawis. 1994. Evaluating the general conditions of a construction contract. *International Journal of Project Management* 12(3):133–136.

Bubshait, K., and I. Manzanera. 1990. Claim management. *International Journal of Project Management* 8(4):222–228.

Carty, G. J. 1995. Construction. *Journal of Construction Engineering and Management* 121(3):319–328.

Central Unit on Purchasing (CUP). 1989. *Contracts and Contract Management for Construction Works*. CUP Guidance No. 12. London: HM Treasury.

———. 1997. *Contract Management*. CUP Guidance No. 61. London: HM Treasury.

Chaney, A. R. 1987. Financial guarantees. *International Journal of Project Management* 5(4):231–236.

Cheung, S. 1999. Critical factors affecting the use of alternative dispute resolution processes in construction. *International Journal of Project Management* 17(3):189–194.

Corbett, E. 2000. FIDIC's new rainbow 1st Edition: An advance. *The International Construction Law Review* 17(2):253–275.

Fenn, P., D. Lowe, and C. Speck. 1997. Conflict and dispute in construction. *Construction Management and Economics* 15:513–518.

The Fédération Internationale des Ingénieurs-Conseils (FIDIC). 1999. *Conditions of Contract for Construction (First Edition)*. Lausanne, Switzerland: FIDIC.

Griffiths, F. 1989. Project contract strategy for 1992 and beyond. *International Journal of Project Management* 7(2):69–83.

Herten, H. J., and W. A. R. Peeters. 1986. Incentive contracting as a project management tool. *International Journal of Project Management* 4(1):34–39.

Hughes, W., and Y. Maeda, Y. 2003. Construction contract policy: Do we mean what we say? FiBRE—Findings in Built and Rural Environments, RICS Foundation, The Royal Institution of Chartered Surveyors, London. www.rics-foundation.org/publish/documents.aspx.

Institution of Chemical Engineers (IChemE). 2003. IChemE Forms of Contract. Rugby, UK: IChemE. www.icheme.org

The Institution of Civil Engineers (ICE), 1995. *The engineering and construction contract: Guidance notes.* London: Thomas Telford.

————. 1995. *The new engineering and construction contract (NEC).* 1995. The Institution of Civil Engineers. London: Thomas Telford.

Jergeas, G. F., and V. G. Cooke. 2000. Law of tender applied to request for proposal process. *Project Management Journal* 28(4):21–34.

Joint Contracts Tribunal (JCT), 2002. JCT contracts. London. The Joint Contracts Tribunal Ltd. www.jctltd.co.uk/contracts.htm.

Kangari, R. 1995. Risk management perceptions and trends of U.S. construction. *Journal of Construction Engineering and Management* 121(4):422–429.

Kumaraswamy, M. M., and K. Yogeswaran. 2003. Substantiation and assessment of claims for extensions of time. *International Journal of Project Management* 21:27–38.

Latham, M. 1994. Constructing the team: Final report of the government/industry review of procurement and contractual arrangements in the UK construction industry. London: The Stationery Office.

Lowe, D. J., P. Fenn, and S. Roberts. 1997. Commercial management: An investigation into the role of the commercial manager within the UK construction industry. CIOB Construction Papers, No. 81, 1–8.

Paterson, F. A., and P. Britton, eds. 2000. *The Construction Act: Time for review.* London: Centre of Construction Law & Management. King's College.

Sadeh, A., D. Dvir, and A. Shenhar. 2000. The role of contract type in the success of R&D defense projects under increasing uncertainty. *Project Management Journal* 31(3):14–22.

Seppala, C. R. 2000. FIDIC's new standard forms of contract: Force majeure, claims, disputes and other clauses. *The International Construction Law Review* 17(2):125–252.

Shadbolt, R. A. 1999. Resolution of construction disputes by dispute review boards. *The International Construction Law Review* 16(1):101–111.

Specific Interest Group on Contracts and Procurement (SIGCP). 1998. *Contract strategy for successful project management.* Norwich, UK: The Association for Project Management.

Smith, C., D. Topping, and C. Benjamin. 1995. Joint ventures. In *The Commercial Project Manager.* J. R. Turner, Maidenhead, UK: McGraw-Hill.

Smith, N. J., and S. H. Wearne. 1993. *Construction contract arrangements in EU countries.* Loughborough, UK: European Construction Institute.

Thomas, H. R., G. R. Smith, and D. J. Cummings. 1995. Have I reached substantial completion? *Journal of Construction Engineering and Management* 121(1):121–129.

Turner, J. R., and R. A. Cochrane. 1993. The goals and methods matrix: coping with projects with ill-defined goals and/or methods of achieving them. *International Journal of Project Management* 11(2): 93–102.

Turner, J. R., and S. J. Simister. 2001. Project contract management and a theory of organisation. *International Journal of Project Management* 19(8):457–464.

Van Dunne, J. 2002. The changing of the guard: Force majeure and frustration in construction contracts: The foreseeability requirement replaced by normative risk allocation. *The International Construction Law Review* 19(2):162–186.

Van Houtte, V. 1999. The role and responsibility of the owner. *The International Construction Law Review* 16(1):59–79.

Veld, J. in't, and W. A. Peeters. 1989. Keeping large projects under control: the importance of contract type selection. *International Journal of Project Management* 7(3):155–162.

Walker, D. H. T., and D. S. Johannes. 2001. Construction industry joint venture behaviour in Hong Kong: Designed for collaborative results?" *International Journal of Project Management* 21:39–49.

Wang, W., K. I. M. Hawwash, and J. G. Perry. 1996. Contract type selector (CTS): A KBS for training young engineers. *International Journal of Project Management* 14(2):95–102.

Ward, S., and C. Chapman. 1994. Choosing contractor payment terms. *International Journal of Project Management* 12(4):216–221.

Wearne, S. H. 1992. Contract administration and project risks. *International Journal of Project Management* 10(1):39–41.

Wearne, S. H. 1999. Contracts for goods and services. In *Project management for the process industries*. G. Lawson, S. Wearne, and P. Iles-Smith. Rugby, UK: Institution of Chemical Engineers.

Wright, D. 1994. A "fair" set of model conditions of contract: Tautology or impossibility? *International Construction Law Review* 11(4):549–555.

# Recommended Further Reading

Specific Interest Group on Contracts and Procurement. 1998. *Contract strategy for successful project management*. Norwich, UK: The Association for Project Management,

Smith, N. J. 1995. Contract strategy. In *Engineering Project Management*. Oxford, UK: Blackwell Science.

Turner, J. R. 1995. *The Commercial Project Manager*. Maidenhead, UK: McGraw-Hill.

# CHAPTER FIFTEEN

# PROJECT CHANGES: SOURCES, IMPACTS, MITIGATION, PRICING, LITIGATION, AND EXCELLENCE

Kenneth G. Cooper, Kimberly Sklar Reichelt

The project was slated to finish in another three years when the first of many change requests came in. The contractor added new staff to accommodate the changes and also worked 60-hour weeks trying to stay on schedule. The new staff needed extra supervision, which was in short supply. Some vendor-supplied design information was late, and the team implemented workarounds to keep things moving. Productivity suffered. Rip-out and rework became a routine "surprise" condition in the build effort. Staff morale worsened and key people left the project and the company. Problems snowballed until, in the end, the project was completed two years late and 50 percent over budget. No one understood how it had gone so wrong.

Was this (A) a naval shipbuilding project, (B) a big civil construction project, (C) a new military aircraft, or (D) the latest big upgrade of a software product? Or was it your latest project? Sadly, the answer could be "all of the above." But why? How is it that such diverse projects could share so consistently the same phenomena?

If there is one thing we should know to expect as project managers and customers, it is that our projects will change. The changes may be to the design of the product, changes from the expected construction conditions, changes in technology, schedule changes, or any of a host of other sources. So what do we do when all those carefully prepared project plans are established and the changes begin?

In this chapter we examine the beneficial and the challenging aspects of project changes. We review what experienced managers believe to be the most effective practices in managing changes and their project impacts. It is not a chapter for the faint of heart, for among the consequences we will examine is a set of impacts that have been responsible for ruining many a project and career. Study after study and survey after survey point to the disturbingly consistent pattern of projects failing to meet targeted objectives, whether they are failures in the product itself or in the cost and time required to get there. How much of a role do changes play in these oft-cited failures? In a word, huge. The ability to handle changes well is a rare and valuable corporate asset. The more common state is one in which changes lead to a persistent pattern of problem-ridden projects.

We have surveyed dozens of managers for their input on best practices and for their descriptions of the most troublesome aspects of managing changes. Among the proactive points of advice are many things that will seem straightforward. Yet they are the practices to which we seem to have the most difficulty adhering in the rush of project execution. In the absence of diligent implementation of these practices, we pay dearly for the consequences, in terms of cost and schedule overruns, lost profits, and ruined relationships. We will examine here how changes can cause projects to spiral out of control, through impacts that are many times the level expected. We will examine the dark side of change impacts, because, at their worst, they become the fodder for major disputes between customer and contractor. And we will describe how many firms have avoided those phenomena by rigorous adherence to a set of practices that have been successful in containing the adverse impacts on projects.

Project change management is arguably the most important under-addressed aspect in all of project management. Why? Do we simply not want to think that the carefully planned effort will be subject to changes (when nearly all of our experience is to the contrary)? Is it such a contentious aspect of project contracting that we are loathe to have the topic infect the euphoria of that exciting new effort (when avoiding it virtually ensures it will hurt the project)? Are we, as some cynics assert, purposefully obfuscating the issue at the outset, with contractors believing that they will "get well on changes" (when the clear experience is that this is rarely achieved)? Regardless of the motivation or mix thereof, one thing is clear: While changes occur with the intent of improving the project, most often we do not do a good job of managing those changes, to the peril of the project.

The situations in which these challenges manifest themselves most acutely are typically in the scenario of a "conventional" contracted project, with a business relation between customer and contractor for the purpose of designing and/or building a product—a building or plant, an electronic system, a ship or aircraft, software, or anything sufficiently complex as to require the coordination of design and build activities. The challenges of handling changes on such projects lie primarily in the hands of the contracted managers. This is not to minimize the importance of the customer's role, for, as we shall see, the customer is pivotal in determining the success or failure of change management and of the project itself. Nor is it to say that the principles described herein do not apply to all manner of projects, even if conducted internally in an organization without a formally contracted customer. We will, however, orient most of this discussion toward the contracted project, and within that, even more specifically to the contracted managers of such projects, on whose shoulders rests the primary burden of managing to typically aggressive cost and schedule targets in an environment so often riddled with change.

## What Changes Are

*"Change is any deviation from the way the work was planned, budgeted or scheduled."*

THOMAS, 2002

Change is a necessary fact of project life. Projects are not just about meeting contractual requirements; they are about achieving the outcomes the end users need. In a world in which markets shift, technology advances, and requirements evolve, projects must be able to accommodate all of these types of changes. The result can be a more capable product that better meets the users' needs. Changes handled well generate long-term business relationships of trust and understanding; handled poorly, however, they can spiral into overruns and disputes. This, then, is the degree of leverage changes have on projects: from the surprisingly successful to the unforeseeably disastrous.

With that much at stake, first we need to know just what changes are. When we think of change, we may be tempted to define it narrowly; in terms of a house, it might be an extra window here, a higher ceiling there. However, we must think more broadly of change, as quoted previously, as "*any* deviation from the way the work was planned, budgeted or scheduled." With this more inclusive definition, changes may come in many forms, such as:

- Design changes
- Work-scope changes
- Late receipt of important technical information
- Excessive delays in design review and approval
- Diversion of key management and technical resources
- Unplanned site conditions
- Inadequately defined specifications or design "baseline"
- Changes in standards and regulations
- Late or inadequate subcontractor performance
- Schedule changes or acceleration
- Superior knowledge
- Technology advances

Whenever a change may have occurred, managers must review the change, plan its execution, consider mitigations, assess its impact, and importantly, communicate with the customer.

## How Changes Impact a Project

To support an informed, proactive, ideally collaborative decision on changes, we need an accurate view of their impacts—and these may be as diverse as projects themselves. There are, however, highly consistent *categories* of impacts, which we describe in this section. No discussion of impacts can occur without first reminding ourselves that changes do not exist simply for the purpose of causing project management problems, overruns, and associated nightmares. Changes exist on nearly every project with an objective to enhance the product of the project, or to accommodate or correct conditions that would otherwise harm the project. While recognizing that these noble purposes exist, let us move on to the ways in which project impacts of the undesirable sort (cost, schedule) occur and what can be done about them.

Logically, analytically, contractually, procedurally, and practically, there are two fundamental categories of change impacts on projects: direct impacts and disruption impacts. Each category may have many alternate labels (direct/hard-core/primary . . . disruption/ ripple/productivity loss/knock-on/impact/cumulative . . .). Basically, however, the practical distinction is this: The visible cost of directly implementing or accommodating a change is *direct* impact; the change's impact on the cost of executing unchanged work (or, indeed, even other changed work) is *disruption*. Among many dozens of management interviews, a universally cited observation is the much greater difficulty of dealing with the latter. Because of that, much of the discussion of change impacts herein will focus on the less well understood phenomenon of disruption. First, however, a brief look at direct impacts follows.

## Direct Impacts

*The direct impact of any given change is likely to have multiple dimensions.* All impacts need to be translated eventually into one measure, of course: cost. But there are three dimensions that are helpful in achieving that translation. (Two excellent sources detailing the direct impacts of many types of changes are Cushman and Butler, 1994, and Hoffar and Tieder, 2002.)

**Added Expenditures/Scope.** This dimension of direct impact can be the most straightforward to document. "Build three, not two." "Buy another generator." "We need to install another two hundred feet of piping." The measures may be dollars of material and/or additional hours (and dollars) of labor.

**Delays.** Two types of delay can directly impact a project. Some sources of delay can be demonstrated to extend the period of performance on the project. For these delays there are accepted categories and formulas for added costs incurred (Hoffar and Tieder FedPubs, 2002).

Other sources of delay may have little or no direct impact on schedules and cost but may still have disruptive consequences. Examples could include late material delivery or delayed access to testing facilities. Even without critical path-delaying impact on the whole project, such conditions should be documented, as they can cause staff productivity impact (disruption).

**Design Uncertainty.** Another dimension of direct impact of changes is design uncertainty. Although not a direct cost source, one of the best early indicators of disruption impact is the degree of the design package affected by changes. The origin of this impact may be explicit design changes, or merely lack of resolution of needed design decisions (such as through delayed design change approval). Monitoring the percentage of the design so impacted, and the duration of the impact, is valuable. It is a strong leading indicator of the magnitude of disruption throughout the project, as it can cause design work to be done out of sequence, reduce productivity, increase rework, and thus cause construction productivity loss as well. This leads us, then, to the subject of disruption impact.

## Disruption

Regardless of the label (cumulative impact, loss of productivity, knock-on effects, ripple, "death by a thousand cuts," secondary impacts, etc.), disruption is the change-induced additional cost of performing work not directly changed. Hence, it is necessarily the added cost (beyond what would have been required otherwise) from lowered productivity or increased rework on the unchanged work, as traceably caused by the change(s). In this section we seek to bring some added clarity to the phenomena to which we will collectively refer in this discussion as "disruption."

***The Challenges of Disruption.*** Let us just acknowledge from the start that disruption is the most difficult and most abused aspect of contractual change pricing. It is precisely because it is the most difficult that it has been the most abused: some contractors use it as a blanket to hide their own problems and cover all cost overruns with "total cost" claims to customer; some customers have taken equally outrageous positions of denying the very existence of disruption and refusing any consideration or compensation for it. Each extreme position is equally absurd. Disruption happens. It is a fact of project life when changes occur. It is a legally recognized set of phenomena that have been established, upheld, and refined over multitudes of court and board decisions of every venue. (For an extensive discussion of the legal background and approaches to cumulative impacts, see Jones, 2001). So why is it so difficult to address, to describe, to analyze, and to quantify? The list of reasons is long.

*The Seven Barriers to Rational Analysis of Disruption*

1. Disruption can be widely separated in space and time from the precipitating event(s), but to be claimed successfully must be causally tied to their source: "Although a change order may directly add, subtract, or change the type of work being performed in one particular area of a construction project, it also may affect other areas of the work that are not addressed by the change order." (Jones, 2001, p. 2) A construction supervisor could be managing a set of impacted work (a) a year after engineering changes have (b) caused more overtime use that (c) lowered productivity, and thus (d) delayed needed work product from (e) a distant part of the organization. The construction supervisor is unlikely even to recognize that they are being impacted by change-induced disruption, let alone be able to quantify it!
2. Disruption impacts can be cumulative across large numbers of individual impacts.
3. Disruption is fundamentally about productivity and rework, which are hard to measure and thus are rarely measured well.

4. The ideal form of damage quantification is to define the amount of impact, including disruption, that would put the injured contractor in the condition the contractor would have been but for the damaging events—a challenging analytical task.
5. Disruption must screen out the effects of other concurrently occurring contributors (such as strikes, difficult labor markets, or mismanagement).
6. Contractor-customer discussions of project cost growth tend to be adversarial, even while the project continues, making efforts to quantify, explain, and mitigate disruption especially challenging.
7. Finally, with all of these difficulties, there is also, quite frankly, a poor track record of rigor in disruption quantification; it is far easier, and usually tempting, for both sides to put all blame on the other without rigorous analysis to back their claims; sloppy logic and analysis may drive assertion ("You caused all our problems!") and counter-assertion ("You mismanaged everything!") (Cooper, 2002).

***Disruption Explained.*** Many hundreds of project managers have described to us various parts of disruption phenomena on hundreds of projects. When we assemble those comments into a cause-effect description, there is quite a high consistency of the kind of disruption phenomena, even as experienced by a wide range of project types. What follows is a description independently offered by managers of aerospace, construction, IT, shipbuilding, electronics systems, and many other projects and programs. Understandably, each has its own peculiarities and variations, but it is the similarity of the descriptions that is striking and that offers the prospect of usable guidelines for explaining project disruption.

The "+ & $\Delta$" shown in Figure 15.1A represent additions and changes to the project's work scope. The direct consequence of changes and additions is to increase the work scope and reduce management's perceived progress on the project. With a lower progress estimate, management's expected hours at completion will grow, and they will increase their staffing requested.

Increasing the staffing request may have the short-term consequence of requiring the use of more overtime until the new hires are brought on board. Sustained high levels of overtime reduce the per hour productivity of staff (see Figure 15.1B). Hiring in a constrained labor market dilutes skills and experience and strains supervision, which further erode productivity and quality (see Figure 15.1C).

Later, the impact of the reduced quality will be felt. The errors created by the fatigued and less experienced staff will have propagated, as subsequent work products build off earlier faulty ones, and thus have been done at lower productivity and quality as well (see Figure 15.1D).

Later still, all the pressures of overrunning the budget and schedule, and finding more and more rework, lead to morale problems, furthering the decline in performance (see Figure 15.1E).

Problems early in the project propagate to downstream work. Change impacts originally isolated in the engineering phase end up affecting construction as well. There another similar set of dynamics is triggered, with the addition of some physical impacts as well, such as crowding (congestion, "trade-stacking"). All these dynamic effects in construction, as they

## FIGURE 15.1A. DIRECT CONSEQUENCE OF CHANGES AND ADDITIONS.

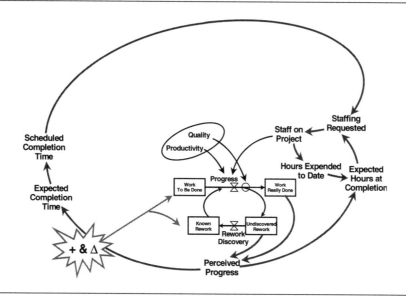

## FIGURE 15.1B. INCREASED STAFFING NEEDS ARE MET WITH HIRING AND OVERTIME.

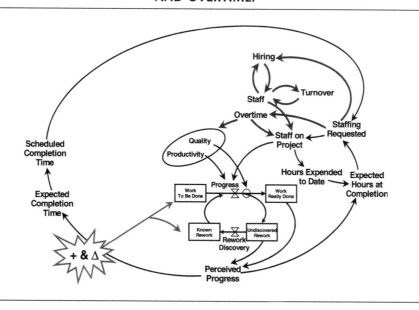

## FIGURE 15.1C. HIRING DILUTES BOTH SUPERVISORY ATTENTION AND OVERALL SKILL LEVELS.

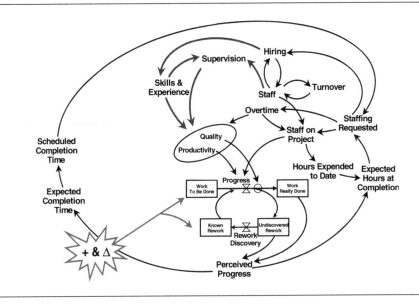

## FIGURE 15.1D. EARLY QUALITY PROBLEMS AND SLOWER PROGRESS CAUSE PROBLEMS LATER.

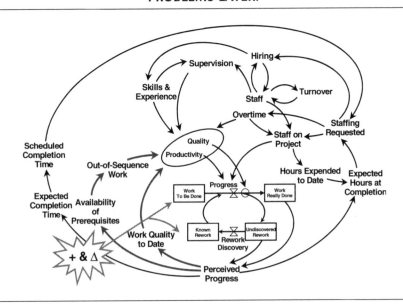

## FIGURE 15.1E. BUILDING PRESSURES EVENTUALLY AFFECT PROGRAM MORALE.

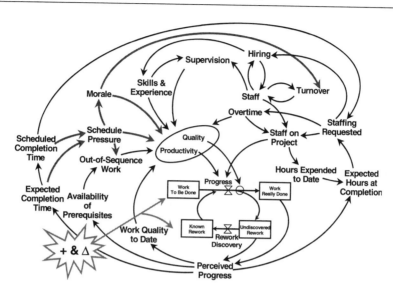

affect labor productivity, can amplify the magnitude of disruption impacts from changes (see Figure 15.1F).

With this causal description of project dynamics, it becomes obvious how problems early in a project can propagate through many stages—as ripple effects, knock-on, . . . Whatever phrase we choose, we really mean disruption.

Now, pause to look back on the assembled diagram (see Figure 15.1F) and note the degree of interconnected *loops* of cause and effect—those circular paths formed by the arrows. Most of these loops are self-reinforcing, and when they are interconnected as they are, the chance for disproportionate cost growth is high when there are many changes or substantial scope growth. Much more scope, for example, generates the need for a more overtime and hiring, diluting productivity and slowing progress versus plans. Work moves more out of sequence, and more rework is generated. Morale suffers, thus worsening productivity more. Staff turnover increases, and so even more overtime and hiring of new staff is needed and so on. It is the self-reinforcing character of those interconnected loops that generates the "cumulative impact" associated with many disruption cases.

With that qualitative description in mind, it is no wonder that the quantification (and responsibility allocation) of disruption remains a challenge. Of one thing we can be sure: When it comes to *anticipating* the additional disruption quantum caused by changes, we are almost universally guilty of underestimation. Some organizations are delighted to receive "extra" compensation of 10 to 50 percent on direct change costs, when in fact the disruption costs can be several *times* that.

FIGURE 15.1F. PROBLEMS IN ENGINEERING LATER FLOW DOWNSTREAM TO AFFECT CONSTRUCTION.

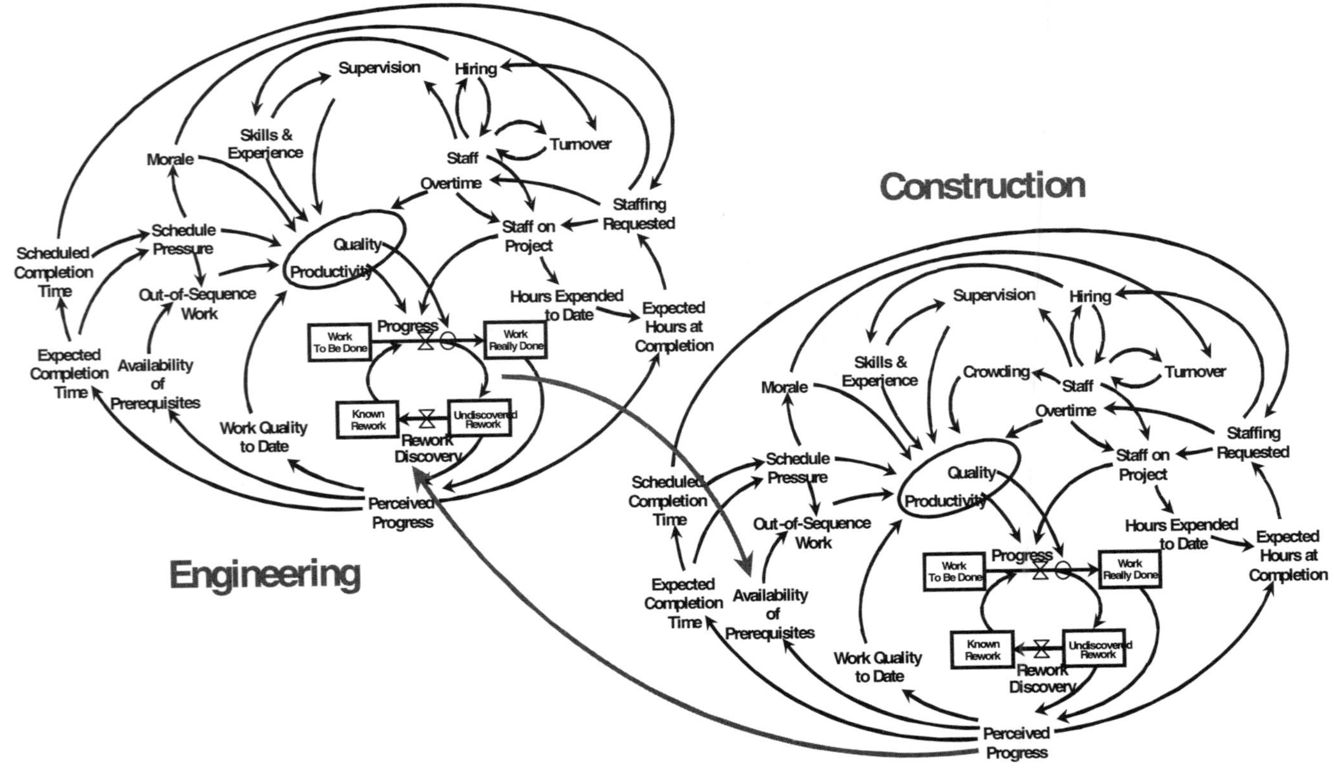

By way of quantitative illustration, we can examine results of several real project simulations. These simulations employ the cause-effect explanation described previously, coded in a model (Cooper, 1994) that is populated by numerical factors drawn from analyses of hundreds of real projects (and thus representative of many projects, but not a model of any single specific project). The model incorporates the most often cited sources of productivity impact from analyses and surveys of hundreds of projects (many excellent reviews and surveys describe sources of project productivity impact. See, for example Schwartzkopf, 1995 and Ibbs and Allen, 1995).

**The Top Ten Causes of Project Productivity Loss.** The following are the ten most prevalent causes of loss of project productivity, in no particular order:

- *Changing work sequence.* Workarounds are frequently necessary, especially when design changes are required and when tight schedules force moving ahead even when necessary information is unavailable, and tasks must be performed out of their ideal or planned sequence.
- *Skill dilution.* Higher staffing in a tight labor market usually means new, less experienced hires are brought into the program.
- *Supervision dilution.* Higher staffing levels, especially with less experienced personnel, can divert supervisors and dilute their effectiveness.
- *Overtime/second shifts.* Overtime or added shifts are frequently used to accelerate work progress; overtime is among the most researched effects on productivity.
- *Rework.* Productivity and rework creation suffer when downstream work products build off of flawed earlier work.
- *Congestion/crowding.* Especially when work is done in constrained spaces, higher staffing levels will hurt work productivity.
- *Late/changing engineering.* Changes or delays in the design will slow work progress and can cause rip-outs in build efforts.
- *Morale.* When program problems mount, the morale and productivity of the workforce may suffer and cause increases in absenteeism and staff turnover.
- *Tools/equipment/materials.* When necessary prerequisites or tools are late, the affected workforce suffers reduced productivity.
- *Schedule pressure.* If the program schedule begins to look difficult to achieve, pressure to regain schedule may cause a "haste makes waste" effect. While output may seem to be accelerated, it is frequently at the cost of increased rework.

**Disruption Quantified.** *A Disrupted Project.* We start with a look at the (simulated) project with no changes. It is a two-million-hour new-design and build project (650,000 hours in engineering, 1.35 million hours in construction direct and support). The construction starts one year after engineering starts and is scheduled for completion in three years (see Figure 15.2). Next we inject growth of scope and associated changes in both engineering and construction that has a *direct* impact of adding 15 percent of those hours to both stages. What additional cost is driven by *directly* adding nearly 300,000 hours to the project?

The graphical view of the impact is displayed in Figures 15.3A to 3C. Figure 15.3A displays the growth caused in engineering and construction staffing versus the no-impact

## FIGURE 15.2. STAFFING IN THE SIMULATED PROJECT WITHOUT CHANGES.

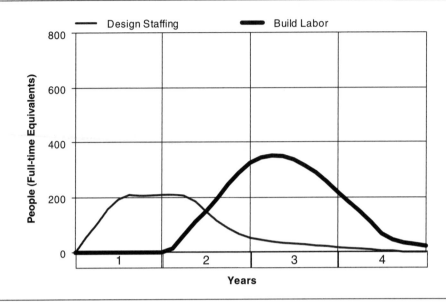

case. Why? Think back to the diagrams in the Figure 15.1 series. The changes lead to significantly higher (and later) engineering effort reworking designs (Figure 15.3B). Later and more engineering changes induce lower construction productivity and thus require more labor. Increased hiring and use of labor reduces skill levels, increases workplace congestion, and more, thus further reducing construction labor productivity (see Figure 15.3C). In the end the total amount of disruption caused through these phenomena is 3.3 times the directly added hours—near one million hours of disruption impact on this project.

*Rule 1: Disruption impacts can appear years after the incident change event and can exceed the direct impact of the change.*

*The Cumulative Character.* Even within a given project, the same change could have different amounts of impact if it is among the first of, or one of few, changes, versus being among the last of many. Some call it "death by a thousand cuts." *Cumulative impact* is the phenomenon of the impact of many changes being greater than the sum of the impact of the individual changes (as discussed previously). In the project simulation model, we tested the injection of multiple levels of changes in order to see the variation in impact. Figure 15.4 illustrates the nonlinearity of total impact, as more increments of exactly the same amount of extra work are successively added to the project. These plots of cumulative construction hours expended show how impacts can increase disproportionately as more and more changed work is added. As noted in Figure 15.5, with the first 5 percent of directly added work, the amount of disruption impact across this project is 2.2 times that of the direct impact. Adding another 5 percent takes that ratio to 2.8. Another 5 percent (the 15 percent-

## FIGURE 15.3. IMPACT OF CHANGES ON SIMULATED PROJECT.

**The Project with the Full Impact of Changes**
(Direct Impact = 15% of original scope)

**Figure 15.3A: Staffing**

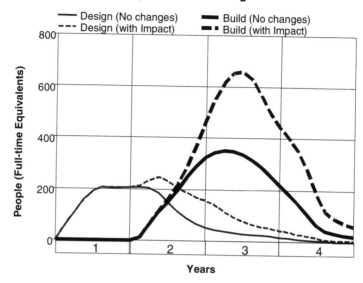

**Figure 15.3B: Design Rework Effort**

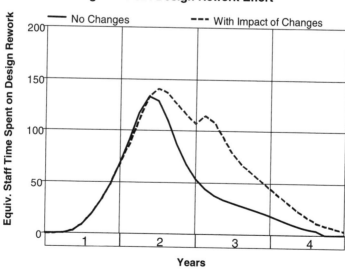

**FIGURE 15.3. (*Continued*)**

**Figure 15.3C: Build Productivity**

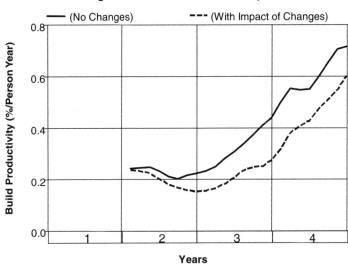

added-scope case we described previously) takes it to 3.3. And yet another 5 percent causes the total disruption impact to grow to 3.7 times the direct impact on this project. The phenomena at work driving these impacts are simply more and more of the self-reinforcing feedback effects described previously.

*Rule 2: Disruption impacts grow disproportionately over more and more changes.*

*Timing Matters.* One of the most consistently cited cautions in our management interviews is the need to resolve early and speedily the content of contemplated changes. Indeed, the timeliness of resolution can make a substantial difference in the impact of changes on project costs. How much? In our simulated project, resolving more rapidly the design issues associated with the changes could cut the disruptive impacts by as much as 40 percent. Figure 15.6A shows the significant improvement (reduction) in disruption impact, as less resolution time is required (moving right to left on the chart). The expanded view in Figure 15.6B shows the timing-induced disruption improvement (again, right to left) for the full range of direct impacts (i.e., from 5 percent to 20 percent of directly added work scope). And recasting these results to the form of display as in Figure 15.5, we can see the reduced range of disruption impact, for each amount of change, with more rapid issue resolution (see Figure 15.6C). The consistent pattern observable here is that cutting the resolution time in half reduces the amount of disruption by 10 to 20 percent, with the biggest percentage improvements in the most disrupted conditions.

## FIGURE 15.4. IMPACT OF CHANGE GROWS WITH GREATER CHANGE MAGNITUDE.

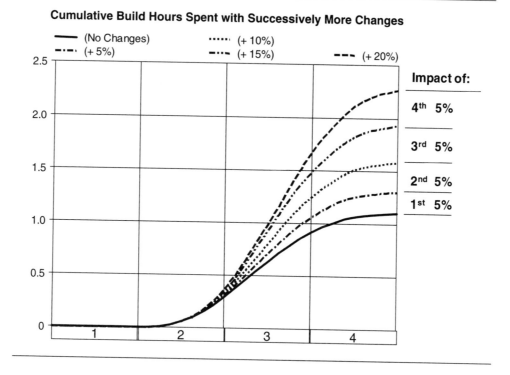

This degree of improvement cannot be surprising. In the extreme, if the changes were known at the project start and resolved instantly, they would not be changes at all; they would, in effect, become part of the project specifications and the known baseline.

*Rule 3: Early issue resolution cuts disruption impacts significantly. Not All Projects Are the Same.*

Surely we can't be saying that all projects are subject to disruption of two, three, or four times the direct impact of changes? Of course not; this is to explain how projects *can* be subject to this degree of impact, but not all projects share the same degree of sensitivity to the many factors active in this simulated project. For example, two of the most significant construction productivity-affecting factors here are (a) the quality of the design package and (b) the workforce skill levels. We tested the very same changes in the very same project, but with no susceptibility to these conditions—that is, if there were no significant design change impact on construction or no dependence on skilled labor (or at least plentiful supply thereof). Eliminating the dependence on skilled labor cuts the disruption impact of the changes from the base ratio of 3.3 (times the direct impact) to 2.2. Eliminating dependence

## FIGURE 15.5. DISRUPTION RATIO INCREASES WITH MORE CHANGE.

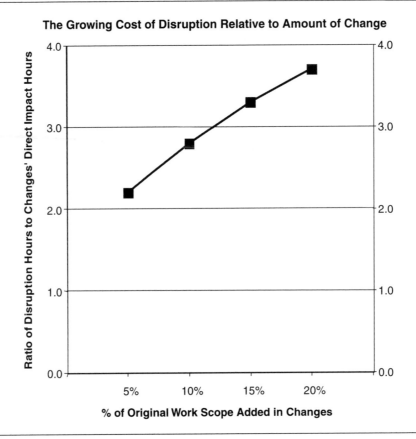

**The Growing Cost of Disruption Relative to Amount of Change**

on design (as in the case, say, of building a completely standardized structure) cuts the ratio to 1.6! Eliminating *both* of these significant conditions reduces the disruption ratio down to "only" 1.2.

This illustrates yet another reason why "every case is different": Any variation in sensitivity to the different productivity-affecting factors on a project alters the amount of disruption caused by changes. Indeed, let's examine the full range of direct impacts tested previously (adding changes with direct impacts totaling 5 percent to 20 percent of original budgets), under the alternate skill and design conditions. Figure 15.7 shows the very different picture for disruption impacts in these scenarios. With no sensitivity to labor skill conditions, the remaining paths of impact yield disruption amounts that are substantially lower throughout all the different levels of change—with impact ratios of 1.4 to 2.7, versus the base conditions of 2.2 to 3.7. With no sensitivity to design conditions, the impact range is even lower—from 1.1 to 1.9. Eliminating sensitivity to *both* skill *and* design brings the range of

# FIGURE 15.6. QUICK RESOLUTION REDUCES CHANGE IMPACT.

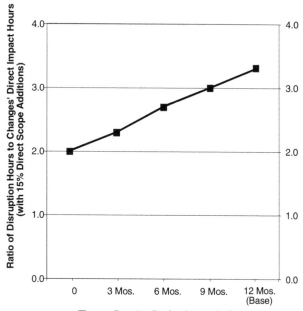

**Figure 15.6A**
**Resolve those design issues quickly...**

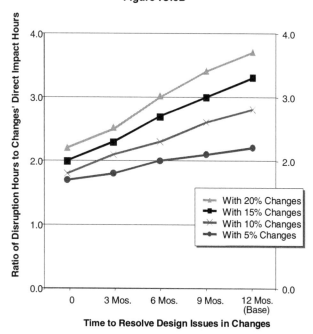

**Figure 15.6B**

**FIGURE 15.6. (Continued)**

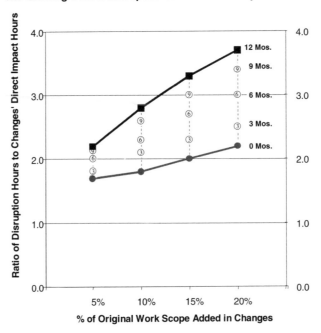

**Figure 15.6C**
**The Growing Cost of Disruption Shrinks with Early Resolution**

the disruption ratio down to a range of 0.8 to 1.0. These reduced ratios of disruption impact stem not only from the eliminated skill and design effects themselves but also because cumulative impacts through other feedback effects (less crowding, less overtime, and so on) also lessen.

*Rule 4: Variations in project conditions drive significantly different disruption impacts.*

*Dominoes, Acceleration, and Mitigation.* Of all the variations in conditions that can cause variation in changes' disruptive impacts, none are more important than the tightness of the schedule. Recall the cumulative cascade of phenomena that can hurt construction productivity—overtime fatigue, skill dilution, out-of-sequence work, congestion, . . . and, especially, late and changing engineering. Each of these phenomena can be significantly aggravated or ameliorated by the tightness of schedule conditions. And, since each can contribute to self-reinforcing cumulative impact, one should expect a significant difference in the domino effect under different schedule conditions. Indeed, this is just what we find in our project simulation analysis results. In this series of analyses, we set identical changes to occur in the

## FIGURE 15.7. DISRUPTION COSTS VARY DEPENDING ON PROJECT CONDITIONS.

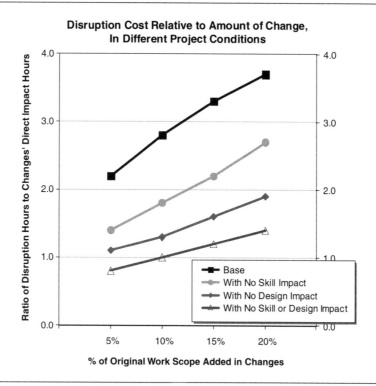

identical project, but under different schedule conditions, ranging from even tighter, with more design-construction overlap, to less tight. Figure 15.8 displays the resulting disruption impact. The even earlier construction start more than doubles the disruption ratio to 7.2 from the base value of 3.3, while the later start of build *drops* the impact ratio 30 percent to 2.3. Combine that later start with an extension in the scheduled completion date, and see a further reduction to 0.7 (i.e., disruption is less than the direct impact).

How tight is your project schedule? Some schedules are intended to be "optimal," a healthy balance between cost and time targets. Other project schedules are acknowledged to be excessively tight—"optimistic," "success-oriented . . ." (If your project schedule is acknowledged to be "loose," you are operating in a rare environment and a distinctly small minority among project managers.) In either case, the addition of changes to the project will tip the schedule toward "tighter" and increase the potential for higher amounts of disruption impact from those changes, as a result of increased hiring and skill dilution, overtime, crowding, and so on.

In a "schedule is supreme" world, it may be viewed initially as career-threatening to talk of schedule extensions. Nevertheless, the addition of work into a tight schedule consti-tutes a *de facto* acceleration and is likely to trigger the kind of disruptive impacts described

## FIGURE 15.8. TIGHTLY SCHEDULED PROJECTS ARE MORE PRONE TO SUFFER DISRUPTION.

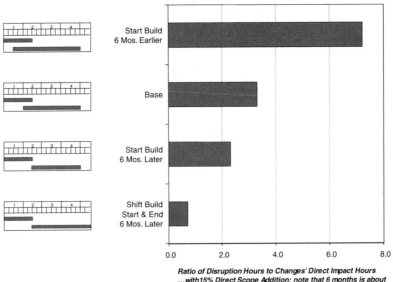

*Proactive Schedule Changes Mitigate Disruption*

*Ratio of Disruption Hours to Changes' Direct Impact Hours
...with 15% Direct Scope Addition; note that 6 months is about
15% of the construction schedules*

previously. Such impacts can grow costs significantly and may jeopardize that earlier schedule anyway.

Perhaps, then, it is only responsible to explore and to offer to the customer/contractor the mitigating option of *reduced* cost impacts with *extended* schedules. Of the cost-mitigating options available, none may be so highly leveraged as schedule relief. In business markets in which even small delays are very costly to the project customer, it may well be worth the much higher price to adhere to tighter schedules, but it is a rare market for which it is "at any price." In this simulated project, the labor disruption costs difference between starting construction six months earlier versus starting and ending six months later is two million hours . . . on a two-million-hour project budget! Indeed, *each successive construction schedule move (start or target completion) that is equal to the percent of direct impact (in this case 15 percent) cuts disruption cost impact successively by 30 to 70 percent.*

Cost-schedule trade-offs, therefore, should be a part of every discussion of significant change impacts.

*Rule 5: Disruption impacts are significantly reduced by less tight project schedules.*

Thus, we have seen five rules of changes' disruption impacts:

*The Five Rules of Changes' Disruption Impacts*

> *Rule 1: Disruption impacts can appear years after the incident change event and can exceed the direct impact of the change.*
>
> *Rule 2: Disruption impacts grow disproportionately over more and more changes.*
>
> *Rule 3: Early issue resolution cuts disruption impacts significantly.*
>
> *Rule 4: Variations in project conditions drive significantly different disruption impacts.*
>
> *Rule 5: Disruption impacts are significantly reduced by less tight project schedules.*

Bear in mind that each time "disruption impact" appears in these rules it means *cost*.

The "impact" means productivity reductions or rework increases that are added project cost, for which someone must pay. Best to resolve the full cost impacts of changes early, for if we don't, it can all go wrong.

## When It All Goes Wrong: Disputes

If anyone should doubt the importance of dealing effectively with changes proactively, one only need talk with an executive or project manager who has been through a major contract dispute. For both customer and contractor (or subcontractor), a contract dispute or claim can be the most management-distracting, relationship-damaging, costly process of their careers. Whether ultimately resolved via negotiations, mediation, arbitration, or trial before a board or court of any jurisdiction, the contract dispute over unresolved costs of project change impacts can consume enormous staff and executive time, decimate customer-contractor goodwill, and cost millions of dollars.

The existence of and toll taken by such disputes is testament to the value of timely resolution of change impact costs during a project. When agreement on those costs cannot be resolved by routine processes, at least one party may choose to initiate a claim process. Processes and venues for such claims will have been spelled out in the contract terms.

Regardless of process or venue, the burden of proof is essentially the same: (1) the contractor must establish entitlement to recover for those "changes"; (2) causation of impact must be established; and (3) the full quantum of that impact must be demonstrated. Very much gone are the days of (successful) total cost claims. Between (a) universal disfavor among courts and (b) the use or threat of use of U.S. federal and state False Claims Acts, contractor claims that overreach (by attempting to recover all costs regardless of demonstrated causation) are likely doomed to failure. Indeed, *any* contractor claim that fails to provide adequate proof of liability, causation, and damage is so doomed. The characteristics of an analysis that are needed for a strong case are that it

- explain variations in productivity and rework;
- assess what would have happened under alternate conditions;
- describe the causation that ties effect to precipitating event(s);
- account for and explain cumulative impact of many individual claimed events;
- explain *why and how* productivity would have been affected;

- account for other concurrent events that influenced project performance;
- permit validation; and
- be testable and auditable.

Given the extreme challenge of explaining and quantifying the disruption elements, it is no wonder that many claims involving disruption are resolved at no more than 30 to 40 percent of the claimed value. This may be less a measure of truth and accuracy than an indicator of the difficulty of explaining and quantifying such claims.

# Management of Change Process: Excellent Practices

With the many adverse impacts that can occur as consequences of change, it is crucial that the management of change be a high-priority aspect of managing projects. Actual practice, however, varies widely. Some project managers see change as something to be avoided. Others plan and behave as though change simply will not happen to them. Still others view change as an opportunity, a chance to recover on bidding mistakes. The best approach is to accept that change is normal and to be expected. On all but the most standard cookie-cutter projects, changes will happen. The key is not to avoid all changes but to anticipate them, be prepared for them, and react to them properly, with trained staff aided by clear guidelines and processes and systems.

In our research for this chapter, we interviewed dozens of top executives, project managers, attorneys, and consultants in the construction, software, and defense industries. The many interviews we conducted highlighted that change management practices vary significantly, as do the resulting project outcomes. Further, when it comes to change management, the experience of the manager is not a good predictor of success; success requires having the right processes in place and following them.

The process areas that are keys to success in handling change are as follows:

- Managing the contract
- Managing the project
- Managing the contractor-customer relationship

## Managing the Contract

*"Most mistakes happen before the shovel hits the ground."*

LANG, 2002

Change planning is critical during the drafting of the contract documents. Contractor and customer should have an open discussion about change, including how much change is reasonable to expect and how, together, changes will be handled.

***Central Contract Review.*** One key element to successful management of contracts is to have central contract review. Firms that do a good job at managing project changes (in terms of pricing, mitigating impacts, recovering damages, and avoiding disputes) have a strong central coordination, if not control, of the process from beginning to end. Starting from before the project begins, *all* contracts should be reviewed centrally by the legal department. While it is fine to craft custom change clauses as needed, central review ensures compliance with key principles (e.g., reservation of rights, consequential damages, damage-delay clauses).

Equally important is knowing when to walk away. Thorough contract review enables parties to make informed decisions regarding the acceptability of a contract and the risks it would bring.

As part of this whole process, the legal department should develop a standard checklist of all the elements that should appear in the contract. Central review will uncover deviations from the norm and enable informed decision about variances. Indeed, the contractor should assign an attorney to each contract and have that attorney support the contract beginning to end. This attorney should be responsible for all the legal aspects of the contract and stay involved throughout, ensuring familiarity with the project when issues arise. In this way, the attorney can serve as an ongoing resource to the project management team, who should themselves be familiar with the contract terms and conditions.

**Specify Planned Conditions.** One the first challenges in dealing with change is getting both sides to agree whether a change has occurred. For this reason, it is important to specify clearly in the contract what the planned conditions and dependencies are (Hatem, 2002). With the baseline conditions clearly specified, the emergence of changed conditions is easier to identify.

**Consider in Advance How to Deal with Disputes.** There are a variety of constructive approaches to resolving disagreements proactively. If both sides can agree in advance on a process with which they are comfortable, much of the tension surrounding change can be alleviated.

- Consider a project "neutral" to review changes. A neutral party who is familiar with the project and knowledgeable about the industry can be invaluable. An unbiased third party can quickly and fairly review proposed changes and their impacts (Shumway, 2002; Sink, 1999).
- Agree how disputes will be escalated. Have an agreed process in place by which disputes are rapidly escalated to the necessary levels of decision-making authority. This prevents issues from lingering too long, which can both breed resentment from both sides and also worsen the impact of the very problem under dispute (Cady, 2002). See the *Disruption* section earlier in the chapter.
- Agree in the contract on whatever alternative dispute resolution (ADR) methods will be employed. In some settings, these can shorten resolution times significantly (Bird, 2002). The two most commonly used methods are mediation (a nonbinding approach that attempts to bring the parties to a settlement with the help of a facilitator) and arbitration (in which a neutral third-party hears their case and provides a ruling that is binding on the parties). ADR can be used to resolve disagreements more quickly and cost-effectively than litigation if both parties are committed. An excellent way to foster this commitment is to establish during the contractual stage a partnering agreement, a pact between con-

tractor and customer to work together proactively. The partnering process is typically kicked off with a contractor-customer retreat and culminates in an agreement describing how the parties will work together, including establishing a fast-track approach to dealing with issues that may arise. These agreements can be effective in fostering proactive communication and setting the right stage for a cooperative working relationship between contractor and customer (Ness, 2002). Consider among the various ADR methods establishing a dispute review board (DRB). If possible, the contractor and customer should pick the DRB as a team. The more traditional we-pick-one-you-pick-one-they-pick-one approach tends to be more adversarial. Instead, work together to pick a team of people who really understand the industry. Involve the team from cradle to grave on the project, and have them invest enough of their time to understand the project thoroughly.

The importance of legal matters continues beyond the contracting. Continue monitoring the legal aspects throughout project execution: Contractors should adhere to notice provisions, be careful about what they sign away, and be clear about what should and should not be in writing.

Finally, ensure that legal talent is seen as part of the team. Project managers often distance themselves from attorneys, viewing their presence as a sign of weakness, that the managers were unable to resolve problems themselves. Instead, Legal should be actively integrated into the project team, and the attorney involved with the project should be viewed as a resource, not as a last resort (Kieve, 2002).

## Managing the Project

*"Know thy project."*

CHIERICHELLA, 2002

***Provide Training.*** Companies that are successful in managing changes make sure the team understands the project and how to identify change. They publish explicit guidelines and standard processes and teach them throughout the company. They train the management team in documentation, letter writing, and handling constructive changes. They provide basic legal education for the project management team. They teach the standard systems that will be employed on each project, evaluate individual lists of responsibilities, and reallocate as deemed appropriate. They provide training on the specifics of the contract and conduct role-playing of possible events (Goff, 2002). They teach key staff not to take verbal instruction from customer representatives without treating it as a possible change (Mountcastle-Walsh, 2002).

***Exercise Discipline.*** Most of the managers and attorneys interviewed emphasized the importance of discipline. Central standards and guidelines may sound bureaucratic and, frankly, boring, but they are helpful in ensuring projects run smoothly. As one executive noted, "new ideas are not needed as much as enforcement of basic 'blocking and tackling' around project change" (Stafford, 2002).

Develop a standard set of processes for projects, train people in the processes, and enforce guidelines to ensure the processes are actually followed. An important element to these processes is to make compliance with the processes as easy as possible. Forms for logging changes, for example, while covering the essentials, should require as little input as possible.

Before the project begins, have a transition session to impart key information from the proposal team to the project team. The proposal manager and project manager should review together all the assumptions used in generating the proposal, including a review of project risks and early warning signs that those risks are developing. Best of all is for the project manager to have been involved in the proposal effort, so the person responsible for executing the project will have been involved in defining it.

Every piece of paper exchanged between customer and contractor should be reviewed for change content. However, even a thorough review of paperwork may miss many changes on a project. Therefore, it is important to monitor constantly for change. If a contractor has confidence in the estimating process employed, then the very fact of overruns in various cost cells might signal that a change occurred. The overrun implies either that there was a misestimation or that something changed—perhaps there were verbal instructions from the customer that never made it into writing, or perhaps the specifications were too vague. Investigate these overruns. Whatever their source, whether changes initiated by the customer, misestimation, or unexpected productivity losses, the contractor should ensure proper diagnosis (Chierichella, 2002).

**Find and Process Changes Promptly.** Contractors need to provide supervisors in the field the right incentive to identify changes. One successful approach is to create a work task category such as "Pending Items," items that have been identified and for which the customer will be or has been notified, but which have not yet been approved. In earned-value progress reporting, supervisors would get credit for these items, such that their performance measures are not penalized as a result of changes. Contractors should provide the customer with timely notice and then ensure they do not execute any work that is neither authorized as a change nor pending (Newman, 2002).

Assign dedicated personnel for handling the change process. Particularly on those jobs with tight budgets (i.e., those most likely to end up with disputes), there is a tendency to cut back on overhead, but it is exactly these projects that are most likely to end up overrunning and have disputes. In the event of a dispute, daily reporting, schedule updating, and disciplined documentation become especially important and can, in fact, promote the resolution of disputes.

### Document Thoroughly

*"One reason that you have disputes is that the parties aren't working with the same factual description of what happened. Having the documentation increases the likelihood of reaching a resolution that both sides can be happy with."*

KRAFTSON, 2002

To increase the likelihood that both sides are working with the same facts, the following are helpful practices (see Hoffar and Tieder FedPubs, 2002, pp. 1–5, and Currie and Sweeney 1994 for more detailed review of documentation practices):

- Prepare daily time and material sheets, and submit reports to the customer every day. Daily reporting accelerates communication and ensures all are well informed on project status (Currie and Sweeney, 1994, p. 218).
- Log time and cost for every change in a separate cost code. Contemporaneous tracking of costs is far better than a re-creation done later. While even real-time reporting is certain to miss some of the impacts of change, it at least provides a more complete picture of the direct costs.
- Document efficiency losses (e.g., when workers are waiting on material, log their time idle). Comment on inefficiencies at job meetings, and record these in meeting minutes. Contemporaneous tracking of productivity problems will ensure that everyone is aware of all relevant problems (perhaps enabling some mitigation ideas), and it provides a good resource for computing disruption.
- Take pictures of the job site and date them. Photographs (or videos) often can relay the story far more clearly than written documentation. For example, a photograph showing workers crowded into a site is better documentation than a report including comments that congestion has been an issue (Currie and Sweeney, 1994, p. 220).
- The contractor should document the customer's performance as well as their own. It is not only the contractor whose performance can drive cost and schedule. Delayed approvals from a customer or late information can impact contractor performance and should be documented and included in reports to the customer (Goff, 2002).

## Managing the Contractor-Customer Relationship

*"If you trust the person on the other side of the table, you can work through just about any problem."*

DEAN, 2002

***Communicate Effectively.*** A key to success is open, honest, and direct communication between the customer and contractor. It is important for the contractor to invite active customer involvement early in the project. Some contractors, for example, require an open exchange with the customer before the project starts to discuss priorities and processes, including the handling of changes.

Reports from contractor to customer should be regular and frequent, include both the good and the bad, and be complete and fair in reporting the true status of the project. It is important that customers be confident that they are hearing about all issues and that the contractor will alert them quickly and be ready to discuss any problem that might arise. In the same spirit of openness, the contractor might consider keeping the regularly updated schedule for the project open to all, including customer and subcontractors. This openness is an important part of developing a partnership.

When a change is identified, the contractor should communicate with the customer quickly. First, the contractor must inform the customer they believe a change has occurred, or is about to. By providing the customer notice (rather than just acting on the change), the customer is afforded the option to avoid the change. If the customer opts to implement the change, the contractor needs to provide a quick, accurate assessment of the impact (and, for changes that involve significant cost or schedule impact, provide mitigation options as well). Indeed, the contractor and customer might assign staff to review changes together before change orders are written and submitted.

Nearly every manager interviewed noted that speed is essential in managing change—speed in identifying changes, estimating direct and, when possible, productivity impacts, (see the *Disruption* section below earlier in the chapter) defining explicitly the cost-schedule trade-offs for the customer, and obtaining customer concurrence. By moving quickly on change issues, much of the impact of the uncertainty that accompanies change can be mitigated (see the *Disruption* section).

### Price Completely

*"Add a door, I can tell you what that costs. Change the door three times, and it's hard to quantify."*

LAX, 2002

When changes are small and uncomplicated, it is usually straightforward to develop an accurate assessment of the total cost of a change. When changes are numerous, large, and/or complex, it is much more difficult. While it is important to resolve issues as quickly as possible, it is wiser to provide an informed estimate than to present a number that may increase dramatically later (Grimes, 1989).

Know the project conditions, and understand how the change fits in before trying to estimate the impact (Schwartzkopf, 1995). It is essential to understand the impacted site, as it is in this environment in which the change will take place (Sanford, 2002). Is labor already crowded into the site where the change will add more people? Has the affected work already started? Are people working in a comfortable environment or are they welding in mid-summer heat?

When providing an estimate, contractors must be realistic, practical, and not overstate the costs. Estimate as much of the costs as possible, minimizing the number of reservations (Allen, 2002). In the estimate, consider additional rework that may occur, as well as any anticipated lost productivity (see the *Disruption* section earlier in the chapter). Industry-standard measures are available, such as MCAA and CII studies, to provide some reasonable benchmarks for estimation. (There have been numerous studies on the impact of change orders and project conditions on labor productivity. See, for example, MCAA, 1994; Thomas and Rayner, 1988; Ibbs and Allen, 1995; Hanna, 1999 et al.; Adrian, 2001). Be cautious, however, as it can be difficult to anticipate the many ways in which a project may be impacted or the degree to which each type of impact will be caused by the changes.

The contractor and customer should review the reasoning for the estimated additional costs. In addition, this is the time to look for offsets that might provide a trade-off in the project scope, or cost-schedule trade-offs that might enable containment of impact costs (see below). With an understanding of the full impact of a change, the parties may even decide to forego it.

## The Importance of Getting it Right

Project changes are inevitable. They are essential in order to improve products, adopt new technologies, adapt to market conditions, and accommodate changed or unexpected circumstances. If we as project managers, executives, and customers know to expect changes, we must learn to manage them far better. We need to execute the basic "blocking and tackling" of change planning and monitoring. We need to evaluate alternatives and trade-offs and review them openly between customer and contractor. We need to anticipate and communicate more completely and candidly the likelihood and nature of disruption impacts, and thus the ultimate cost effects on the project. Companies who do so see substantial performance improvement. The failure to do so results in project overruns and failures, damaged customer-contractor relations, resource-consuming disputes, ruined careers, and lost profits and shareholder value.

## Acknowledgments

Many executives, consultants, and attorneys generously gave us their time and experience-honed ideas in the dozens of interviews conducted in support of this chapter. Although only a few have been quoted and cited, many others helped shape the ideas and content herein. To all of them we owe our gratitude. In addition, without the contribution of colleagues in our firm and our clients over many years, the insights into and analyses of change impacts summarized here would not have been possible. Of most recent significant help have been Tom Kelly and Sharon Els, who helped us conduct so many of the interviews; Hua Yang, who conducted all of the simulation analyses reported herein; Sheri Dreier and Jane Hemingway, who prepared the graphics for this chapter; and Doris Walsh, our patient and ever-productive assistant, who (many times over) converted bits of notes into the assembled text and charts for the manuscript.

## References

Adrian, J. 2001. *Jim Adrian's Construction Productivity Newsletter*. see, for example, Change orders: How they affect time and Cost (Vol. 18, No. 2); The impact of temperature on productivity" (Vol. 19,

No. 5),; and The impact of the loss of learning on construction productivity (Vol. 19, No. 6); among many others.

Allen, S. 2002. Stephen Allen, Washington Group; Boise, Idaho. Personal communication. June 18, 2002.

Bird, K. 2002. Karl Bird, USAF, Wright-Patterson Air Force Base. Personal communication. July 29, 2002

Cady, J. 2002. Jim Cady, Granite Construction, Inc.; Watsonville, California. Personal communication. July 3, 2002

Chierichella, J. 2002. John Chierichella, Fried, Frank, Harris, Shriver & Jacobson; Washington, D.C. Personal communication. August 7, 2002

Cooper, K. 1994 The $2000 hour: How managers influence project performance through the rework cycle. *Project Management Journal.* (March 1994).

Cooper, K.G. and K. S. Reichelt (2002) Quantifying project disruption with simulation. San Antonio: Project Management Institute

Currie, O. A., and N. J. Sweeney. 1994. Prelitigation advice. In *Construction Change Order Claims*, Chap. 12, 215–237, ed. R. F. Cushman and. S. D. Butler. New York: Wiley.

Cushman, R. F. and S. Butler, eds. 1994. *Construction change order claims.* New York: Wiley.

Dean, W. 2002. William Dean, The Clark Construction Group; Bethesda, Maryland. Personal communication. October 3, 2002

Goff, C. M. 2002. Colleen Mullen Goff, Zachry Construction Corporation; San Antonio, TX. Personal communication. July 23, 2002.

Grimes, J. E. 1989. *Construction paperwork: An efficient management system*, 159. Kingston, MA: R. S. Means Company, Inc.

Hanna, A. S., J. S. Russell, E. V. Nordheim, and M. J. Bruggink. 1999. Impact of change orders on labor efficiency for electrical construction. *Journal of Construction Engineering and Management—ASCE* 125(4):224–232

Hatem, D. 2002. David Hatem, Donovan Hatem, Boston. Personal communication. July 24, 2002.

Hoffar, Julian F. and Tieder, John B. 2002. *Proving construction contract damages.* Federal Publications, Inc. Washington, D.C.

Ibbs, C. W. and W. E. Allen. 1995. Quantitative impacts of project change. Source Document 108. Construction Industry Institute; Austin, TX pages 1–46.

Jones, R. M. 2002. Lost productivity: Claims for the cumulative impact of multiple change orders. *Public Contract Law Journal* 31: (Fall, 1).

Kieve, L. 2002. Loren Kieve, Quinn Emmanuel; San Francisco. Personal communication. September 5, 2002.

Kraftson, D. J. 2002. Daniel J. Kraftson, Jenkens & Gilchrist; Washington D.C. Personal communication. July 15, 2002.

Lang, R. 2002. Roger Lang, Turner Construction; New York. Personal communication. June 14, 2002.

Lax, P. 2002. Paul Lax, Lax & Stevens; Los Angeles. Personal communication. June 18, 2002.

MCAA. 1994. Change orders, overtime, and productivity. Mechanical Contractors Association of America; Rockville, MD. Publication M3.

Mountcastle-Walsh, H. 2002. Harriet Mountcastle-Walsh, Honeywell International; Columbia, MD. Personal communication. July 18, 2002.

Myers, J. M. 1994 Changes resulting from delays. In *Construction Change Order Claims*, ed. R. F. Cushman and S. D. Butler. 215–237. New York: Wiley.

Ness, A. 2002. Andy Ness, Thelen Reid & Priest; Washington D.C. Personal communication. August 8, 2002.

Newman, J. 2002. Joe Newman, Bechtel Corporation; San Francisco. Personal communication. September 26, 2002.

Sanford, J. 2002. Jim Sanford, Northrop Corporation; Los Angeles. Personal communication. July 25, 2002

Schwartzkopf, W. 1995. Calculating lost labor productivity in construction claims," 125–130. Frederick, MD: Aspen Law and Business.

Shumway, R. 2002. Ron Shumway, KPMG; San Francisco. Personal communication. August 7, 2002.

Sink, C. M. 1999. Ten ways to improve the contract claims process. *Water Environment & Technology* (April).

Stafford, T. 2002. Trevor Stafford, Fluor Corporation; Aliso Viejo, CA. Personal communication. July 29, 2002

Thomas, H. R., and K. A. Rayner. 1988. *The effects of scheduled overtime and shift schedule on construction craft productivity*. Austin, TX: Construction Industry Institute, Source Document 98.

Thomas, M. E. 2002. Mary Edith Thomas, Harris Corporation; Melbourne, FL. Personal communication. July 9, 2002

# INDEX